Using Vis-NIR Spectroscopy for Predicting Quality Compounds in Foods

Using Vis-NIR Spectroscopy for Predicting Quality Compounds in Foods

Editors

Mercedes Del Río Celestino
Rafael Font Villa

MDPI • Basel • Beijing • Wuhan • Barcelona • Belgrade • Manchester • Tokyo • Cluj • Tianjin

Editors

Mercedes Del Río Celestino
Agri-Food Laboratory
CAGPDS
Córdoba
Spain

Rafael Font Villa
Agri-Food Laboratory
CAGPDS
Córdoba
Spain

Editorial Office
MDPI
St. Alban-Anlage 66
4052 Basel, Switzerland

This is a reprint of articles from the Special Issue published online in the open access journal *Sensors* (ISSN 1424-8220) (available at: www.mdpi.com/journal/sensors/special_issues/NIR-Foods).

For citation purposes, cite each article independently as indicated on the article page online and as indicated below:

LastName, A.A.; LastName, B.B.; LastName, C.C. Article Title. *Journal Name* **Year**, *Volume Number*, Page Range.

ISBN 978-3-0365-7501-8 (Hbk)
ISBN 978-3-0365-7500-1 (PDF)

Cover image courtesy of Mercedes Del Río Celestino

Contents

About the Editors

Mercedes Del Río Celestino

Mercedes Del Río Celestino, Ph.D. in Biological Sciences (2000), has been working in different institutes and research centers in Spain (IAS-CSIC, Cordoba) and Belgium (Université Libre de Bruxelles), and has been awarded a contract from the "Ramon y Cajal" Spanish post-doc program. She was a permanent researcher in the Department of Plant Breeding and Biotechnology at the IFAPA (Almería, Spain) from 2009 to 2018. She is currently a research scientist at the Agri-Food Laboratory of Cordoba (Spain). She has published over 80 peer-reviewed scientific papers and various book chapters in the area of Plant Breeding. For several years, Dr. Del Río Celestino has been studying the genetic control of the fatty acids of the Ethiopian mustard seed and how to increase the added value of the fruit of zucchini (Cucurbita pepo subsp. pepo) through nutritional quality. Her achievements include obtaining materials from Ethiopian mustard with different profiles of the fatty acid composition of the seed adapted to the semi-arid conditions of Southern Spain and the obtaining of the first TILLING platform in zucchini fruit. Later, after the toxic waste spill of the Aznalcóllar mine in 1999, she became a part of the CSIC expert Group and a member of the Bioremediation Network for monitoring and recovering the ecosystem altered by metalloids. She has been a pioneer in genotoxicity and cytotoxicity studies in complex biological matrices (horticultural products) in order to quickly and economically determine its toxicity. Her research also focuses on the developments of chemometric and NIR spectroscopy for determining quality components in horticultural products.

Rafael Font Villa

Rafael Font Villa, Ph.D. in Biological Sciences (2003) has been working in Spain (IAS-CSIC, Cordoba) and has obtained a pre-doctoral mobility fellowship for short stays in different research centers in Reino Unido (SAK, Aberdeen), Italy (University of Genoa), and Belgium (Universite Libre de Bruxelles). He was a permanent researcher in the Department of Food Science and Health, IFAPA Center La Mojonera (Almería, Spain) from 2009 to 2018. He is currently a research scientist at the Agri-Food Laboratory of Cordoba (Spain). He has published over 75 peer-reviewed scientific papers and various book chapters in the the research areas of Plant Breeding and Postharvest. After the toxic waste spill of the Aznalcollar mine in 1999, he became a pioneer for using chemometric and spectroscopic methods to determine metalloids in different matrices (soil, plant, animals) and quality components in horticultural products. For several years, Dr. Font has also studied the genetic control of quality components (glucosinolates, fiber, fatty acids) of the Brassica seed and how to increase the added value of the fruit of zucchini (Cucurbita pepo subsp. pepo) through nutritional quality. He has worked on genotoxicity and cytotoxicity studies in complex biological matrices (horticultural products) for quickly and economically determining its toxicity. His research also focuses on postharvest technologies such as controlled ripening, edible coating, temperature management, and chemical treatment methods that are potential tools to reduce fruit and vegetable postharvest losses.

Preface to "Using Vis-NIR Spectroscopy for Predicting Quality Compounds in Foods"

The development of affordable and more reliable methods of controlling food quality and managing crops is imperative in order to maximize productivity and profitability and to minimize the environmental impacts of agriculture.

For several years, visible and near-infrared (VIS–NIR) spectroscopy has contributed to improving the control of food quality by providing the possibility to probe the internal quality of fresh fruits, vegetables, cereals, and other edibles.

We aimed to provide the readership with a comprehensive summary of present state-of-the-art NIR spectroscopy, current development trends, and future possibilities. We also believe that by doing so, we will be able to provide an acceptable chance for all contributors to make their results and methodologies more visible, as well as to highlight their current achievements in their respective fields which have been made possible by the use of NIR spectroscopy. These articles cover a wide range of topics related to NIR spectroscopy in a broad sense.

We would like to thank all of the authors and co-authors for their contributions, as well as all of the reviewers for their time and effort in carefully analyzing the submissions. Last but not least, we would like to express our gratitude to the *Sensor* journal's editorial office for their cooperation in preparing this Special Issue.

Mercedes Del Río Celestino and Rafael Font Villa
Editors

sensors

MDPI

Editorial

Using Vis-NIR Spectroscopy for Predicting Quality Compounds in Foods

Mercedes del Río Celestino * and Rafael Font

Agri-Food Laboratory, CAGPDS, Avd. Menéndez Pidal, s/n, 14080 Córdoba, Spain; rafaelm.font@juntadeandalucia.es

* Correspondence: mercedes.rio.celestino@juntadeandalucia.es

Citation: del Río Celestino, M.; Font, R. Using Vis-NIR Spectroscopy for Predicting Quality Compounds in Foods. *Sensors* **2022**, *22*, 4845. https://doi.org/10.3390/s22134845

Received: 16 June 2022
Accepted: 22 June 2022
Published: 27 June 2022

Publisher's Note: MDPI stays neutral with regard to jurisdictional claims in published maps and institutional affiliations.

Over the past four decades, near-infrared reflectance spectroscopy (NIRS) has become one of the most attractive and used technique for analysis as it allows for fast and simultaneous qualitative and quantitative characterization of a wide variety of food samples [1]. NIR spectroscopy is also essential in various other fields, e.g., pharmaceuticals [2], petrochemicals [3], textiles [4], cosmetics [5], medical applications [6], and chemicals such as polymers [7].

The high level of interest in NIR spectroscopy among scientific and professional sectors demonstrates its relevance. We hope that this Special Issue's scope facilitates the interchange of ideas and thereby aids in expanding the frontiers of this field of knowledge. Furthermore, we aim to provide readers with a comprehensive summary of present state-of-the-art NIR spectroscopy, trends in development, and future possibilities. We believe that by doing so, we will be able to provide a chance for all contributors to make their results and methodologies more visible, as well as to highlight current achievements in their respective fields made possible by the use of NIR spectroscopy.

This Special Issue has had a resoundingly enthusiastic response, with several submissions from academics and professional spectroscopists, resulting in a collection of 13 papers, including one exhaustive review paper [8–20]. The articles submitted represent the variety of the discussed field well, covering a wide range of topics related to NIR spectroscopy. The majority of the papers concentrate on applied qualitative and quantitative analysis in a variety of fields.

New progress has been made in improving food quality thanks to the first investigation. Accordingly, it was determined that the use of variable selection algorithms provided a better performance in predicting the amount of organophosphorus pesticide residues in tomatoes using NIRS than the use of all spectral data [8].

The feasibility of measuring physicochemical quality parameters of mangetout pods by means of VIS-NIRS has also been demonstrated. The results revealed that the models allow for an accurate quantification of protein and total polyphenol content and a rough screening method of the samples for color parameters (c^* and h^*), firmness, ascorbic acid content and pH [9].

In addition, despite the advantages of NIR nondestructive measurement, there is a lack of basic studies comparatively evaluating various forms of sampling with and without minimal processing. The analyses conducted in this Special Issue have showed that Vis-NIR spectroscopy could be used as a quick method to assess the abundance of chemical compounds (soluble solids content, saccharose (Pol), fiber, Pol of cane, and total recoverable sugars) of sugarcane. Moreover, the performance of the models on defibrated cane and raw juice samples were similar, but defibrated cane samples involve less preparation as they do not require juice extraction [10].

For the first time, this research shows the applicability of NIR spectroscopy to assess volatile phenol contents (guaiacol, 4-methyl-guaiacol, eugenol, syringol 4-methyl-syringol and 4-allyl-syringol) and confirms the ability of this technique to quantify compounds that contribute to the sensory quality of aged wine spirits [11].

NIRS technology can be a powerful tool to ensure the quality of food products and prevent fraud. From the results obtained, it can be concluded that NIRS together with artificial neural networks allow for the accurate prediction of almost all sensory parameters selected for an exhaustive characterization of dry-cured beef meat—cecina—quality. It would be possible to substitute the sensory panel with a faster, reliable, nondestructive and cheaper instrumental technique that may be implemented on site [12].

In addition, this Special Issue showed that NIRS is a feasible and useful tool for screening purposes, and it has the potential to predict most of the fatty acids of freeze-dried beef [13].

Moreover, a comprehensive review of the state of the art in research and the actual potential of NIRS for the analysis of olive oil has been included. It can be concluded that the four most common physicochemical parameters that define the quality of olive oils, namely free acidity, peroxide value, K232, and K270, can be measured using NIRS with high precision. In addition, NIRS is suitable for the nutritional labeling of olive oil because of its great performance in predicting the total fat, total saturated fatty acid, monounsaturated fatty acid, and polyunsaturated fatty acid contents in olive oils [14].

Likewise, the potential of hyperspectral imaging can be also recognized on the basis of the articles collected in this Special Issue [15–17]. Hyperspectral imaging (his) emerges as a non-destructive and rapid analytical tool for assessing food quality, safety, and authenticity. This technology can not only identify the physical chemistry characteristics of a substance through spectroscopic analysis, but also simultaneously obtains information about the spatial distribution of certain components through image analysis [21]. In this Special Issue, we present the possibility of rapidly inspecting and detecting *Escherichia coli* and *Salmonella typhimurium* on the surface of food processing facilities, which is a major global public health problem [22], via fluorescence hyperspectral imaging and various discriminant analysis techniques [15].

This Special Issue aims to investigate the potential of combining the spectral and spatial features of HSI data with the aid of deep-learning approaches for the pixel-wise classification of food products (sweet products and salmon fillets). The results demonstrated that spectral pre-processing techniques prior to convolutional neural network model's development can enhance the classification performance. This work will open the door for more research in the area of practical applications in food industry [16].

Important information is generated for the agrifood industry thanks to the new data provided in this Special Issue. Hyperspectral imaging technology has been used to develop a method for diagnosing the soil plant analysis development (SPAD) value and mapping the spatial distribution of chlorophyll in leaves located at different positions during the growth season of pepper plants. The results show that hyperspectral imaging is a very promising technology and has great potential for the intuitive monitoring of crop growth, laying the foundation for the development of hyperspectral field dynamic monitoring sensors [17].

The growing applicability and importance of portable NIR spectrometers is reflected by several articles, opening a new window for the utilization of these types of instruments in the analysis and monitoring of the composition of foods. In this context, the ability of a micro-near-infrared portable instrument to predict vitamin C in both whole and pureed Kakadu plum fruit samples was demonstrated [18].

In this regard, the use of MicroNIR as a tool for estimating dry matter and reducing sugars of fresh potato in a warehouses by directly measuring the tubers without chemical treatment and destruction of samples has been demonstrated. The efficiency of such automation techniques optimizes the management of industrial processing, guaranteeing the quality of the potato tubers during in-line processing [19].

In this work, we also focused on the development of a real-time and simple methodology to quantify the macronutrients (fat, raw protein and carbohydrates) in breast milk using a portable NIRS instrument. Notably, the implementation of this procedure requires the use of low-cost and handheld NIRS instruments where expert personnel are not required for analyzing samples, facilitating the quality-control procedure in the feeding of newborns in neonatology units [20].

It should be noted that these contributions accurately reflect the diversity and dynamism of current NIR spectroscopy development trends.

This Special Issue is accessible through the following link: https://www.mdpi.com/journal/sensors/special_issues/NIR-Foods (accessed on 24 June 2022). We would like to thank all of the authors and co-authors for their contributions, as well as all of the reviewers for their time and effort in carefully analyzing the submissions. Last but not least, we would like to express our gratitude to the editorial office of *Sensors* for their cooperation in preparing this Special Issue.

Author Contributions: Conceptualization, M.d.R.C. and R.F.; writing—review and editing, M.d.R.C. All authors have read and agreed to the published version of the manuscript.

Funding: This research received no external funding.

Conflicts of Interest: The authors declare no conflict of interest.

References

1. Shenk, J.S.; Workman, J.J.; Westerhaus, M.O. Application of NIR Spectroscopy to Agricultural Products. In *Handbook of Near-Infrared Analysis*, 2nd ed.; Burns, D.A., Ciurczak, E.W., Eds.; Marcel Dekker, Inc.: New York, NY, USA, 2001.
2. Jamrógiewicz, M. Application of the near-infrared spectroscopy in the pharmaceutical technology. *J. Pharm. Biomed. Anal.* **2012**, *66*, 1–10. [CrossRef]
3. Workman, J., Jr. A Brief Review of near Infrared in Petroleum Product Analysis. *J. Near Infrared Spectrosc.* **1996**, *4*, 69–74. [CrossRef]
4. Cleve, E.; Bach, E.; Schollmeyer, E. Using chemometric methods and NIR spectrophotometry in the textile industry. *Anal. Chim. Acta* **2000**, *420*, 163–167. [CrossRef]
5. Blanco, M.; Alcalá, M.; Planells, J.; Mulero, R. Quality control of cosmetic mixtures by NIR spectroscopy. *Anal. Bioanal. Chem.* **2007**, *389*, 1577–1583. [CrossRef] [PubMed]
6. Ferrari, M.; Norris, K.H.; Sowa, M.G. Medical near Infrared Spectroscopy 35 Years after the Discovery. *J. Near Infrared Spectrosc.* **2012**, *20*, vii–ix. [CrossRef]
7. Heigl, N.; Petter, C.H.; Rainer, M.; Najam-ul-Haq, M.; Vallant, R.M.; Bakry, R.; Bonn, G.K.; Huck, C.W. Near Infrared Spectroscopy for Polymer Research, Quality Control and Reaction Monitoring. *J. Near Infrared Spectrosc.* **2007**, *15*, 269–282. [CrossRef]
8. Nazarloo, A.S.; Sharabiani, V.R.; Gilandeh, Y.A.; Taghinezhad, E.; Szymanek, M. Evaluation of Different Models for Non-Destructive Detection of Tomato Pesticide Residues Based on Near-Infrared Spectroscopy. *Sensors* **2021**, *21*, 3032. [CrossRef]
9. García-García, M.D.C.; Martín-Expósito, E.; Font, I.; Martínez-García, B.dC.; Fernández, J.A.; Valenzuela, J.L.; Gómez, P.; Del Río-Celestino, M. Determination of Quality Parameters in Mangetout (*Pisum sativum* L. ssp. *arvense*) by Using Vis/Near-Infrared Reflectance Spectroscopy. *Sensors* **2022**, *22*, 4113. [CrossRef]
10. Corrêdo, L.d.P.; Maldaner, L.F.; Bazame, H.C.; Molin, J.P. Evaluation of Minimum Preparation Sampling Strategies for Sugarcane Quality Prediction by vis-NIR Spectroscopy. *Sensors* **2021**, *21*, 2195. [CrossRef]
11. Anjos, O.; Caldeira, I.; Fernandes, T.A.; Pedro, S.I.; Vitória, C.; Oliveira-Alves, S.; Catarino, S.; Canas, S. PLS-R Calibration Models for Wine Spirit Volatile Phenols Prediction by Near-Infrared Spectroscopy. *Sensors* **2022**, *22*, 286. [CrossRef]
12. Revilla, I.; Vivar-Quintana, A.M.; González-Martín, M.I.; Hernández-Jiménez, M.; Martínez-Martín, I.; Hernández-Ramos, P. NIR Spectroscopy for Discriminating and Predicting the Sensory Profile of Dry-Cured Beef "Cecina". *Sensors* **2020**, *20*, 6892. [CrossRef]
13. Ripoll, G.; Failla, S.; Panea, B.; Hocquette, J.-F.; Dunner, S.; Olleta, J.L.; Christensen, M.; Ertbjerg, P.; Richardson, I.; Contò, M.; et al. Near-Infrared Reflectance Spectroscopy for Predicting the Phospholipid Fraction and the Total Fatty Acid Composition of Freeze-Dried Beef. *Sensors* **2021**, *21*, 4230. [CrossRef] [PubMed]
14. García Martín, J.F. Potential of Near-Infrared Spectroscopy for the Determination of Olive Oil Quality. *Sensors* **2022**, *22*, 2831. [CrossRef] [PubMed]
15. Lee, A.; Park, S.; Yoo, J.; Kang, J.; Lim, J.; Seo, Y.; Kim, B.; Kim, G. Detecting Bacterial Biofilms Using Fluorescence Hyperspectral Imaging and Various Discriminant Analyses. *Sensors* **2021**, *21*, 2213. [CrossRef] [PubMed]
16. Zhu, H.; Gowen, A.; Feng, H.; Yu, K.; Xu, J.-L. Deep Spectral-Spatial Features of Near Infrared Hyperspectral Images for Pixel-Wise Classification of Food Products. *Sensors* **2020**, *20*, 5322. [CrossRef]
17. Yuan, Z.; Ye, Y.; Wei, L.; Yang, X.; Huang, C. Study on the Optimization of Hyperspectral Characteristic Bands Combined with Monitoring and Visualization of Pepper Leaf SPAD Value. *Sensors* **2022**, *22*, 183. [CrossRef]
18. Bobasa, E.; Phan, A.D.T.; Netzel, M.; Smyth, H.E.; Sultanbawa, Y.; Cozzolino, D. The Use of a Micro Near Infrared Portable Instrument to Predict Bioactive Compounds in a Wild Harvested Fruit—Kakadu Plum (*Terminalia ferdinandiana*). *Sensors* **2021**, *21*, 1413. [CrossRef]

19. Escuredo, O.; Meno, L.; Rodríguez-Flores, M.S.; Seijo, M.C. Rapid Estimation of Potato Quality Parameters by a Portable Near-Infrared Spectroscopy Device. *Sensors* **2021**, *21*, 8222. [CrossRef]
20. Melendreras, C.; Forcada, S.; Fernández-Sánchez, M.L.; Fernández-Colomer, B.; Costa-Fernández, J.M.; López, A.; Ferrero, F.; Soldado, A. Near-Infrared Sensors for Onsite and Noninvasive Quantification of Macronutrients in Breast Milk. *Sensors* **2022**, *22*, 1311. [CrossRef]
21. Feng, C.-H.; Makino, Y.; Oshita, S.; Martín, J.F.G. Hyperspectral imaging and multispectral imaging as the novel techniques for detecting defects in raw and processed meat products: Current state-of-the-art research advances. *Food Control* **2018**, *84*, 165–176. [CrossRef]
22. Srey, S.; Jahid, I.K.; Ha, S.D. Biofilm formation in food industries: A food safety concern. *Food Control* **2013**, *31*, 572–585. [CrossRef]

sensors

MDPI

Article

Evaluation of Different Models for Non-Destructive Detection of Tomato Pesticide Residues Based on Near-Infrared Spectroscopy

Araz Soltani Nazarloo [1], Vali Rasooli Sharabiani [1,*], Yousef Abbaspour Gilandeh [1], Ebrahim Taghinezhad [2] and Mariusz Szymanek [3]

[1] Department of Biosystem Engineering, Faculty of Agriculture and Natural Resources, University of Mohaghegh Ardabili, Ardabil 56199-11367, Iran; arazsoltani@uma.ac.ir (A.S.N.); abbaspour@uma.ac.ir (Y.A.G.)

[2] Department of Agricultural Engineering and Technology, Moghan College of Agriculture and Natural Resources, University of Mohaghegh Ardabili, Ardabil 56199-11367, Iran; e.taghinezhad@uma.ac.ir

[3] Department of Agricultural, Forest and Transport Machinery, University of Life Sciences in Lublin, Street Głęboka 28, 20-612 Lublin, Poland; mariusz.szymanek@up.lublin.pl

* Correspondence: vrasooli@uma.ac.ir; Tel.: +98-914-451-9089

Citation: Nazarloo, A.S.; Sharabiani, V.R.; Gilandeh, Y.A.; Taghinezhad, E.; Szymanek, M. Evaluation of Different Models for Non-Destructive Detection of Tomato Pesticide Residues Based on Near-Infrared Spectroscopy. *Sensors* **2021**, *21*, 3032. https://doi.org/10.3390/s21093032

Academic Editors: Mercedes Del Río Celestino and Federico Angelini

Received: 24 March 2021
Accepted: 23 April 2021
Published: 26 April 2021

Abstract: In this study, the possibility of non-destructive detection of tomato pesticide residues was investigated using Vis/NIRS and prediction models such as PLSR and ANN. First, Vis/NIR spectral data from 180 samples of non-pesticide tomatoes (used as a control treatment) and samples impregnated with pesticide with a concentration of 2 L per 1000 L between 350–1100 nm were recorded by a spectroradiometer. Then, they were divided into two parts: Calibration data (70%) and prediction data (30%). Next, the prediction performance of PLSR and ANN models after processing was compared with 10 spectral preprocessing methods. Spectral data obtained from spectroscopy were used as input and pesticide values obtained by gas chromatography method were used as output data. Data dimension reduction methods (principal component analysis (PCA), Random frog (RF), and Successive prediction algorithm (SPA)) were used to select the number of main variables. According to the values obtained for root-mean-square error (RMSE) and correlation coefficient (R) of the calibration and prediction data, it was found that the combined model SPA-ANN has the best performance (RC = 0.988, RP = 0.982, RMSEC = 0.141, RMSEP = 0.166). The investigational consequences obtained can be a reference for the development of internal content of agricultural products, based on NIR spectroscopy.

Keywords: pesticide residues; spectroscopy; PLS; soft computing; algorithm

1. Introduction

Tomato (*Solanum lycopersicum*) is one of the most widely used crops in the world, which is rich in antioxidants such as carotenoids, total phenols, vitamin E, and vitamin C [1]. Related empirical studies have shown that vitamin C affects the human immune system and prevents diseases such as Alzheimer's [2]. In addition, the prevention of illnesses by fruits and vegetables also depends on antioxidants [3].

Tomatoes need intensive pest management due to their low resistance to pests and diseases. The need to use pesticides can leave harmful residues in the product. Organophosphorus pesticides can be stable for a considerable time even after washing and cooking in the product if used without observing its pre-harvest interval [4–6].

Today, many countries have restricted the use of pesticides, requiring the pesticide maximum residue limit (MRL) in food [7], and the amount is specified for each crop [8–11]. Currently, there are several methods for determining the concentration of pesticides, including GC, HPLC, thin layer chromatography, and capillary electrophoresis [12]. However, due to time constraints and high costs, it is not possible to use these methods to control all products [13].

Many studies are currently underway to develop safe, rapid, reliable, and low-cost methods for determining pesticide residues that can prevent the use of organic solvents and reduce operator exposure to toxic substances. Spectroscopy-based methods are a potential method that can solve the problems mentioned above.

NIRS is suitable non-destructive method for quantitative and qualitative analysis in agriculture, chemistry, medicine, and other sciences [14–18]. This technique is faster and cheaper than conventional methods and environmentally friendly and can usually be used without the need to prepare samples [19–22]. This technology is based on the absorption of radiation in the infrared region near the electromagnetic spectrum, which can be used to control the quality of food products [23–25]. Furthermore, in some studies, this technology has been used to detect pesticide residues in agricultural products [26–28].

Jun et al. [29] examined cadmium residue in tomato leaves using hyperspectral imaging. In this method, WT and LSSVR were used to choose the best wavelength and create a detection model. The best prediction performance for the detection of cadmium (Cd) content in tomato leaves was obtained using the second derivative preprocessing method.

Chen et al. [30] used NIRS to determine organophosphate chemicals. PLSR was used to create the prediction models. The best prediction result was obtained using PLSR with MSC and the first derivative as the preprocessing method.

Fen et al. [31] used NIRS and ANN for non-destructive detection of a common pesticide on the Longan surface. The results showed that the correct diagnosis ratio was 93%.

Jiang et al. [32] combined deep learning and machine vision to predict the pesticide. The consequences showed that when the training epoch is 10, the precision of the test set detection will be 90.09% and the average picture bandwidth detection precision will be 95.35%.

Wei et al. [33] offered a technique for removing residues of pesticide in apple juice. This technique can precisely identify and classify data about residues of pesticide in apples.

Soltani et al. [18] used NIRS technology with multivariate regression analysis to predict pesticide residues in tomato. The best prediction results were obtained using the PLS model based on the smoothing + moving average method (Rcv = 0.92, RMSECV = 4.25).

Xue et al. [34] used the PSO algorithm to predict dichlorvos residue on the orange surface by Vis-NIR spectroscopy. The PSO-PLS model was able to predict the dichlorvos residue with a correlation coefficient of 0.8732. They have stated that the selection of wavelengths through a PSO algorithm increases the ability to predict when using the PLS model.

According to previous studies, the NIRS can be used to predict pesticide residues from other crops. To the best of our knowledge, there is no research to determine the organophosphorus pesticides and their prediction methods in tomatoes.

Therefore, in this paper, we use NIRS and chemometric methods to create a prediction model without destruction to detect the tomato pesticide residues. Spectral data obtained with a spectroradiometer and reference data obtained by a gas chromatography equipment were used as input and output of the models used in this study, respectively. PCA, SPA, and RF algorithms were utilized to select the variable as input for artificial neural network (ANN) and PLSR. First, all spectral data without dimension reduction and then spectral data obtained from variable selection algorithms were used to predict the amount of pesticide in tomatoes. Then 8 combined modes (PLS, ANN, PCA-ANN, RF-ANN, SPA-ANN, PLS-PCA, PLS-RF, and PLS-SPA) were developed for pesticides residues prediction. The use of several algorithms for variable selection to predict organophosphorus pesticide in tomatoes has not been evaluated in previous research. New progress can be made in improving food quality by this investigation.

2. Materials and Methods

2.1. Sample Preparation

180 samples of tomatoes (Queen) were randomly harvested from a greenhouse where almost all their produce was uniform in size and stored until 5 °C until use. Pest control

in tomatoes was non-chemical from the beginning of planting to the harvest stage. In order to achieve different pesticide residual concentrations, the samples were infected with Profenofos 40% (EC 40%) (C11H15BrClO3PS) with a Pre-Harvest Interval (PHI) of 14 days. Therefore, the solution of Profenofos pesticide with a concentration of 2 per 1000 L of water was prepared and sprayed on the samples. The samples were divided into 6 categories: The first group (P0) was used without any spraying as control and non-pesticide samples; second group two hours (P-2H); third group two days (P-2D); the fourth group is the same as the third category, except that it was washed after spraying (P-2D-W); the fifth group for one week (P-1W) and the sixth group for two weeks (P-2W) were subjected to VIS/NIR spectroscopy after spraying with the prepared solution. All samples reached equilibrium temperature in the laboratory before completing the measurements.

2.2. Vis/NIR Spectroscopy

Vis/NIR spectroscopy tests was performed using a PS-100 spectroradiometer (Apogee Instruments, INC., Logan, UT, USA) with CCD detector, 2048 pixels, 1 nm resolution and halogen-tungsten light source in the wavelength range of 350–1100 nm. Prior to spectroscopy, black and white (reference) spectra were first defined and stored. In this way, first by turning off the light source, the dark spectrum was taken, then in the light source mode, a standard Teflon disk with the ability to reflect above 97 in the range of 300 to 1700 nm was used to achieve the reference spectrum. For each tomato sample from 4 different points of each sample with 8 scans, within the spectral range of the equipment used, spectroscopy was performed with software Spectra-Wiz Spectrometer OS v5.33 (c) 2014 and the data were recorded after averaging. To find the spectral regions in the pesticide solution a quartz cell and two single-stranded fiber optics P400-2-VIS-NIR was used (Figure 1) [35]. Reference measurements were performed one day after spectroscopic analysis [18].

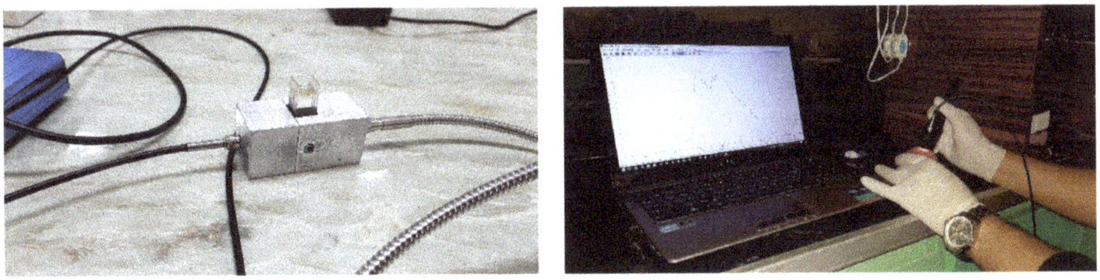

Figure 1. Measurement of Vis/NIR spectra of tomato samples in reflection mode and pesticide in passing mode.

2.3. Reference Measurements

After Vis/NIR spectroscopy, all tomatoes were prepared frozen to measure profenofos by gas chromatographic reference method (Agilent 5977A Series GC/MSD—Santa Clara, CA 95051, USA). To determine the retention time of the peak of the diagram obtained for Profenofos pesticide, the Profenofos standard material (95%) prepared from Agricultural Exir Company was injected into the chromatograph. For this purpose, sample preparation was performed according to the British standard BS EN 15662 [36,37]. First, 10 g of the homogenized sample was poured into a 50 mL centrifuge falcon. Then 10 mL of ethyl acetate, 1.9 mL of distilled water and 5 g of nitrogen sulfate were added and stirred for 1 min. It was then centrifuged at 5000 rpm for 5 min and 6 mL of the extract formed on top of the falcon was transferred to another glass falcon. It was shaken for 1 min and centrifuged at 5000 rpm for 5 min. Then 4 mL of the upper extract of glass was poured into another falcon and 50 μL of ethyl acetate was added. After filtration, 1 μL of extract was injected into the equipment. The run conditions of the gas chromatography equipment are fully described in Table 1.

Table 1. GC run conditions.

Analytical Column	HP-5 ms Ultra Inert 30 m × 250 μm, 0.25 μm (p/n 19091S-433UI)
Injection volume	1 μL
Injection mode	Spitless
Inlet temperature	280 °C
Liner	UI, split less, single taper, glass wool (p/n 5190-2293)
Plated seal kit	Gold Seal, Ultra Inert, with washer (p/n 5190-6144)
Carrier gas	Helium, constant flow, 1 mL/min
Oven program	60 °C for 1 min then 40 °C/min to 170 °C then 10 °C/min to 310 °C then hold for 2 min
Transfer line temperature	280 °C

2.4. Remove the Outlier Data

The Monte Carlo cross-validation method was used to remove outliers. This method can simultaneously detect spectral outliers and reference data [38]. Initially, the data were randomly divided into two categories: Calibration set (70%) and prediction set (30%). Then, PLS models were got with full cross-validation. When the RMSECV is minimized, the best number of PC of the model is achieved. Next, the statistical characteristic parameters of each model and the cumulative value of the sum of squares of predicted residual errors of each sample were determined [39,40]. In this paper, outlier data (20 samples) have been deleted by the method mentioned and the amount of R of the model has been improved from 0.8113 to 0.8609 after their removal. Table 2 shows the reference values (mean, standard deviation, and range) for the profenofos content (mg kg^{-1}) in the tomato samples used in this study. As can be seen, the values ranged from n.d (Not detected) to 42.9 mg/kg.

Table 2. Reference values (mean, standard deviation (SD) and range) for profenofos content (mg/kg).

	Profenofos (mg/kg)			
	Number	Range	Mean	Standard Deviation
calibration	112	n.d. *–42.9	14.0	10.1
validation	48	n.d.–34.0	13.7	8.9

* Not detected.

2.5. Variable Selection Method

2.5.1. Random frog (RF) Algorithm

The RF algorithm is generally used in the set of meta-heuristic algorithms. This algorithm is a useful wavelength selection method that calculates the probability of selection for each variable [40]. In short, the random frog algorithm consists of three steps [41,42]: (1) The random initialization of a subset of variable V_0 containing the variables Q; (2) creating a subset of the variable V^* including the variable Q^*; accepting V^* as V_1 with a certain probability and considering $V_0 = V_1$; the above procedure is repeated until the end of N and (3) calculating the probability of selecting each variable that can be used as a measure of the importance of the variable. The schematic of the algorithm is shown in Figure 2.

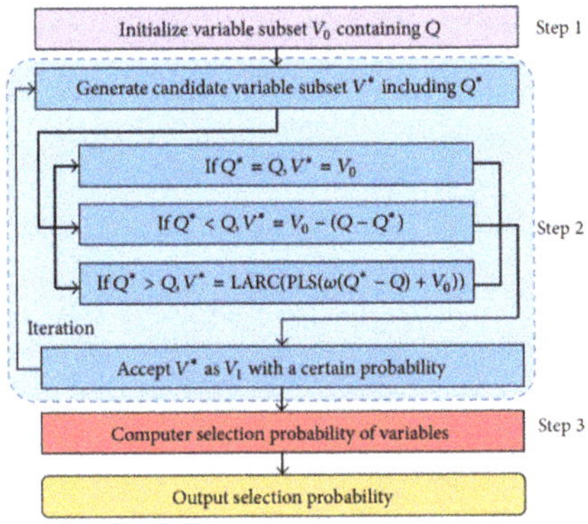

Figure 2. Flowchart of a random frog algorithm.

Figure 3 shows the appropriate wavelengths attained by the RF algorithm. In order to have the large part of impressible data in the main spectrum, the selection threshold was determined experimentally by 20% trial and error method and the wavelengths above this selection threshold were selected as the number of characteristic wavelengths. Therefore, 28 wavelengths above the dotted line were used as the final wavelengths to predict pesticide residues in tomatoes.

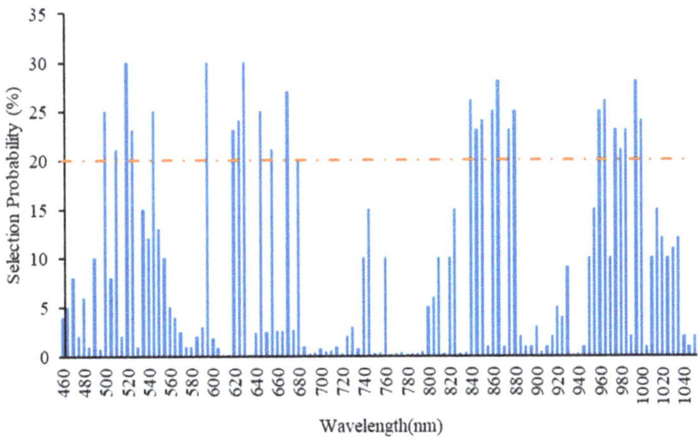

Figure 3. The result of extracting useful wavelengths using the RF algorithm.

2.5.2. SPA

SPA is a forward selection method that uses simple operations in a vector space to minimize the linearity of variables. The useful variable can be selected in spectral data analysis for multivariate calibration using this new method. This technique is widely used in optimizing specific spectral wavelengths that evaluate variable subsets based on RMSEC [43]. According to the change curve of RMSEC in relation to the number of wavelengths, it was determined that by selecting 14 characteristic wavelengths, the value of RMSE attained a lowest value of 0.141 (Figure 4). Thus, 14 effective wavelengths were applied as input to the prediction model. The selected characteristic wavelength distributions across the whole spectrum are shown in Figure 5. Wavelengths close to 650–700, 750–800 and 960–1000 were chosen to build the model. These wavelengths were in some cases like the wavelengths of the RF algorithm.

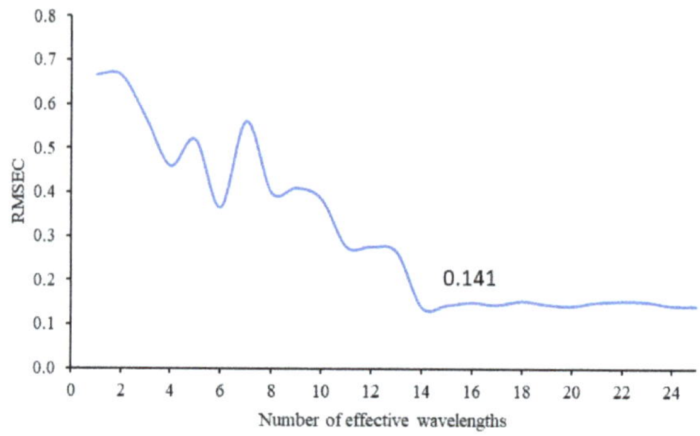

Figure 4. Change in RMSEC and Number of effective wavelengths.

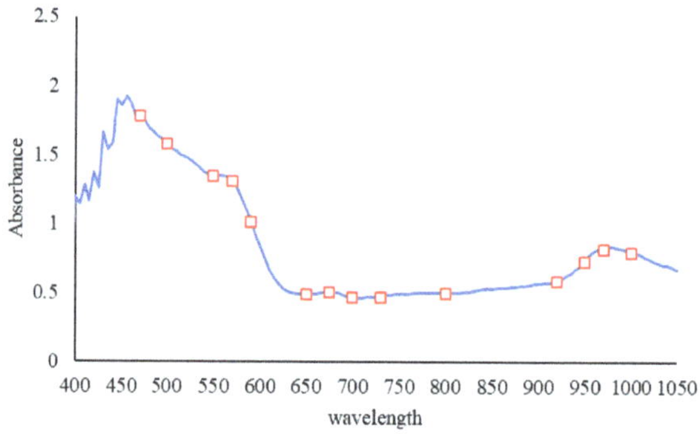

Figure 5. Selected bands by SPA.

2.5.3. PCA

PCA is one of the most widely used multivariate statistical methods in chemistry [44,45]. The corresponding mathematical model for PCA is based on the decomposition of matrix X into score matrix $n \times A$ (T) and loading matrix $N \times A$ (P) as Equation (1):

$$X = TP\prime + F = \sum_{a=1}^{A} t_a p'_a + F \quad (1)$$

where X is the spectral data matrix, T is the score matrix for X, P is the loading matrix for X, F is the residual or model error matrix, t_a is the sample score vector on each PC for X, and p_a is the variable loading vector on each PC for X. In this study, the share of the first principal component (PC1), the second principal component (PC2), the third principal component (PC3) and the fourth component were 55%, 18%, 8%, and 6%, respectively. In total, the cumulative share rate of these four components reached 87.00%. To avoid under-fitting of the prediction model due to lack of components, and to prevent over-fitting due to information of redundant components, finally 14 main components were selected as input to the prediction model of the amount of pesticide residues in tomatoes.

2.6. Prediction Models

2.6.1. PLSR

PLSR is a method for relating two matrices X (predictor) and Y (response), by a linear multivariate model, which also models the structure of X and Y [46]. It works well for analyzing large, noisy, and collinear data. In this model, by increasing the number of variables and related observations, the accuracy of the model parameters improves [47].

This method, the least squares solution, is applied to several orthogonal components that are a linear combination of independent variables and are created alternately with the aim of maximizing the covariance of the linear transformation of independent variables and dependent variables. It is very important to select the main factor when using PLSR for regression analysis. Wrong selection of the number of main factors causes the model to under-fitting or over-fitting, thus reducing the model prediction accuracy [48]. In this study, the mentioned method in the wavelength range of 300–1100 nm was used for modeling and analysis of spectral data. The fully cross-validation method was used to enhance the selection and the number of main factors RF, SPA, and PCA were 28, 14, and 14, respectively.

2.6.2. BP-ANN

BP-ANN, a multilayer feed-forward neural network trained by the post-propagation error algorithm, is today the most widely used reductive neural network [40,49]. In this paper, a BP feed-forward neural network with one and two hidden layers was modeled. "tansig", "logsig", and "purlin" were used in the hidden and output layers as transfer functions. The training function used in this model was "trainlm" and the maximum number of repetitions was 3000. The optimal number of hidden layer neurons for RF-BP, PCA-BP, and SPA-BP combined models was obtained by trial-and-error method, 8, 12, and 14, respectively.

2.7. Model Validation

Validation methods are important to assess calibration precision and avoid data over-fitting. The predictive power of a calibration model can be evaluated by the R, RMSEP and RMSEC between the predicted value and the measured value in the validation set [50]. In this research, we used R and RMSEC-RMSEP values to evaluate the accuracy and overall strength of the model, respectively. These indicators are defined as follows:

$$R = \sqrt{\frac{\sum_{i=1}^{n}(\hat{y}_i - y_i)^2}{\sum_{i=1}^{n}(\hat{y}_i - y_{mean})^2}} \tag{2}$$

$$RMSEC = \sqrt{\frac{1}{n_c}\sum_{i=1}^{n_c}(\hat{y}_i - y_i)^2} \tag{3}$$

$$RMSECV = RMSEP = \sqrt{\frac{1}{n_p}\sum_{i=1}^{n_p}(\hat{y}_i - y_i)^2} \tag{4}$$

$\hat{y}_i$: Predicted value of ith observation.
y_i: Measured value of ith observation.
y_{mean}: Mean of the prediction or calibration set.
n, n_c, n_p: The number of observations in the data set, calibration and prediction set, respectively.

In general, a good model should have higher correlation coefficients, lower RMSEC, lower RMSEP [51,52].

3. Results and Discussion

Pre-Processing Spectra

Due to the presence of noise in the initial and final parts of the diagram of absorption spectra of tomato samples with different concentrations of pesticides, the spectrum range from 460–1050 nm was considered (Figure 6). The following 10 spectral preprocessing methods were applied to stabilize the models: Moving average, gaussian filter, median filter, S-Golay, Maximum normalize, derivative-S-Golay, SNV, MSC, (Gaussian filter) + (median filter), Normalize + Gaussian.

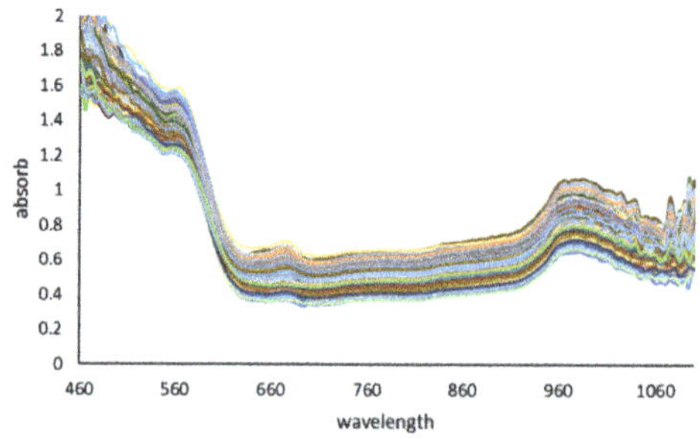

Figure 6. Absorption spectra of tomato samples with different concentrations of pesticides.

The residual reference values of the pesticide obtained by the GC-MS destructive test are between 42.9–"n.d" percent. Moreover, according to the prediction results of the combined models, the values of pesticide residues in the calibration and prediction data were between "n.d" up to 62.75%. The spectral diagram for tomatoes with different concentrations of pesticides is shown in Figure 6. In the diagram, the peak points in the visible and infrared region are closely visible. The peak points between 650–700, 750–800, and 960–1000 are related to the absorption of red pigments, the second and third overtone vibrations of OH and the first and second overtone vibrations of OH are related to water absorption. The results of PLS models obtained with different preprocessing methods to predict the Profenofos pesticide residues in tomato samples were shown in Table 3. Most of the developed calibration models had an acceptable ability to predict pesticide residues in samples with an RCV above 0.8. However, the best prediction results were obtained using the PLS model based on the Smoothing + moving average method (Rcv = 0.92, RMSECV = 4.25). Hence, this model was selected for further analysis. Shan et al. (2020), Soltani et al. (2021), Yi et al. (2010) and Sharabiani et al. (2019) also used the method used in this study to predict the amount of soil atrazine uptake, residual pesticides in strawberries, the amount of nitrogen in orange leaves and the amount of wheat protein, respectively, and achieved acceptable results [16,18,53,54].

Table 3. Results of different preprocessing methods for predicting Profenofos residues.

Pre-Processing	$RMSE_{CV}$	R_{CV}	LV
No preprocessing	5.7129	0.8609	15
Smoothing-moving average	4.2562	0.9254	13
Smoothing-gaussian filter	4.2680	0.9251	14
Smoothing-median filter	5.2481	0.8847	13
Smoothing	4.1379	0.9295	15
Maximum normalize	5.5788	0.8679	11
1derivative (S-Golay)	7.6328	0.7522	15
SNV	6.8656	0.7978	13
MSC	7.1441	0.7828	15
(Smoothing-Gaussian) + (smoothing median)	7.0276	0.7778	11
Normalize + Gaussian	5.9218	0.8490	10

Figure 7 shows the correlation diagrams of the predicted values versus the main values of the models used.

The use of NIRS technology in the detection of pesticide residues in fruits and vegetables, as well as their qualitative prediction, provides the researcher with a myriad of spectral data for analysis. Large amounts of spectral data complicate analysis, prediction errors, as well as over-fitting and under-fitting correlation curves. As a result, we need to reduce the data dimension. In this paper, it was found that the combined models used to predict the amount of Profenofos pesticide residues in tomato based on RF, SPA, and

PCA can achieve the same performance using only a few characteristic spectra, and in some cases achieve better performance than the mode based on all spectral data (Figure 7). Some variables selected using the frog (28 wavelength) and SPA (14 wavelengths) algorithms were similar and the rest of the characteristic wavelengths were close to each other. Table 4 shows the results of model evaluation indicators. According to the results obtained in ANN-based combined models, using SPA algorithm with values of Rc = 0.989, Rp = 0.982, RMSEC = 0.141 and RMSEP = 0.166 and using total spectral data with values of Rc = 0.86, Rp = 0.81, RMSEC = 0.521 and RMSEP = 0.561, respectively, had the best and worst performance in predicting Profenofos pesticide in tomatoes. Also, in PLS-based combined models, the modes of using SPA, RF, PCA, and total spectral data had the best and worst performance in predicting, respectively. In general, according to the results obtained in terms of validation parameters, the best model proposed in this paper is the SPA-ANN model.

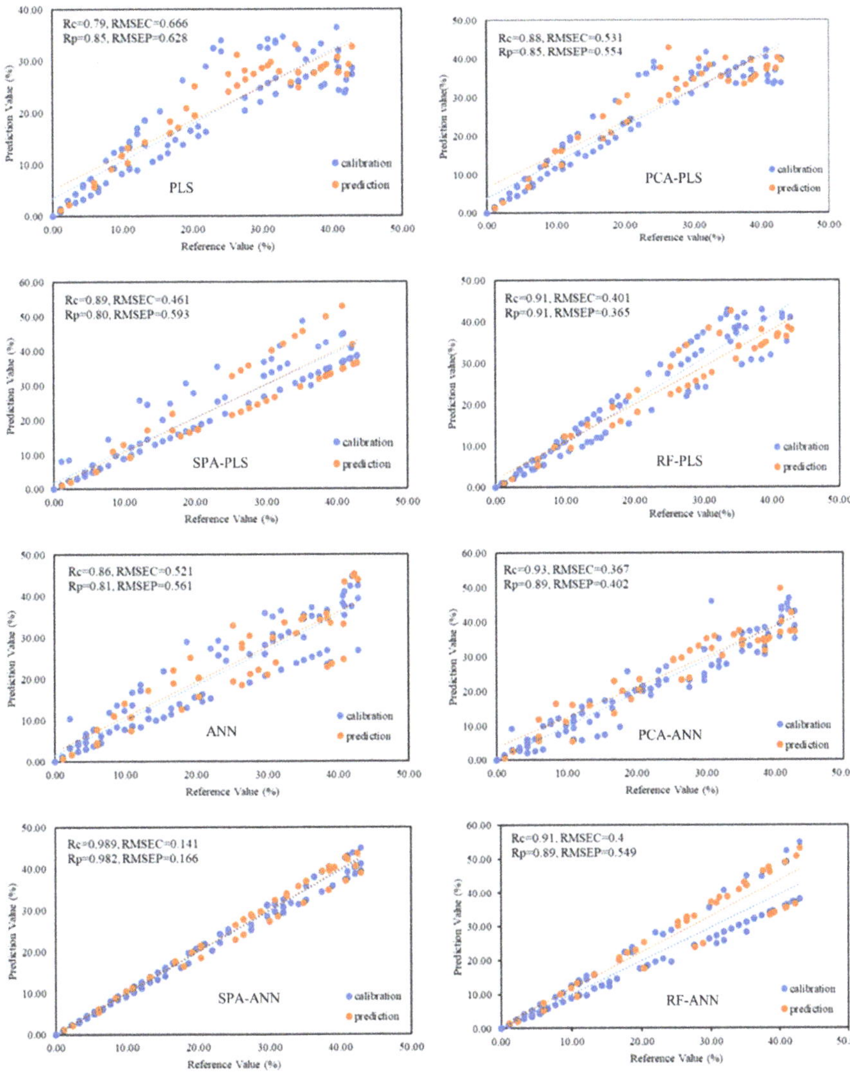

Figure 7. Correlation diagrams of the predicted values versus the main values of the models used.

Table 4. Results of validation parameters of combined models.

Combined Models	Validation Parameters			
	Rc	Rp	RMSEC	RMSEP
PLS	0.79	0.85	0.66	0.62
PCA-PLS	0.88	0.85	0.53	0.55
SPA-PLS	0.89	0.80	0.46	0.59
RF-PLS	0.91	0.91	0.40	0.36
ANN	0.86	0.81	0.52	0.56
PCA-ANN	0.93	0.89	0.36	0.40
SPA-ANN	0.98	0.98	0.14	0.16
RF-ANN	0.91	0.89	0.40	0.54

In a similar study the feasibility of using NIRS to detect the soluble solids content (SSC) of Malus micro malus Makino were studied using SPA, SVR, PLSR, and BP-ANN. The comparison studies confirmed that the optimal fusion model of SPA-SVR had the best performance (R_C = 0.9629, R_P = 0.9029, RMSEC = 0.199, RMSEP = 0.271) [42]. In other study, a new method of variable interval selection based on random frog (RF), known as Interval Selection based on Random Frog (ISRF), is developed. The results show that the proposed method is very efficient to find the best interval variables and improve the model's prediction performance and interpretation [55]. The results of various studies show that the use of soft computing has been an effective method in the qualitative diagnosis of products. This is evidence of the confirmation of the results obtained from our study.

4. Conclusions

In this paper, a rapid and non-destructive near-infrared method was used to predict the profenofos pesticide residues in tomatoes. Spectral data obtained with a spectroradiometer and reference data obtained by a gas chromatography equipment were used as input and output of the models used in this research, respectively. PCA, SPA, and RF algorithms were used to select the variable as input for artificial neural network (ANN) and PLSR. First, all spectral data without dimension reduction and then spectral data obtained from variable selection algorithms were used to predict the amount of pesticide in tomatoes. Afterwards, 8 combined modes (pls, ANN, PCA-ANN, RF-ANN, SPA-ANN, PLS-PCA, PLS-RF, and PLS-SPA) were developed for prediction. Finally, the prediction accuracy of different combined models was compared and the best case was introduced. Based on what was said in the previous sections of the paper, it was found that it is possible to predict the amount of pesticide residues in tomatoes using the spectrum in the range of 460–1050 nm. Accordingly, it was determined that the use of variable selection methods had a better performance in predicting the amount of pesticide residues than the use of all spectral data. Finally, according to the results of the validation parameters of the combined models used, the SPA-ANN combined model with values of Rc = 0.989, Rp = 0.982, RMSEC = 0.141 and RMSEP = 0.166 had the best performance in predicting Profenofos pesticide in tomatoes.

At the end of the article, it can be mentioned that NIRS technology, in addition to advantages such as a non-destructive method, low cost measurement, high speed and online uses in the processes of quality determination. However, this method has some disadvantages, which are: NIRS requires chemometric techniques to provide the reference data for calibration and validation of experimental results, and the large number of samples with large variations to extract accurate information.

Author Contributions: Methodology, V.R.S., A.S.N. and Y.A.G.; formal analysis, V.R.S., E.T. and M.S.; investigation, V.R.S. and Y.A.G.; experiment conduction, V.R.S. and Y.A.G.; statistical analysis, V.R.S. and A.S.N.; Writing—original draft preparation, V.R.S., A.S.N., E.T., Y.A.G. and M.S. writing—review and editing, A.S.N., V.R.S., Y.A.G., E.T. and M.S. All authors have read and agreed to the published version of the manuscript.

Funding: This research received no external funding.

Institutional Review Board Statement: Not applicable.

Informed Consent Statement: Not applicable.

Data Availability Statement: Data is contained within the article.

Conflicts of Interest: The authors declare no conflict of interest.

Abbreviations

MRL	Maximum Residue Limit
EU	European Union
VIS/NIRS	Visible/Near InfraRed Spectroscopy
PCA	Principal Component Analysis
PLSR	Partial Least Squares Regression
ANN	Artificial Neural Network
R	Regression Coefficient
RMSE	Root Mean Square Error
GC	Gas Chromatography
HPLC	High-Performance Liquid Chromatography
WT	Wavelet Transform
LSSVR	Least-Square Support Vector Machine Regression
PSO	Particle Swarm Optimization
RPD	Residual Prediction Deviation
PLS-DA	Partial Least Squares-Discriminant Analysis
PHI	Pre-Harvest Interval
LOD	Limit of detection

References

1. Zhang, L.; Chen, F.; Zhang, P.; Lai, S.; Yang, H. Influence of rice bran wax coating on the physicochemical properties and pectin nanostructure of cherry tomatoes. *Food Bioprocess Technol.* **2017**, *10*, 349–357. [CrossRef]
2. Sánchez-Moreno, C.; Cano, M.P.; De Ancos, B.; Plaza, L.; Olmedilla, B.; Granado, F.; Martín, A. Effect of orange juice intake on vitamin C concentrations and biomarkers of antioxidant status in humans. *Am. J. Clin. Nutr.* **2003**, *78*, 454–460. [CrossRef] [PubMed]
3. Odriozola-Serrano, I.; Soliva-Fortuny, R.; Martín-Belloso, O. Effect of minimal processing on bioactive compounds and color attributes of fresh-cut tomatoes. *LWT Food Sci. Technol.* **2008**, *41*, 217–226. [CrossRef]
4. Agnihotri, N. *Pesticide Safety Evaluation and Monitoring, Published All India Co-Ordinated Research Project on Pesticide Residues Division of Agricultural Chemicals*; Indian Agricultural Research Institute: New Delhi, India, 1999.
5. Chung, S.W.; Chen, B.L. Determination of organochlorine pesticide residues in fatty foods: A critical review on the analytical methods and their testing capabilities. *J. Chromatogr. A* **2011**, *1218*, 5555–5567. [CrossRef]
6. Kabir, K.H.; Rahman, M.A.; Ahmed, M.S.; Prodhan, M.D.H.; Akon, M.W. Pesticide analytical research at BARI related to quality, residue, and maximum residue limit. In Proceedings of the Workshop on Maximum Residue Limits of Pesticides in Agricultural Commodities and Food in Bangladesh, Dhaka, Bangladesh, 1 April 2007.
7. Cho, T.; Kim, B.; Jo, S.; Kang, H.; Choi, B.; Kim, M. Pesticide residue monitoring in Korean agricultural products, 2003–2005. *Food Addit. Contam. B* **2009**, *2*, 27–37. [CrossRef]
8. Barnat, S.; Boisset, M.; Casse, F.; Catteau, M.; Lecerf, J.-M.; Veschambre, D.; Périquet, A. Pesticide residues intake of French adults under increased consumption of fresh fruits and vegetables—A theoretical study. *J. Environ. Sci. Heal. B* **2010**, *45*, 102–107. [CrossRef]
9. Knežević, Z.; Serdar, M. Screening of fresh fruit and vegetables for pesticide residues on Croatian market. *Food Control* **2009**, *20*, 419–422. [CrossRef]
10. Claeys, W.L.; Schmit, J.-F.; Bragard, C.; Maghuin-Rogister, G.; Pussemier, L.; Schiffers, B. Exposure of several Belgian consumer groups to pesticide residues through fresh fruit and vegetable consumption. *Food Control* **2011**, *22*, 508–516. [CrossRef]
11. Omirou, M.; Vryzas, Z.; Papadopoulou-Mourkidou, E.; Economou, A. Dissipation rates of iprodione and thiacloprid during tomato production in greenhouse. *Food Chem.* **2009**, *116*, 499–504. [CrossRef]
12. Gambacorta, G.; Faccia, M.; Lamacchia, C.; Di Luccia, A.; La Notte, E. Pesticide residues in tomato grown in open field. *Food Control.* **2005**, *16*, 629–632. [CrossRef]
13. Luypaert, J.; Zhang, M.; Massart, D. Feasibility study for the use of near infrared spectroscopy in the qualitative and quantitative analysis of green tea, Camellia sinensis (L.). *Anal. Chim. Acta* **2003**, *478*, 303–312. [CrossRef]
14. Peng, Y.; Lu, R. Analysis of spatially resolved hyperspectral scattering images for assessing apple fruit firmness and soluble solids content. *Postharvest Biol. Technol.* **2008**, *48*, 52–62. [CrossRef]

15. Peng, Y.; Zhang, J.; Wu, J.; Hang, H. Hyperspectral scattering profiles for prediction of the microbial spoilage of beef. In *Sensing for Agriculture and Food Quality and Safety*; International Society for Optics and Photonics: Bellingham, WA, USA, 2009.
16. Sharabiani, V.R.; Nazarloo, A.S.; Taghinezhad, E. Prediction of Protein Content of Winter Wheat by Canopy of Near Infrared Spectroscopy (NIRS), using Partial Least Squares Regression (PLSR) and Artificial Neural Network (ANN) models. *Yüzüncü Yıl Üniversitesi Tarım Bilimleri Derg.* **2019**, *29*, 43–51. [CrossRef]
17. Soltani, A.; Noguchi, N. Multivariate analyzing and artificial neural networks for prediction of protein content in winter wheat using spectral characteristics. *Sci. Bus. Soc.* **2018**, *3*, 153–157.
18. Nazarloo, A.S.; Sharabiani, V.R.; Gilandeh, Y.A.; Taghinezhad, E.; Szymanek, M.; Sprawka, M. Feasibility of using VIS/NIR spectroscopy and multivariate analysis for pesticide residue detection in tomatoes. *Processes* **2021**, *9*, 196. [CrossRef]
19. Jamshidi, B.; Mohajerani, E.; Jamshidi, J.; Minaei, S.; Sharifi, A.; Malvajerdi, A.S. Non-destructive detection of pesticide residues in cucumber using visible/near-infrared spectroscopy. *Food Addit. Contam. A* **2015**, *32*, 857–863. [CrossRef]
20. Sánchez, M.-T.; Flores-Rojas, K.; Guerrero, J.E.; Garrido-Varo, A.; Pérez-Marín, D. Measurement of pesticide residues in peppers by near-infrared reflectance spectroscopy. *Pest Manag. Sci.* **2010**, *66*, 580–586. [CrossRef]
21. Jamshidi, B. Non-destructive safety assessment of agricultural products using Vis/NIR spectroscopy. *NIR News* **2017**, *28*, 4–8. [CrossRef]
22. Yazici, A.; Tiryaki, G.Y.; Ayvaz, H. Determination of pesticide residual levels in strawberry (Fragaria) by near-infrared spectroscopy. *J. Sci. Food Agric.* **2020**, *100*, 1980–1989. [CrossRef] [PubMed]
23. Hu, L.; Yin, C.; Ma, S.; Liu, Z. Vis-NIR spectroscopy combined with wavelengths selection by PSO optimization algorithm for simultaneous determination of four quality parameters and classification of soy sauce. *Food Anal. Methods* **2019**, *12*, 633–643. [CrossRef]
24. Quiñones, M.d.C.S.; M artínez, L.A.O.; Herrera, S.M.G.; Quiñones, O.M.R.; Lared, R.F.G. Near-Infrared Spectroscopy (NIRS) applied to legume analysis: A Review. *Spectroscopy* **2018**, *8*.
25. Williams, P. *Near Infrared Technology: Getting the Best Out of Light*; African Sun Media: Stellenbosch, South Africa, 2019.
26. Saranwong, I.; Kawano, S. Rapid determination of fungicide contaminated on tomato surfaces using the DESIR-NIR: A system for ppm-order concentration. *J. Near Infrared Spectrosc.* **2005**, *13*, 169–175. [CrossRef]
27. Teye, E.; Huang, X.-Y.; Afoakwa, N. Review on the potential use of near infrared spectroscopy (NIRS) for the measurement of chemical residues in food. *Am. J. Food Sci. Technol.* **2013**, *1*, 1–8.
28. Watanabe, E.; Kobara, Y.; Baba, K.; Eun, H. Determination of seven neonicotinoid insecticides in cucumber and eggplant by water-based extraction and high-performance liquid chromatography. *Anal. Lett.* **2015**, *48*, 213–220. [CrossRef]
29. Jun, S.; Xin, Z.; Xiaohong, W.; Bing, L.; Chunxia, D.; Jifeng, S. Research and analysis of cadmium residue in tomato leaves based on WT-LSSVR and Vis-NIR hyper-spectral imaging. *Spectrochimica Acta A Mol. Biomol. Spectrosc.* **2019**, *212*, 215–221. [CrossRef]
30. Chen, J.; Peng, Y.; Li, Y.; Wang, W.; Wu, J. A method for determining organophosphorus pesticide concentration based on near-infrared spectroscopy. *Trans. ASABE* **2011**, *54*, 1025–1030. [CrossRef]
31. Fen, D.; Tiansheng, H.; Kun, Z.; Ya, H. Nondestructive detection of pesticide residue on longan surface based on near infrared spectroscopy. In Proceedings of the 2010 International Conference on Intelligent Computation Technology and Automation, Institute of Electrical and Electronics Engineers (IEEE), Changsha, China, 11–12 May 2010; Volume 2, pp. 781–783.
32. Jiang, B.; He, J.; Yang, S.; Fu, H.; Li, T.; Song, H.; He, D. Fusion of machine vision technology and AlexNet-CNNs deep learning network for the detection of postharvest apple pesticide residues. *Artif. Intell. Agric.* **2019**, *1*, 1–8. [CrossRef]
33. Wei, Z. Determination of organophosphorus pesticides in cider by GC-MS coupled with cloud point extraction. *Sci. Technol. Food Ind.* **2017**, *23*, 225–231.
34. Xue, L.; Cai, J.; Li, J.; Liu, M. Application of Particle Swarm Optimization (PSO) algorithm to determine dichlorvos residue on the surface of navel orange with Vis-NIR spectroscopy. *Procedia Eng.* **2012**, *29*, 4124–4128. [CrossRef]
35. González-Martín, M.; Revilla, I.; Vivar-Quintana, A.; Salcedo, E.B. Pesticide residues in propolis from Spain and Chile. An approach using near infrared spectroscopy. *Talanta* **2017**, *165*, 533–539. [CrossRef] [PubMed]
36. Anastassiades, M.; Lehotay, S.J.; Štajnbaher, D.; Schenck, F.J. Fast, and easy multiresidue method employing acetonitrile extraction/partitioning and "dispersive solid-phase extraction" for the determination of pesticide residues in produce. *J. AOAC Int.* **2003**, *86*, 412–431. [CrossRef] [PubMed]
37. Ciscato, C.; Barbosa, C.; Gebara, A. Analysis of Pesticide Residues in Mango by GC/MS/MS with Bond Elut QuEChERS EN Kits. Available online: https://www.agilent.com/cs/library/applications/5991-6054EN.pdf (accessed on 26 April 2021).
38. Xiao-yu, Y.; Gui-shan, L.; Jia-xing, D.; Ya-bin, C.; Meng-meng, F.; Chao, M.; Jian-guo, H. A rapid evaluation of vc content on lingwu long jujube using hyperspectral technique. *Spectrosc. Spectr. Anal.* **2019**, *39*, 230–234.
39. Liu, Z.; Cai, W.; Shao, X. Outlier detection in near-infrared spectroscopic analysis by using Monte Carlo cross-validation. *Sci. China Ser. B Chem.* **2008**, *51*, 751–759. [CrossRef]
40. Gao, Q.; Wang, M.; Guo, Y.; Zhao, X.; He, D. Comparative analysis of non-destructive prediction model of soluble solids content for malus micro malus makino based on near-infrared spectroscopy. *IEEE Access* **2019**, *7*, 128064–128075. [CrossRef]
41. Maldonado, A.I.L.; Rodriguez-Fuentes, H.; Contreras, J.A.V. *Hyperspectral Imaging in Agriculture, Food and Environment*; BoD–Books on Demand: Norderstedt, Germany, 2018.
42. Zhao, Y.-R.; Yu, K.-Q.; He, Y. Hyperspectral imaging coupled with random frog and calibration models for assessment of total soluble solids in mulberries. *J. Anal. Methods Chem.* **2015**, *2015*. [CrossRef] [PubMed]

43. Luo, W.; Du, Y.-Z.; Zhang, H.-L. Discrimination of varieties of cabbage with near infrared spectra based on principal component analysis and successive projections algorithm. *Guang Pu Xue Yu Guang Pu Fen Xi = Guang Pu* **2016**, *36*, 3536–3541. [PubMed]
44. Wang, Y.; Sun, F.; Li, X. Compound dimensionality reduction based multi-dynamic kernel principal component analysis monitoring method for batch process with large-scale data sets. *J. Intell. Fuzzy Syst.* **2020**, *38*, 471–480. [CrossRef]
45. Brereton, R. Principal component analysis: The method. In *Chemometrics. Data Analysis for the Laboratory and Chemical Plant*; R.G. Brereton, Ed.; John Wiley and Sons, Ltd.: Chichester, UK, 2003; pp. 191–223.
46. Miller, J.; Miller, J.C. *Statistics and Chemometrics for Analytical Chemistry*; Pearson Education: London, UK, 2018.
47. Wold, S.; Sjöström, M.; Eriksson, L. PLS-regression: A basic tool of chemometrics. *Chemom. Intell. Lab. Syst.* **2001**, *58*, 109–130. [CrossRef]
48. Nturambirwe, J.F.I.; Nieuwoudt, H.H.; Perold, W.J.; Opara, U.L. Non-destructive measurement of internal quality of apple fruit by a contactless NIR spectrometer with genetic algorithm model optimization. *Sci. Afr.* **2019**, 3, e00051. [CrossRef]
49. Khoshnoudi-Nia, S.; Moosavi-Nasab, M. Comparison of various chemometric analysis for rapid prediction of thiobarbituric acid reactive substances in rainbow trout fillets by hyperspectral imaging technique. *Food Sci. Nutr.* **2019**, *7*, 1875–1883. [CrossRef] [PubMed]
50. Wang, H.; Peng, J.; Xie, C.; Bao, Y.; He, Y. Fruit quality evaluation using spectroscopy technology: A review. *Sensors* **2015**, *15*, 11889–11927. [CrossRef] [PubMed]
51. Lin, S.; Huang, X. Advances in computer science, environment, ecoinformatics, and education, Part III. In Proceedings of the International Conference, CSEE 2011, Wuhan, China, 21–22 August 2011.
52. Shao, Y.; He, Y. Visible/Near infrared spectroscopy and chemometrics for the prediction of trace element (Fe and Zn) levels in rice leaf. *Sensors* **2013**, *13*, 1872–1883. [CrossRef] [PubMed]
53. Shan, R.; Chen, Y.; Meng, L.; Li, H.; Zhao, Z.; Gao, M.; Sun, X. Rapid prediction of atrazine sorption in soil using visible near-infrared spectroscopy. *Spectrochim. Acta A Mol. Biomol. Spectrosc.* **2020**, *224*, 117455. [CrossRef]
54. Yi, S.; Deng, L.; He, S.; Zheng, Y.; Zhang, X. Research on nitrogen content of leaf of Jincheng orange cultivar using visible near infrared spectroscopy model. *J. Fruit Sci.* **2010**, *27*, 13–17.
55. Sun, J.; Yang, W.; Feng, M.; Liu, Q.; Kubar, M.S. An efficient variable selection method based on random frog for the multivariate calibration of NIR spectra. *RSC Adv.* **2020**, *10*, 16245–16253. [CrossRef]

sensors

MDPI

Article

Determination of Quality Parameters in Mangetout (*Pisum sativum* L. ssp. *arvense*) by Using Vis/Near-Infrared Reflectance Spectroscopy

María del Carmen García-García [1,*], Emilio Martín-Expósito [1], Isabel Font [2], Bárbara del Carmen Martínez-García [3], Juan A. Fernández [4], Juan Luis Valenzuela [5], Pedro Gómez [6] and Mercedes del Río-Celestino [7,*]

1 Department of Agro-Food Engineering and Technology, IFAPA Centro La Mojonera, CAGPDS, 04745 Almería, Spain; emilio.martin.exposito@juntadeandalucia.es
2 ETSIIT, Campus Aynadamar, University of Granada, 18071 Granada, Spain; isabelfont@correo.ugr.es
3 ESI, Higher Engineering School, University of Almería, 04120 Almería, Spain; barbaramgbm@gmail.com
4 Department of Agronomical Engineering, Technical University of Cartagena, 30203 Murcia, Spain; juan.fernandez@upct.es
5 Department of Biology and Geology, Higher Engineering School, University of Almería, 04120 Almería, Spain; jvalenzu@ual.es
6 Department of Plant Breeding and Biotechnology, IFAPA Centro La Mojonera, CAGPDS, 04745 Almería, Spain; pedro.gomez.j@juntadeandalucia.es
7 Agri-Food Laboratory, CAGPDS, Avda, Menéndez Pidal, s/n, 14080 Córdoba, Spain
* Correspondence: mariac.garcia.g@juntadeandalucia.es (M.d.C.G.-G.); mercedes.rio.celestino@juntadeandalucia.es (M.d.R.-C.)

Citation: García-García, M.d.C.; Martín-Expósito, E.; Font, I.; Martínez-García, B.d.C.; Fernández, J.A.; Valenzuela, J.L.; Gómez, P.; Río-Celestino, M.d. Determination of Quality Parameters in Mangetout (*Pisum sativum* L. ssp. *arvense*) by Using Vis/Near-Infrared Reflectance Spectroscopy. *Sensors* **2022**, *22*, 4113. https://doi.org/10.3390/s22114113

Academic Editor: Luca Fiorani

Received: 7 April 2022
Accepted: 26 May 2022
Published: 28 May 2022

Abstract: *Pisum sativum* L. ssp. *arvense*, is colloquially called *tirabeque* or mangetout because it is eaten whole; its pods are recognized as a delicatessen in cooking due to its crunch on the palate and high sweetness. Furthermore, this legume is an important source of protein and antioxidant compounds. Quality control in this species requires the analysis of a large number of samples using costly and laborious conventional methods. For this reason, a non-chemical and rapid technique as near-infrared reflectance spectroscopy (NIRS) was explored to determine its physicochemical quality (color, firmness, total soluble solids, pH, total polyphenols, ascorbic acid and protein content). Pod samples from different cultivars and grown under different fertigation treatments were added to the NIRS analysis to increase spectral and chemical variability in the calibration set. Modified partial least squares regression was used for obtaining the calibration models of these parameters. The coefficients of determination in the external validation ranged from 0.50 to 0.88. The RPD (standard deviation to standard error of prediction ratio) and RER (standard deviation to range) were variable for quality parameters and showed values that were characteristic of equations suitable for quantitative prediction and screening purposes, except for the total soluble solid calibration model.

Keywords: mangetout; pea pod; near-infrared reflectance spectroscopy; quality parameters

1. Introduction

Vegetable proteins are appearing as a sustainable source for human consumption [1]. Demand for protein is likely to increase significantly over the next few decades to keep pace with a growing population, which is projected to reach nearly ten billion by 2050 [2]. The trend of animal protein consumption is increasing in recent decades [3,4], with production of animal source foods responsible for a significant proportion of global greenhouse gas (GHG) emissions, water consumption and land use [5]. However, the proportion of protein consumption that the World Health Organization recommends is 75% vegetable and 25% animal [6]. In this context, legumes, including soybeans, peanuts, beans, peas, fava beans and lentils, among others, have a higher protein content than most plant foods and about twice the protein content of cereals [7]. The high protein content of legumes may be related to their association with nitrogen-fixing bacteria in their roots, which converts the unusable nitrogen into ammonium that is used for protein synthesis [8].

At present, the consumer demands new products on the supermarket shelves and is also attracted by local markets and products. A segment of the population considers itself a green consumer [9], in its different variants, and values healthy and quality foods. Legumes, for all the above exposed, satisfy the current market trends [10].

Several species have been the subject of research for the diversification of vegetables in the agricultural system of the province of Almería (Southeast Spain), with more than 32,000 hectares of greenhouses [11], more than 60% of cultivated vegetables belonging to the *Solanaceae* family. The species tested to diversify these horticultural crops are sweet cucumber, berries, pitahaya, passion fruit, fig tree and a wide range of legumes, among some of them *Pisum sativum* L. ssp. *arvense*, colloquially called *tirabeque* or mangetout [12]. This species is recognized as a delicatessen in cooking due to its crocanti on the palate and high sweetness. Whole mangetout pods are cooked and eaten, this being possible by the absence of "parchment" in the pod walls, hence its pod is indehiscent. The external appearance of pods, particularly their color, is also of great importance when considering the fruits destined for fresh products.

Previous studies have also revealed the nutritional potential of mangetout, not only for its protein content but also for its content of total soluble solids and antioxidant compounds such as polyphenols, ascorbic acid, fiber, phytoprostanes and phytofurans [12–14].

Overall, the methodology used for the determination of phenolic compounds and ascorbic acid content is based on spectrophotometric and chromatographic techniques; however, these techniques require expensive equipment and usually use hazardous and pollutant reagents [15,16]. Another relevant method includes colorimetric and titration measurements, since it represents a relatively simple method for measuring total phenolic compounds and ascorbic acid content, respectively.

The need to carry out screening in breeding programs, quality controls, traceability studies and/or obtaining rapid information for labelling in a large number of samples using conventional methods, leads to high costs, labour input and delays in the rapid decision making. For this reason a non-chemical (producing no chemical waste) and rapid technique, near-infrared reflectance spectroscopy (NIRS), which has been successfully applied in various fields from life sciences to environmental issues, is explored here to screen quality in mangetout pods [17]. Near-infrared spectroscopy is a technique that uses the radiation absorbed by a set of samples in the region from 780 to 2500 nm (near-infrared region-NIR spectroscopy in combination with chemometric analyses can be used for analysis of numerous components (protein, carbohydrates, carotenoid, minerals, glucosinolates, phenolics) and parameters of the sample (firmness, Brix, acidity, color) to be analyzed [18–23]. NIRS depends on the number and type of C-H, N-H and O-H bonds in the material being analyzed, then spectral features are combined with reliable compositional or functional analyses of the material in a predictive statistical model. This model is then used to predict the composition of new or unknown samples [24].

Recently, the use of NIRS models for predicting the quality of vegetables has been reported, several of which have addressed zucchini [19,20], pepper, rocket leaves, blackberries [16,21,22] and Ethiopian mustard leaves [23], among others. The seed quality of various legume species has also been analyzed using NIRS such as lentils [25], chickpeas [26] and pea accessions from different germplasm collections [27,28]. Other studies have focused on predicting the sensory quality and maturity of peas [29,30] using NIRS. To the best of our knowledge, there is no research that predicted the quality in mangetout pods.

NIRS calibration models have been developed using a variety of linear regression approaches, including modified partial least squares regression (MPLS). The modified partial least squares (MPLS) is an improved version of traditional PLS that was developed by Shenk and Westerhaus [31]. The MPLS procedure copes more effectively with non-analyte interference in multicomponent determinations. This regression approach is a soft-modeling method for generating predictive models when the factors are many and very collinear. It allows us to develop a model that is then evaluated on external samples to estimate the predictive ability of the model. The mathematical procedure's end goal is to decrease the large amount of spectral data points (1050 data points from 400 to 2500 nm wavelength range, every 2 nm) and remove the correlation presented by neighboring wavelengths. As a result, the model developed only takes into account the most significant factors, with the "noise" encapsulated in the less important factors, hence the accuracy of NIRS analysis is improved.

At present, the purpose of the producers and the Andalusian Administrations involved in the cultivation of mangetout is to apply for a "Protected Geographical Indication" (PGI) for the Dalías Valley (Almería, Southeast Spain). This European Indication distinguishes the quality attributes of the products grown in a certain region, and the NIRS technique is a suitable tool that could contribute quickly and accurately to verify the quality of the productions.

The objective of this paper was to investigate the feasibility for measuring physico-chemical quality parameters (color, firmness, total soluble solids, pH, total polyphenols, ascorbic acid and protein content) of mangetout pods by means of VIS-NIRS. For this purpose, different cultivars of mangetout grown under organic cultivation and two fertigation regimes were tested to generate the highest variability for the development of NIRS prediction models.

2. Material and Methods

2.1. Plant Material

The vegetal material consisted of a local landrace (germplasm maintained by local growers in Almería Province, Southeast Spain) and 7 commercial cultivars of mangetout (Figure 1, Table 1).

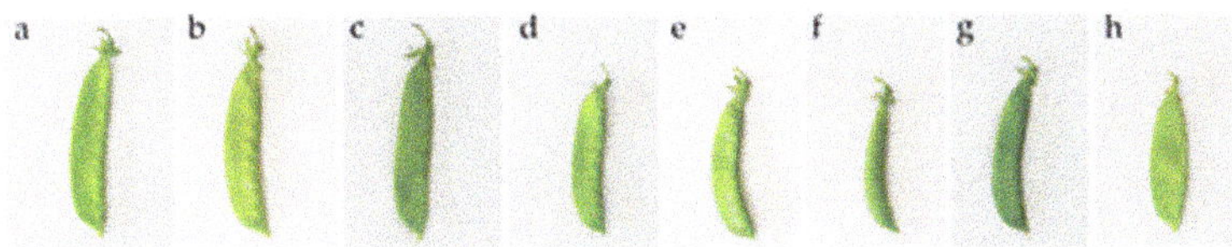

Figure 1. Pea pods of the different cultivars of mangetout analyzed. From left to right: Local landrace (**a**), AR-24007 (**b**), Capuchino (**c**), Tirabeque IS (**d**), Tirabí (**e**), Pea Zuccola (**f**), Pea Delikata (**g**) and Bamby (**h**).

Table 1. Cultivars, companies and growth habit of mangetout used in this study.

Cultivars	Companies	Growth Habit
Local landrace	Growers production	Indeterminate climbing
AR-24007	Ramiro Arnedo	Indeterminate climbing
Capuchino	Batlle	Indeterminate climbing
Tirabeque IS	Intersemillas	Indeterminate climbing
Tirabí	Fitó	Indeterminate climbing
Pea Zuccola	Tozer	Determinate climbing
Pea Delikata	Tozer	Determinate climbing
Bamby	Gautier	Deterninate postrate

Edible pods of *Pisum sativum* L. spp. *arvense* (tirabeque or mangetout) were grown in an organic greenhouse of 800 m^2, at Instituto de Investigacion y Formacion Agraria y Pesquera (IFAPA) Center "La Mojonera" (36°48′ N, 2°41′ W; altitude 142 m). The crop (Figure 2) was carried out according to European ecological regulations [32]. The crop cycle took place between October 2020 and March 2021. Two treatments, T100 (100% fertigation treatment) and T50 (50% of water and fertilizers applied), were arranged in a randomized complete block design with 3 replicates, for each cultivar and fertigation treatment, being the planting density of 4 plants per m^2 [13]. T100 consisted of water and fertilizer provided according to fertigation management. The fertigation treatments allowed us to have a larger number of samples with physicochemical variability (different qualities) to develop NIRS predictive models.

Figure 2. Detail of flower, leaves (**left up**) and pod (**left down**) of mangetout. Panoramic view of field trial (**right**).

A random monitoring of disease and pest symptoms was conducted weekly. In T100, the consumption of irrigation water was 100 L m^{-2}, applying ecological fertilizers so that the average nutrient solution reached 2.3 mS cm^{-1}. Pods were harvested when reached standard commercial sizes.

2.2. *Physicochemical Parameters*

The parameters considered to assess the physical quality in mangetout fruit were firmness and skin color, whereas the parameters of chemical quality were total soluble solids content, pH, total vitamin C, total polyphenol content and protein content. All these characters were determined on the fruit of fresh mangetout except the protein content. For each cultivar (8), treatment (2) and replicate (3), three samples were used. Each sample was composed of 5 pods from 3 plants selected at random, which were then averaged (n = 144).

2.2.1. Firmness

Texturometer XTPlus (Texture Analyzer, Surrey, UK) was used to obtain pod firmness (Figure 3). Shear force was measured by the Warner-Bratzler test. The pod was cut perpendicular with a Warner-Bratzler blade at 1 mm s^{-1} during 5 s. The result was expressed in Newton (N).

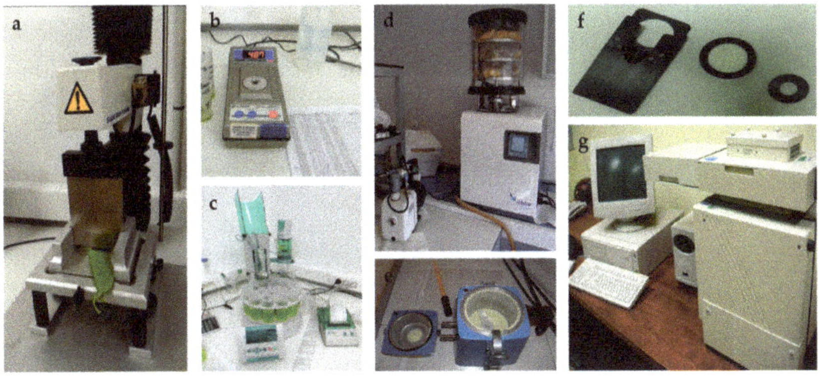

Figure 3. Detail of Texturometer XTPlus Texture Analyzer (**a**); Smart-1 digital refractometer (**b**); Automatic Metrohm 862 Titrosampler (**c**); Freeze-drying equipment (**d**); Mill (**e**); NIRS sample holder (**f**); Spectrometer Model 6500 Foss-NIRSystems (**g**).

2.2.2. Color

CM-700d Konica Minolta portable colorimeter was used. Chroma and Hue angle were measured externally, in two different pod locations, in the central plane.

2.2.3. Total Soluble Solids and pH

The soluble solid content (TSS) of the pods was obtained through measurement with a Smart-1 digital refractometer (Atago, Japan) (Figure 3), and the previous sample was homogenized for 30 s at 700 Braun CombiMax. The result was expressed in Brix. The pH was obtained by automatic Metrohm 862 Titrosampler (Metrohm, Riverview, Florida, USA) (Figure 3)

2.2.4. Total Polyphenol Content

In total, 10 g of the pods was homogenized with 10 mL of ethanol in PT3100 Polytron (Littau, Switzerland) and then centrifuged for 10 min at 4 °C in J2-21M/E Beckman (Fullerton, CA, USA). The pellet was resuspended in 10 mL 70% methanol in water (*v*/*v*) and centrifuged again. Finally, the supernatant was diluted with 25 mL of 70% methanol. This extract was used to determine the TPC according to the Folin–Ciocalteu procedure [33]. In total, 200 µL of the extract, 1 mL of Folin–Ciocalteu solution (diluted 1:10 in water) and 800 µL of Na_2CO_3 (7.5%) were mixed vigorously, then the mixture was incubated in the dark at room temperature. After 1 h, absorbance at 765 nm was determined on ThermoSpectronic (Thermo Fisher Scientific, Waltham, MA, USA. The quantification of TPC was expressed in Gallic acid equivalents (mg GAE kg^{-1} Fw).

2.2.5. Vitamin C

The reference values for ascorbic acid content (AAC) were obtained using the iodine titration method by means of an automatic Metrohm 862 Titrosampler [34]. In total, 5 g of sample juice was mixed with distilled deionized water until reaching 50 g of final weight, mixing with 2 mL of glyoxal solution (40%). We proceeded to a brief stirring briefly and 5 min of rest. Once 5 mL of sulfuric acid (25%) was added, it was titrated with iodine (0.01 mol L^{-1}) to the end point (EP1). Pure ascorbic acid (AA) was used as an external standard to determine the linearity of the method. For each standard solution, valuations were performed in triplicate. The values of the regression equation and the regression coefficient (r^2 = 0.9998) were obtained. The ascorbic acid content was expressed as mg 100 g^{-1} fresh weight (fw).

2.2.6. Protein Content

The nitrogen (N) content of the dried and ground pod samples was determined by the Kjeldahl method using a distillation apparatus (k314, Büchi Labortechnik GmbH, Essen, Germany) and then converted to protein content by multiplying it by 6.25. The protein content was expressed as g 100 g^{-1} dry weight (dw).

2.3. *Statistical Analysis*

Analysis of variance (ANOVA) was used to compare differences among treatments for total marketable yield. Previously, normality and homoscedasticity were tested using the Shapiro–Wilk and Levenne tests, respectively. For these analyses, Fisher's least significant difference (LSD) test was used to compare the treatments, using the 5% level of significance. Data were analyzed using the Statistical Package for the Social Sciences (SPSS) 24.0 software package (LEAD Technologies, Inc., Chicago, IL, USA).

2.4. *VIS-NIRS Analysis*

Six replicate spectra were recorded for each sample (n = 144) and the average of the spectra was calculated. The samples were lyophilized using freeze-drying equipment (Telstar LyoQuest, Terrassa, Spain), then ground in a mill (Janke & Kunkel, model A10, IKA®-Labortechnik) for about 20 s to pass through a 0.5 mm screen and stored at −80 °C until analysis. The samples were freeze-dried to eliminate the strong absorbance of water in the infrared spectral region, which overlaps with important bands of nutritional compounds that are present in low concentration. Samples were placed in the NIRS sample holder (3 cm diameter) until it was $\frac{3}{4}$ full (weight ≅ 3.50 g) and were scanned (Spectrometer Model 6500 Foss-NIRSystems, Inc., Silver Spring, MD, USA). Their NIR spectra were acquired over a wavelength range from 400 to 2500 nm (VIS + NIR regions) at 2 nm intervals.

Principal component analysis (PCA) was used to detect and remove possible spectral outliers (spectra with a standardized Mahalanobis distance (H) from the mean spectrum of the population greater than 3) [35].

Then, laboratory values were added to the spectra files. The reference values were plotted as the dependent variable and the predicted NIRS values plotted as the independent variable. The raw optical data (as log 1/R, being R = reflectance) or first or second derivatives of the log 1/R data, with several combinations of derivative (gap) sizes and segment (smoothing) were used to develop calibration equations [36,37]. Modified partial least squares was used as regression method to correlate the spectral information (raw optical and the different spectral treatments) of the samples and the quality components. The applied pre-treatments to correct baseline offset due to spectral dispersion effects (differences in particle size between samples) were standard normal variate and detrending (SNV-DT) transformations.

2.5. *Cross-Validation*

Cross-validation is an internal validation method [38] and is useful because all samples can be used to perform the calibration equation without the need to maintain separate calibration sets and validation [39]. The method involves dividing the calibration set into M segments (six) and calibrating M times, each time assessing a different part of the set of calibration (1/M) [40]. This number was proposed by WinISI software (Infrasoft International, Port Matilda, PA, USA), five groups being used as the calibration set and then tested on the remaining samples, performing a validation. This process continued until each group of the six was used as a validation group. WinISI software uses principal component analysis as a tool for selecting samples (spectra) to establish the calibration and validation groups. Thus, both groups comprised samples representative of the whole spectral variability of the population with similar mean and standard deviations for each trait.

Thus, cross-validation was conducted on the calibration set to establish the optimum number of terms to be used in building the calibration equations and to identify spectral (H) or chemical (T) outliers. "T" outliers are samples with high residuals when predicted by the model build in the cross-validation. T values of greater than 2.5 are considered significant and those NIR analyses which have large T values may possibly be outliers. The H outlier identifies a sample that is spectrally different from other samples in the population and has a standardized H value of greater than 3.0. The outlier elimination pass was set to allow the software to remove outliers twice before completing the final calibration [41].

The performances of the different calibration equations obtained were determined from cross-validation. Thus, the prediction ability of the equations obtained for each quality component was determined on the basis of two mathematical relationships, which are the standard error of cross-validation (SECV) [42] to standard deviation (SD) ratio (RPD = relative percent difference).

2.6. External Validation

To evaluate the precision and accuracy of the equations obtained in the calibration models, an external validation procedure in 30 independent samples was completed. Thus, having ordered the sample set by spectral distance using the CENTER algorithm (Winisi), the 30 samples forming the validation set were selected by taking approximately 1 of every 5 samples in the final 144 sample set. The calibration set thus comprised the remaining 114 samples.

The statistical methods applied in this study included the coefficient of determination calculated in cross-validation (R^2 CV) and external validation (R^2 V), the root mean square error of calibration (RMSEC), the root mean square error of cross-validation (RMSECV) and the root mean square error of prediction (RMSEP). Moreover, the ratio of prediction to deviation (RPD), which indicated the correlations between the SD of the standard wet chemical analyzed data and prediction data by NIRS model (RMSECV or RMSEP) [42], was applied to estimate the prediction ability of the model.

NIR models can be classified depending on the R^2 value from the external validation [43] as: models ($0.26 < R^2$ v < 0.49) with a low correlation; models ($0.50 < R^2$ v < 0.64) that can be used to discriminate between low and high values of the samples; models ($0.65 < R^2$ v < 0.81) that can be used for rough predictions of samples; models ($0.82 < R^2$ v < 0.90) with good correlations; and models (R^2 v > 0.90) with excellent precision.

The RPD statistic demonstrates how well the calibration model predicts data. The RPD value >3 is desirable for excellent calibration equations, while equations with an RPD <1.5 are unsuitable, according to the guideline used for defining performance calibrations [43]. With regard to the range error ratio (RER), values in the 4 to 8 range indicate the ability to discriminate between high and low values, and RER values from 8 to 12 establish the ability to predict quantitative data [44,45].

The mathematical expressions of these statistics are as follows:

$$\mathrm{RPD} = SD\left\langle \left[\left(\sum_{i=1}^{n} (y_i - \hat{y}_i)^2 \right) (N - K - 1)^{-1} \right]^{1/2} \right\rangle^{-1}$$

where y_i = lab reference value for the ith sample; $\hat{y}$ = NIR measured value; N = number of samples; K = number of wavelengths used in an equation; and SD = standard deviation.

The coefficient of determination in the cross-validation (R^2):

$$R^2 = \left(\sum_{i=1}^{n} (\hat{y} - \overline{y})^2 \right) \left(\sum_{i=1}^{n} (y_i - \overline{y})^2 \right)^{-1}$$

where $\hat{y}$ = NIR measured value; $\overline{y}$= mean "y" value for all samples; y_i = lab reference value for the ith sample.

$$\mathrm{RER} = range\left\langle \left[\left(\sum_{i=1}^{n} (y_i - \hat{y}_i)^2 \right) (N - K - 1)^{-1} \right]^{1/2} \right\rangle^{-1}$$

where y_i = lab reference value for the ith sample; $\hat{y}$ = NIR measured value; N = number of samples; and K = number of wavelengths used in an equation.

3. Results and Discussion

3.1. Marketable Yield

Figure 4 shows the total marketable yield of the diverse varieties in response to different fertigation treatments which ranged from 0.54 to 2.49 kg m^{-2}. Significant differences were found between the different cultivars of mangetout, the most productive being the varietal types of the indeterminate climbing growth plant, corresponding to the local Landrace (T50 2.49 kg m^{-2}, T100 2.44 kg m^{-2}), AR-24009 (T50 2.22 kg m^{-2}, T100 2.05 kg m^{-2}) and Tirabeque IS (T50 1.76 kg m^{-2}, T100 2.05 kg m^{-2}), followed by the varieties Tirabí (T50 1.57 kg m^{-2}, T100 1.58 kg m^{-2}), Pea Zuccola (T50 1.42 kg m^{-2}, T100 1.34 kg m^{-2}), Capuchino (T50 1.37 kg m^{-2}, T100 1.60 kg m^{-2}), Pea Delikata (T50 1.37 kg m^{-2}, T100 1.34 kg m^{-2}) and lastly the Bamby variety which shows the prostrate growth (T50 0.57 kg m^{-2}, T100 0.54 kg m^{-2}).

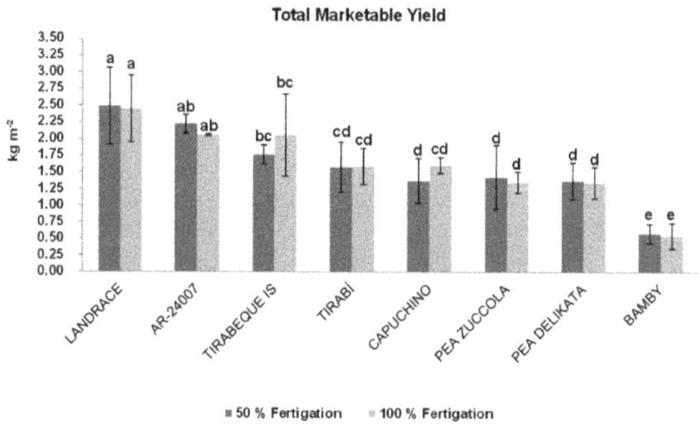

Figure 4. Total marketable yield (kg m^{-2}) of the different varieties of mangetout under different fertigation treatments (T50 and T100). Bars with different lowercase letters were significantly different at $p < 0.05$ (Tukey's multiple range test).

The production data obtained in the field trial for most of cultivars were higher than those described previously in mangetout by García-García [46] (0.55–0.65 kg m^{-2}), and similar to those indicated by Estrada and Ibáñez [47] (1.5–2 kg m^{-2}) in Mediterranean greenhouse conditions.

Increasingly, the use of organic production regulations [48] as well as appropriate fertigation management play an important role in enhancing crop quality and economizing water [13] according to the Sustainable Development Goals (SDGs) by 2030. In this regard, previous studies have showed that yield and quality of snap pods can be significantly affected by different compositions of fertilization [49,50] and by different doses of water in the fertigation solution [13,51]. In order to obtain the highest possible physicochemical variability to develop NIRS predictive models, two fertigation treatments and different mangetout cultivars were used.

3.2. Physicochemical Profiles

The samples analyzed varied in all variables as shown by the range and coefficient of variation (CV) of the calibration set (Table 2). The highest values for the CV were observed for C* chromatic value, firmness, ascorbic acid content and total polyphenol content (>20%), possibly due to the different fertigation treatments and varieties used.

Table 2. Mean, range and standard deviation (n = 144) for quality parameters of the mangetout samples used in this study.

Parameters	Mean	Range	SD
C*	27.87	15.20–35.58	6.15
h*	109.46	105.13–112.91	1.49
Firmness (N)	43.62	20.59–67.52	12.74
TSS (Brix)	7.53	6.08–8.85	0.65
pH	6.80	5.99–7.28	0.27
Protein (g 100 g^{-1} dw)	23.48	11.50–29.75	3.02
AAC (mg 100 g^{-1} fw)	43.82	19.75–68.86	10.82
TPC (mg GAE kg^{-1} fw)	389.09	202.30–685.05	111.52

Based on the results of this study, the chromatic parameters (C* and h*) varied from 15.20 to 35.58 and 105.13 to 112.91, respectively. The h* values correspond to the color green. Green color of fresh pods is one the key factors for deciding the commercial acceptance of snap bean as a fresh vegetable. Similar results have been previously found in snap pods with values ranged from 107 to 111 for h* parameter, but a narrow variation range (27 to 33) for C* chromatic parameter [13,52,53].

Texture is a quality attribute in mangetout fruits very important for consumers since its singular quality of edible crunchy pod is highly appreciated. From our study, the firmness values in mangetout pods ranged from 20.59 to 67.52 N. Although information is lacking for the comparison of firmness with other mangetout cultivars from the literature, our previous research results showed that mangetout "Tirabí" showed values included in the range mentioned above [13].

Vitamin C is essential in both plants and animals. The main suppliers of this vitamin in the diet are fruits and vegetables [54]. Legumes are considered an important source of vitamins, especially rich in ascorbic acid content in the pods [55]. Considerable variation was found for AAC which ranged from 19.75 to 68.86 mg AA 100 g^{-1} fw in mangetout pods. Previous studies revealed AA content within the range of AAC showed in this work for three pea varieties (26 to 31 mg AA 100 g^{-1}) [56]. Our findings are also in agreement with those of Rickman et al. [57] and Avilés and Cruz [58], who described AA values of 40 and 27 mg 100 g^{-1} fw in peas and pea pods, respectively. Mangetout pods can be considered a rich source of vitamin C, since orange and lemon contain 30–50 mg of ascorbic acid 100 g^{-1} fw [54].

The pH of foods is an important parameter related to the taste perceived by consumers. In our study, the pH ranged from 5.99 to 8.85. The values obtained agree with previous studies on legumes [59–61], but lower than those obtained in French bean pods (5.84–5.96) by Segura et al. [62].

TSS is another taste quality determinant [63], and cultivars with higher TSS have higher taste quality. Mangetout pods are rich in TSS content (6.29–8.83 Brix) in comparison with other legume pods; thus, cowpea accessions from different Mediterranean countries showed lower sweetness (range 5.07–7.57 Brix) [55] in relation to our results.

On the other hand, the fresh mangetout pod TSS content in our study was lower compared to those previously reported in the scientific literature [12]. This previous work revealed that the TSS of fresh pods ranged from 9.1 to 11.3 under specific fertigation treatments demonstrating that the environmental factors such as available water had a highly significant effect on this quality parameter. According to the Brix reference values of the main greenhouse vegetables, the mangetout pods analyzed showed a higher sweetness than California green pepper fruits (4.03–6.31 Brix) and similar to red California pepper fruits (7.37–8.85 Brix) [64].

The presence of polyphenols in plants is very varied, depending on the plant species, variety, part of the plant, growing conditions, etc. More than 8000 phenolic compounds with a very varied structure have been identified from simple molecules, such as phenolic acids and complex polymers of high molecular mass such as tannins [65]. In our study, mangetout exhibited higher total polyphenol content (202.30 to 685.05 mg GAE kg^{-1} fw) that those reported for other snap pods, such as the French bean with 300 mg GAE kg^{-1} fw [46]. Our results agree with those of Devi et al. [66] who found a wide variation range (126.3–1286.3 mg GAE kg^{-1} fw) in pea pods from 22 different genotypes. On the other hand, the consumer increasingly appreciates fruits with antioxidant properties due to the health benefits. A source of phenolic compounds is identified as a chemopreventive agent since it eliminates free radicals and has a preventive effect on degenerative diseases, among others [67]. Mangetout pods have a high potential to be used in the development of functional foods or nutraceutical products and unlike pea pods they would not require any processing as the whole pod is edible.

Our results showed a wide variability for protein content (11.50–29.75 g 100 g^{-1} dw) and our results agree with those of Hood-Niefer et al. [68] (24.4 to 27.5 g 100 g^{-1} dw), but are higher than the results obtained by Mateos-Aparicio et al. [69] (10.8 $\pm$ 0.3 g 100 g^{-1} dw) in pea pods. Overall, in the pea, both the seeds (20.5–22.6%) and pods (13.37%) are a rich source of protein [70]. A diet rich in vegetable protein is increasingly important nowadays due to its health benefits and thus it is recommended that people reduce their consumption of animal protein. In addition, pea pods have protein-denaturing properties that show anti-inflammatory effects and anti-cholinesterase activity because of the strong antidiabetic properties of peas [71].

3.3. VIS-NIRS Analysis

3.3.1. Raw Spectra on Mangetout

Raw spectra of the calibration set samples are shown in Figure 5. A remarkable variability in the VIS region (400–850 nm) absorbance spectra was observed because of pigments. The peak around 640–700 nm illustrated the color transition of pea pods correlated with the chlorophyll content that absorbs radiation in this region [72].

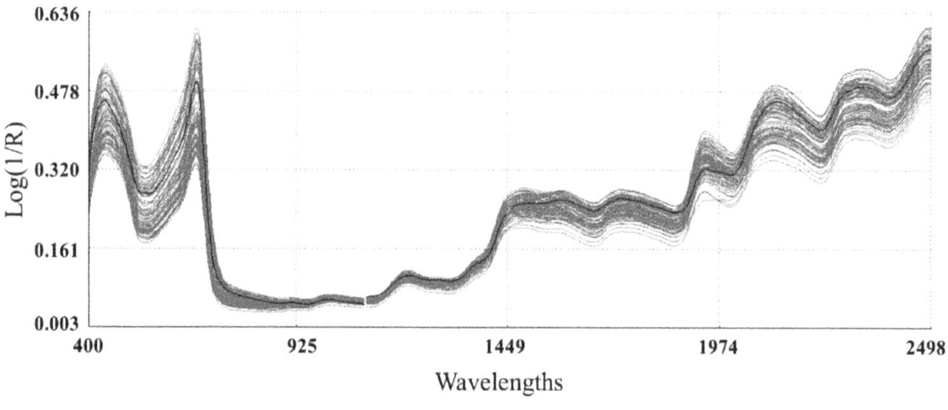

Figure 5. Raw spectra for dried mangetout samples.

Absorption bands in the region from 1300 to 2000 nm have been assigned to the third overtones of C-N (amines); C=O (ketones, amino acids); and C-O (long-chain fatty acids, phenols). From 2200 to 2400 nm, absorptions of C-N (primary amines) and C-O (alcohols) have been assigned to the third overtones of these compounds, while in the same region, C-H (asymmetrical deformation) and C-O (symmetrical vibrations) have been assigned to the second overtones of these molecules. Finally, the second overtones of C-H deformation and C-N (amides) have been reported in the 2400–2500 nm region [73].

3.3.2. Second Derivative Spectra of Mangetout

The second derivative and SNT-DT (standard normal variate and de-trending) algorithms to the raw spectra led to a substantial correction (Figure 6) of the baseline shift produced by differences in path length and particle size. The increase in the complexity of the derivative spectra resulted in a clear separation between peaks which overlap in the raw spectra.

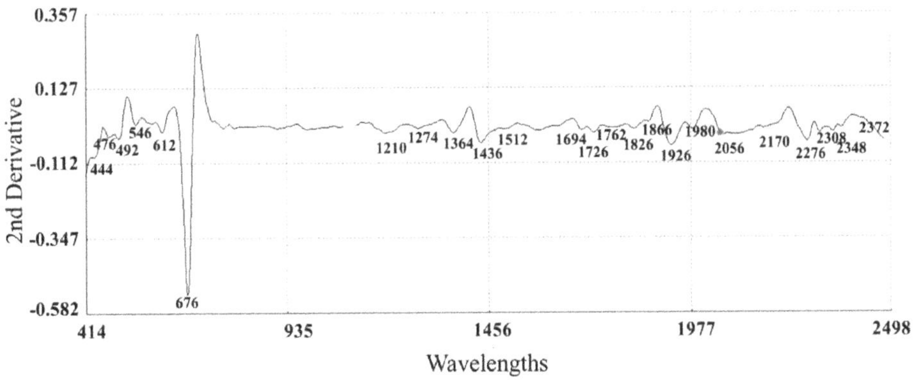

Figure 6. Second derivative spectra (2, 5, 5, 2; SNV + DT) of the raw optical data for mangetout samples in the range of 400 to 2500 nm, together with the most relevant absorption bands.

Absorption maxima bands (λmax) were observed between 400 and 700 nm (at 444, 546 nm and 670 nm) in the spectra attributed to pod pigments that absorb in visible region (Figure 6). From all pigments that can be found in plants, chlorophylls are used for photosynthesis ("a" and "b"), which absorb preferentially violet-blue light (400–500 nm) and red light (600–700 nm), respectively [74].

Pigment–protein complex molecules could be responsible for some of the traits that determine the VIS region at longer wavelengths. Thus, binding proteins in chlorophyll *a/b* absorb in the 498–568 nm range [74] and red absorbing pigments, particularly chlorophyll, give the fruit its green color [75,76].

In the region NIRS of the spectra, peaks at 1208 nm (attributed to a CH second overtone), 1726 nm and 1762 nm (assigned to CH first overtone), 2308 nm and 2348 nm (attributed to CH stretch and deformation in a CH_2 group) were detected which are related to lipids [77,78]. Other peaks located at 1210 nm corresponded to absorption by OH groups in carbohydrates [79,80]. Other peaks at 1512, 2056 and 2174 nm related to protein, specifically to NH stretch, NH stretch and amide II, and amide I and amide III, respectively [78]. The last significant peaks were observed at 1436 and 2270 nm, these wavelengths corresponding to the deformation of the OH + CO cellulose groups [79].

3.3.3. Calibration Development

Tables 3 and 4 show the summary of the statistics obtained from calibration, cross-validation and external validation models in mangetout samples, respectively. The full available visible region and near-infrared region (400–2500 nm) were used.

The coefficients of determination (R^2) achieved in calibration were higher than those found in external validation models for mangetout, as expected. The coefficient of determination for cross-validation (R^2 CV), oscillated between 0.55 for pH to 0.92 for protein (Table 3), whereas RPDcv values ranges from 1.50 for pH to 3.45 for protein.

Based on the R^2 values of the external validation, the models were as follows [39]: models that can be used to discriminate between low and high values of the samples (0.50 m< R^2 < 0.64), in our work the models developed for AAC and TSS; models that can be used for rough predictions of samples (0.65 < R^2 < 0.81), in our case the calibrations achieved for C* and h* color parameters, firmness and pH; and models with good correlations (0.83 < R^2 < 0.90), these values corresponding with models obtained for total polyphenol content and protein.

The SEP values of the validation were lower than their respective SD, which indicates that NIRS is able to determine these traits in mangetout.

Table 3. Calibration and cross-validation statistics of quality compounds for mangetout.

Parameters	Range	[1] SD	[2] R^2	[3] SEC	[4] R^2 CV	[5] SECV	[6] RPDcv	[7] Treatment	[8] Cv
C*	15.20–35.58	6.35	0.87	2.24	0.81	2.78	2.28	2,5,5,2	0.22
h*	106.41–112.10	1.41	0.80	0.62	0.71	0.75	1.88	1,4,4,1	0.01
Firmness (N)	21.75–67.52	10.09	0.71	5.46	0.71	5.93	1.70	1,4,4,1	0.21
[9] TSS (Brix)	6.29–8.83	0.65	0.93	0.18	0.68	0.39	1.66	1,4,4,1	0.08
pH	6.01–7.28	0.27	0.60	0.17	0.55	0.18	1.50	1,4,4,1	0.04
Protein (g 100 g^{-1} dw)	15.69–29.75	2.80	0.97	0.48	0.92	0.81	3.45	2,5,5,2	0.13
[10] AAC (mg 100 g^{-1} fw)	19.75–64.40	10.89	0.79	5.02	0.56	7.16	1.52	1,4,4,1	0.24
[11] TPC (mg GAE kg^{-1} fw)	239.28–670.30	101.91	0.93	27.01	0.86	39.08	2.61	1,4,4,1	0.28

[1] SD: Standard deviation; [2] R^2: Coefficient of determination in calibration; [3] SEC: Standard error in calibration; [4] R^2: Coefficient of determination in cross-validation; [5] SECV: Standard error of cross-validation; [6] RPDcv: Ratio of the standard deviation to standard error of cross-validation; [7] Mathematical treatment; [8] Coefficient of variation; [9] TSS: Total soluble solids; [10] AAC: Ascorbic acid content; [11] TPC: Total polyphenol content.

Table 4. Reference values and external validation statistics of the NIRS calibrations for quality compounds in mangetout.

	Reference Values (n = 30)			External Validation			
Parameters	Range	Mean	[1] SD	[2] Rv2	[3] SEP	[4] RPDp	[5] RER
C*	15.20–34.89	25.50	7.33	0.78	3.34	2.19	5.89
H*	107.40–111.71	109.71	1.24	0.68	0.56	2.00	6.95
Firmness (N)	24.45–67.20	40.48	12.51	0.65	7.34	1.70	5.96
[6] TSS (Brix)	6.29–8.76	7.54	0.69	0.52	0.51	1.35	4.84
pH	6.22–7.20	6.83	0.22	0.50	0.14	1.57	7.00
Protein (g 100 g^{-1} dw)	17.22–29.5	24.95	2.18	0.88	0.68	3.20	14.89
[7] AAC (mg 100 g^{-1} fw)	22.71–63.47	45.69	8.82	0.50	8.82	1.50	7.03
[8] TPC (mg GAE kg^{-1} fw)	250.89–570.21	360.89	80.37	0.84	29.46	2.72	10.84

TSS: Total soluble solids; AAC: Ascorbic acid content; TPC: Total polyphenol content; [1] SD: Standard deviation; [2] Rv^2: Coefficient of determination in external validation; [3] SEP: Standard error of prediction corrected for bias; [4] RPDp: Ratio of the standard deviation to standard error of prediction (performance); [5] RER: Ratio of the range to standard error of prediction (performance); [6] TSS: Total soluble solids; [7] AAC: Ascorbic acid content; [8] TPC: Total polyphenol content.

According to the guideline used for defining performance calibrations [43] when this ratio is greater than 3, the calibration equation is very significant, and this was reached in our study for protein content; if RPD values range between 2.5 < RPD < 3, predictive models are considered very good, in our case corresponding to the TPC model; while RPD range between 1.5 < RPD < 2.5 predictive models are appropriate for screening purposes, which was achieved for AAC, pH, firmness and C* and h* color parameter models.

Figure 7 shows the relationship between the predicted reflectance spectroscopy in the near infrared (NIRS) and reference values for all parameters (color parameters (chroma * and hue angle), firmness, total soluble solids, pH, protein content, ascorbic acid and total polyphenol content) in the mangetout validation set samples.

In reference to RER (ratio of the range to standard error of prediction) coefficients, this dimensionless parameter is also used to evaluate the predictive ability of NIRS equations, in this work ranged from 4.84–14.89.

Prediction models for C*, h*, firmness, TSS, pH, AAC showed RER values within the range from 4 to 8, which suggest the possibility of discriminating between high and low values; while RER values in the range of 8 to 12 represent the possibility of predicting quantitative data [44,45] which was achieved for protein and TPC predictive models.

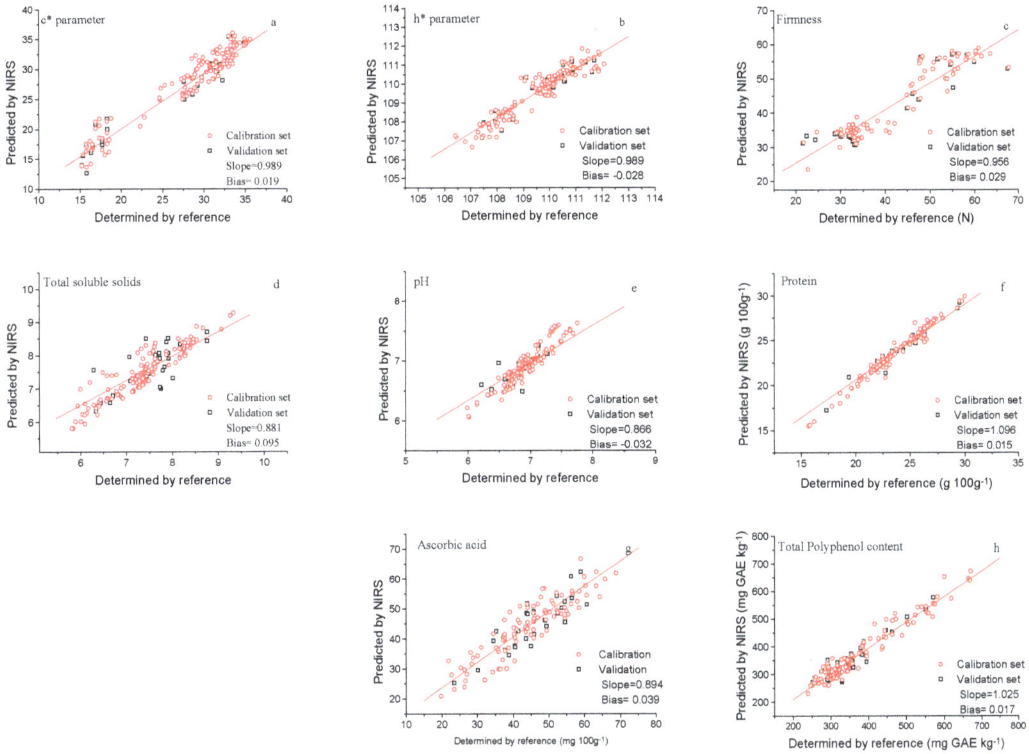

Figure 7. Predicted versus reference values for calibration and external validation for all parameters: c* (**a**); h* (**b**); firmness (**c**); total soluble solids (**d**); pH (**e**); protein content (**f**); ascorbic acid (**g**); total polyphenol content in the mangetout (**h**).

Previous works have demonstrated the validity of the NIRS technique in evaluating the accuracy of pea single seed protein with R^2 = 0.94 and RPD = 3.7 in external validation [28], and also for predicting soybean single seed protein content with $Rval^2$ = 0.84 and RPDval = 2.28 values [81].

The estimation of protein and total polyphenol content in common beans (*Phaseolus vulgaris* L.) by NIRS has also been previously assayed by several authors reaching significantly good results in general. Thus, the high R^2 obtained ranged from 0.91–0.94 and RPD values above 3.5 [82–85]. Other authors supported the validity of the NIRS technique in similar approaches, with R^2 and RPD values for firmness of 0.61 and 1.7, respectively, in soybean single seed [82]. Wang [86] used NIRS to predict the total polyphenol content in ground faba bean (*Vicia faba* L.), with an R^2 of 0.79, RMSECV of 0.40 and RPD of 2.20, and also for the determination of protein in ground faba bean seed powder with an R^2 of 0.94.

It should be noted that the prediction accuracies in all of the above-mentioned studies were comparable to those reported for mangetout in this study. To our knowledge, this is the first article dealing with the use of NIRS to predict pod quality traits in mangetout.

Modified PLS regression was employed to reduce the spectral information of the mangetout samples by creating a much smaller number of new orthogonal variables (factors) which retain the essential information needed to predict the composition of the samples.

3.3.4. Modified Partial Least Squares Loadings for Quality Equations

The scores of the best models for all quality parameters were plotted by their first MPLS loadings (Figure 8) to identify those areas within the spectral range where variance had influenced the model fitting, to a lesser or greater degree, as well as the direction (negative or positive).

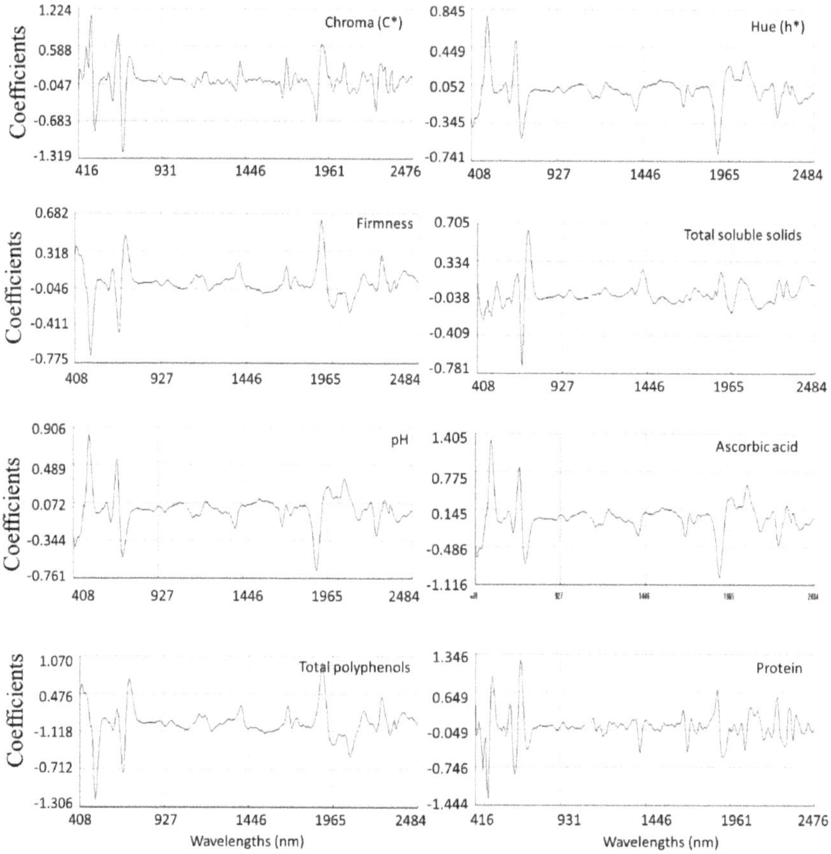

Figure 8. Modified partial least squares (MPLS) loading of the optimal calibrations for physicochemical compounds measured by NIRS.

The region of the spectrum which most influenced the fitting of the model was the visible segment between 480 to 700 nm. Thus, the contribution of chlorophyll (672 nm) showed the highest weight on first MPLS loading [75] (Figure 8). Other chromophores absorbing at 496 and 512 nm also participated in the equations. With respect to NIR region, previous studies have shown the contribution of this region to predict color parameters for species such as fresh *Ginkgo biloba* leaves [87], green-leafy species [88] or *Sassafras tzumu* [89]. Some plant chemical compounds (e.g., phenolics and flavonoids) respond to the stress and environmental changes and correlate in a secondary way with the color parameters. The characteristic bands of phenolics and flavonoids can be detected in wavelength regions from 1415 nm to 1512 nm, 1650 to 1750 nm and from 1955 to 2035 nm in the MPLS loadings for the color parameters (Figure 8) [90]. Furthermore, the color is caused by the reflection of helicoidally stacked cellulose microfibrils that form multilayers in the cell walls of the epicarp [91]. Thus, the wavelengths at 1932 nm (O-H stretching plus O-H deformation) could be related to the cellulose of the pod tissues which can be observed in the MPLS loadings of the optimal calibrations for the color parameters (Figure 8). Others main NIR contributions were those at 2284 nm (C-H stretching plus C-H deformation), 2300 nm (stretching–bending of CH—CH2 bonds and C—O bonds) and 2348 nm (C-H combination of methylene groups) [77]. In addition, absorption bands in the NIR region that influenced the fitting of the models were found at 1212, 1388, 1412 and 1990 nm (associated with glucides and water absorptions), and the region around 2072 nm (N—H bonds) associated with protein.

4. Conclusions

This work has showed that genetic variability exists for the quality parameters analyzed in mangetout cultivars. Many of the traits analyzed are of economic interest (color, firmness, protein content and antioxidant compounds). These new understandings could be useful in selecting parents for breeding programs aimed at enhancing physicochemical parameters that respond to the new trends market.

Moreover, the result of the present investigation explores the potential of NIRS to simultaneously determine eight quality traits in mangetout, as an alternative to reference methods. The measurements with the reference methods of most of these parameters are expensive, have laborious protocols and require a long analysis time. Utilizing NIRS, every 2 min, we can analyze all the quality parameters of a sample. The results reveal that the models allow an accurate quantification of protein and TPC and a rough screening method of the samples for color parameters (c^* and h^*), firmness, AAC and pH.

The inclusion future of mangetout cultivars from different geographical origins and segregant populations in the calibration models will allow us to increase the robustness of the equations for these parameters.

The performance of the calibration model for TSS was lower than that obtained for the other quality parameters in this work. The low variability among mangetout cultivars used in this work (6.29–8.76 Brix) could be based on the lower accuracy of the calibration model for TSS. An increase in both the number of samples and trait variation can be crucial factors for improving the accuracy of this calibration model.

It is interesting to focus attention on firmness (shear force). Pod firmness is an excellent indicator of pod quality, but its quantification is time consuming and not easily measured. Pods must be harvested before they become tough and develop poor culinary acceptance, even if it means sacrificing maximum yield. From this point of view, the use of NIRS instead of a texturometer could be clearly advantageous.

Spectral ranges associated with the absorbance of chromophores, carbohydrates, water and protein were used by MPLS regression for the model fitting of quality equations in mangetout.

Author Contributions: M.d.R.-C. and J.A.F. conceived and designed the experiments; B.d.C.M.-G. and E.M.-E. performed the field trials and laboratory analysis; J.L.V. and P.G. performed supervision; I.F.: MPLS regression analysis; M.d.C.G.-G.: NIRS analysis and wrote the paper. All authors have read and agreed to the published version of the manuscript.

Funding: This study was carried out with the support of the Projects 'Innovación en cultivos protegidos: biodiversidad y bioeconomía aplicada' (PP.TRA.TRA2019.003) and 'Nuevas alternativas de cultivos a la producción hortícola protegida' (PP.AVA.AVA2019.023). FEDER cofund 80% 'Andalucía se mueve con Europa'.

Institutional Review Board Statement: Not applicable.

Informed Consent Statement: Not applicable.

Data Availability Statement: Data sharing not applicable.

Conflicts of Interest: The authors declare no conflict of interest.

References

1. Rivero-Pino, F.; Espejo-Carpio, F.J.; Guadix, E.M. Identification of dipeptidyl peptidase IV (DPP-IV) inhibitory peptides from vegetable protein sources. *Food Chem.* **2021**, *354*, 129473. [CrossRef] [PubMed]
2. Semba, R.D.; Ramsing, R.; Rahman, N.; Kraemer, K.; Bloem, M.W. Legumes as a sustainable source of protein in human diets. *Glob. Food Secur.* **2021**, *28*, 100520. [CrossRef]
3. Andreoli, V.; Bagliani, M.; Corsi, A.; Frontuto, V. Drivers of Protein Consumption: A Cross-Country Analysis. *Sustainability* **2021**, *13*, 7399. [CrossRef]
4. Quesada, D.; Gómez, G. ¿Proteínas de origen vegetal o de origen animal?: Una mirada a su impacto sobre la salud y el medio ambiente. *Nutr. Clin. Metab.* **2019**, *2*, 79–86. [CrossRef]
5. Henchion, M.; De Backer, C.J.S.; Hudders, L. Ethical and sustainable aspects of meat production; consumer perceptions and system credibility. In *New Aspects of Meat Quality*; Woodhead Publishing: Sawston, UK, 2017; pp. 649–666.
6. Organización Mundial de la Salud. Informe de Expertos Independientes Sobre Dieta, Nutrición y Prevención de Enfermedades Crónicas. OMS. Available online: http://apps.who.int/iris/bitstream/handle/10665/42755/WHO_TRS_916_spa.pdf;jsessionid=317C195443FD49BB85BA3FDE262C4AF7?sequence=1 (accessed on 27 March 2022).
7. Leonard, E. Cultivating good health. In *Grains and Legumes Nutrition Council*; Cadillac: Adelaide, Australia, 2012; pp. 5–15. ISSN 1039-6217.
8. Maphosa, Y.; Jideani, V.A. The role of legumes in human nutrition. *Funct. Food-Improv. Health Through Adequate Food* **2017**, *1*, 13.
9. Salazar Cubides, C.A. Prototipo Ideal del Consumidor Verde. 2018. Available online: https://repositoriocrai.ucompensar.edu.co/bitstream/handle/compensar/2607/PFG_Prototipoidealdelconsumidorverde.pdf?sequence=1&isAllowed=y (accessed on 27 March 2022).
10. Troncoso-Pantoja, C. Comidas tradicionales y alimentación saludable: El ejemplo de la dieta mediterránea. *Horiz. Méd. Lima* **2019**, *19*, 72–77. [CrossRef]
11. Servicio de Estudios Agroalimentarios de Cajamar. *Análisis de la Campaña Hortofrutícola de Almería. Campaña 2020/2021*; Cajamar Caja Rural: Almeria, Spain, 2021.
12. García-García, M.C.; del Río Celestino, M.; Gil-Izquierdo, Á.; Egea-Gilabert, C.; Galano, J.M.; Durand, T.; Oger, C.; Fernández, J.; Ferreres, F.; Domínguez-Perles, R. The value of legume foods as a dietary source of phytoprostanes and phytofurans is dependent on species, variety, and growing conditions. *Eur. J. Lipid Sci. Technol.* **2019**, *121*, 1800484. [CrossRef]
13. García-García, M.D.C.; Font, R.; Gómez, P.; Valenzuela, J.L.; Fernández, J.A.; Río-Celestino, M. Effects of Fertigation Management on the Quality of Organic Legumes Grown in Protected Cultivation. *Horticulturae* **2021**, *7*, 28. [CrossRef]
14. El-Seifi, S.K.; Hassan, M.A.; El-Bassiouny, R.E.I.; Elwan, M.W.M.; Nasef, I.N. Effect of maturity stage on physical and chemical characteristics and determination of harvest time of sugar snap pea pods. *J. Plant Prod.* **2014**, *5*, 305–314. [CrossRef]
15. Nicolaï, B.M.; Beullens, K.; Bobelyn, E.; Peirs, A.; Saeys, W.; Theron, K.I.; Lammertyn, J. Nondestructive measurement of fruit and vegetable quality by means of NIR spectroscopy: A review. *Postharvest Biol. Technol.* **2007**, *46*, 99–118. [CrossRef]
16. Toledo-Martín, E.M.; García-García, M.C.; Font, R.; Rojas, J.M.M.; Gómez, P.; Navarro, M.S.; Del Río-Celestino, M. Application of visible/near-infrared reflectance spectroscopy for predicting internal and external quality in pepper. *J. Sci. Food Agric.* **2015**, *96*, 3114–3125. [CrossRef] [PubMed]
17. Shepherd, K.D.; Walsh, M.G. Infrared Spectroscopy-Enabling an Evidence-Based Diagnostic Surveillance Approach to Agricultural and Environmental Management in Developing Countries. *J. Near Infrared Spectrosc.* **2007**, *15*, 1–19. [CrossRef]
18. Malley, D.F.; Williams, P.C.; Hauser, B.; Hall, J. Prediction of organic C, N, and P in freshwater sediments using near-infrared reflectance spectroscopy. In *Near Infrared Spectroscopy: The Future Waves*; Davies, A.M.C., Williams, P.C., Eds.; NIR Publications: Chichester, UK, 1996; pp. 691–699.
19. Martínez-Valdivieso, D.; Font, R.; Fernández-Bedmar, Z.; Merinas-Amo, T.; Gómez, P.; Alonso-Moraga, Á.; Del Río-Celestino, M. Role of Zucchini and Its Distinctive Components in the Modulation of Degenerative Processes: Genotoxicity, Anti-Genotoxicity, Cytotoxicity and Apoptotic Effects. *Nutrients* **2017**, *9*, 755. [CrossRef] [PubMed]
20. Pomares-Viciana, T.; Martínez-Valdivieso, D.; Font, R.; Gómez, P.; Del Río-Celestino, M. Characterisation and prediction of carbohydrate content in zucchini fruit using near infrared spectroscopy. *J. Sci. Food Agric.* **2017**, *98*, 1703–1711. [CrossRef]
21. Toledo-Martín, E.M.; Font, R.; Obregón-Cano, S.; De Haro-Bailón, A.; Villatoro-Pulido, M.; Del Río-Celestino, M. Rapid and Cost-Effective Quantification of Glucosinolates and Total Phenolic Content in Rocket Leaves by Visible/Near-Infrared Spectroscopy. *Molecules* **2017**, *22*, 851. [CrossRef]
22. Toledo-Martín, E.M.; García-García, M.C.; Font, R.; Moreno-Rojas, J.M.; Salinas-Navarro, M.; Gómez, P.; Del Río-Celestino, M. Quantification of Total Phenolic and Carotenoid Content in Blackberries (*Rubus Fructicosus* L.) Using Near Infrared Spectroscopy (NIRS) and Multivariate Analysis. *Molecules* **2018**, *23*, 3191. [CrossRef]
23. Martínez-Valdivieso, D.; Font, R.; Del Río-Celestino, M. Prediction of Agro-Morphological and Nutritional Traits in Ethiopian Mustard Leaves (*Brassica Carinata* A. Braun) by Visible-Near-Infrared Spectroscopy. *Foods* **2019**, *8*, 6. [CrossRef]
24. Ozaki, Y. Near-infrared spectroscopy-its versatility in analytical chemistry. *Anal. Sci.* **2012**, *28*, 545–563. [CrossRef]
25. Hang, J. Genome-Wide Association Study of Seed Protein and Amino acid Contents in Cultivated Lentils as Determined by Near-Infrared Reflectance Spectroscopy. 2022. Available online: http://hdl.handle.net/1993/36250 (accessed on 27 March 2022).

26. Font, R.; del Río-Celestino, M.; Luna, D.; Gil, J.; de Haro-Bailón, A. Rapid and Cost-Effective Assessment of the Neutral and Acid Detergent Fiber Fractions of Chickpea (*Cicer arietinum* L.) by Combining Modified PLS and Visible with Near-Infrared Spectroscopy. *Agronomy* **2021**, *11*, 666. [CrossRef]
27. Jung, C.S.; Kim, B.J.; Kwon, Y.C.; Han, W.Y.; Kwack, Y.H. Analysis of Protein and Moisture Contents in Pea (*Pisum sativum* L) Using Near-Infrared Reflectance Spectroscopy. *Korean J. Crop Sci.* **1998**, *43*, 101–104.
28. Hacisalihoglu, G.; Freeman, J.; Armstrong, P.R.; Seabourn, B.W.; Porter, L.D.; Settles, A.M.; Gustin, J.L. Protein, weight, and oil prediction by single-seed near-infrared spectroscopy for selection of seed quality and yield traits in pea (*Pisum sativum*). *J. Sci. Food Agric.* **2020**, *100*, 3488–3497. [CrossRef] [PubMed]
29. Martens, M.; Martens, H. Near-infrared reflectance determination of sensory quality of peas. *Appl. Spectrosc.* **1986**, *40*, 303–310. [CrossRef]
30. Digman, M.F.; Runge, W.M. The utility of a near-infrared spectrometer to predict the maturity of green peas (*Pisum sativum*). *Comput. Electron. Agric.* **2021**, *193*, 106643. [CrossRef]
31. Shenk, J.S.; Westerhaus, M.O. Population structuring of near infrared spectra and modified partial least squares regression. *Crop Sci.* **1991**, *31*, 1548–1555. [CrossRef]
32. Europea, C. Reglamento (CE) nº 889/2008 de la Comisión, de 5 de septiembre de 2008, por el que se establecen disposiciones de aplicación del Reglamento (CE) nº 834/2007 del Consejo sobre producción y etiquetado de los productos ecológicos, con respecto a la producción ecológica, su etiquetado y su control. *D. Of. Unión Eur.* **2008**, *18*, 1–84.
33. Kähkönen, M.P.; Hopia, A.I.; Vuorela, H.J.; Rauha, J.P.; Pihlaja, K.; Kujala, T.S. Antioxidant activity of plant extracts containing phenolic compounds. *J. Agric. Food Chem.* **1999**, *47*, 3954–3962. [CrossRef]
34. Suntornsuk, L.; Gritsanapun, W.; Nilkamhank, S.; Paochom, A. Quantitation of vitamin C content in herbal juice using direct titration. *J. Pharm. Biomed. Anal.* **2001**, *28*, 849–855. [CrossRef]
35. Ferrer-Gallego, R.; Hernández-Hierro, J.M.; Rivas-Gonzalo, J.C.; Escribano-Bailón, M.T. A comparative study to distinguish the vineyard of origin by NIRS using entire grapes, skins and seeds. *J. Sci. Food Agric. 2013 93*, 967–972. [CrossRef]
36. Hruschka, W.R. *Data analysis: Wavelength selection methods, In Near-Infrared Technology in the Agricultural Food Industries*; Williams, P., Norris, K., Eds.; American Association of Cereal Chemists, Inc.: St. Paul, MN, USA, 1987; pp. 35–55.
37. Shenk, J.S.; Workman, J.J.; Westerhaus, M.O. Application of NIR spectroscopy to agricultural products. In *Handbook of NearInfrared Analysis*; Burns, D.A., Ciurczak, E.W., Eds.; Marcel Dekker: New York, NY, USA, 1992; pp. 383–431.
38. Stone, M.; Roy, J. The linear complementarity problem. *Stat. Soc.* **1974**, *B*, 111–133.
39. Foley, W.J.; McIlwe, A.; Lawler, I.; Aragones, L.; Woolnough, A.P.; Berding, N. Ecological applications of near infrared reflectance spectroscopy—A tool for rapid, cost-effective prediction of the composition of plant and animal tissues and aspects of animal performance. *Oecologia* **1998**, *116*, 293–305. [CrossRef]
40. Martens, H.; Naes, T. *Multivariate Calibration*; John Wiley & Sons: Hoboken, NJ, USA, 1992; p. 440.
41. *NIRS 2, Routine Analysis Manual. NIR Systems User's Manual*; Infrasoft International: Port Matilda, PA, USA, 1995.
42. Shenk, J.S.; Westerhaus, M.O. Calibration the ISI way. In *Near Infrared Spectroscopy: The Future Waves*; Davies, A.M.C., Williams, P.C., Eds.; NIR Publications: Chichester, UK, 1996; pp. 198–202.
43. Williams, P.; Norris, K. *Near-Infrared Technology in the Agricultural and Food Industries*; Minnesota American Association: St. Paul, MN, USA, 2001.
44. Barbin, D.F.; Kaminishikawahara, C.M.; Lourenco Soares, A.; Mizubuti, I.Y.; Grespan, M.; Shimokomaki, M.; Hirook, E.Y. Prediction of chicken quality attributes by near infrared spectroscopy. *Food Chem.* **2015**, *168*, 554–560. [CrossRef] [PubMed]
45. Millmier, A.; Lorimor, J.; Hurburgh, C.; Fulhage, C.; Hattey, J.; Zhang, H. Near infrared sensing of manure nutrients. *Trans. ASAE* **2000**, *43*, 903–908. [CrossRef]
46. García-García, M.C. Reducción de la Fertirrigación en Cultivo Ecológico de Leguminosas Protegidas y su Efecto Sobre la Cosecha. Master's Thesis, Universidad de Almería, Almería, Spain, 2017.
47. Estrada, J.M.; Ibáñez, J. Cultivo del tirabeque. *Hortic* **1985**, *2019*, 11–23.
48. Regulation (EU) 2018/848 of the European Parliament and of the Council of 30 May 2018 on organic production and labelling of organic products and repealing Council Regulation (EC) No 834/2007. *Off. J. Eur. Union* **2018**, *150*, 1–92.
49. Salinas-Ramírez, N.; Escalante-Estrada, J.; Rodríguez-González, M.; Sosa-Montes, E. Yield and nutritional quality of snap bean in terms of biofertilization. *Trop. Subtrop. Agroecosyst.* **2011**, *13*, 347–355.
50. Tantawy, A.; Abdel-Mawgoud, A.M.R.; Habib, H.; Hafez, M. Growth, Productivity and Pod Quality Responses of Green Bean Plants *Phaseolus vulgaris* to Foliar Application of Nutrients and Pollen Extracts. *Res. J. Agric. Biol. Sci.* **2009**, *5*, 1032–1038.
51. Mattar, M.A.; Soliman, S.S.; Al-Obeed, R.S. Effects of various quantities of three irrigation water types on yield and fruit quality of 'succary'date palm. *Agronomy* **2021**, *11*, 796. [CrossRef]
52. Ram, H.; Hedau, N.K.; Chaudhari, G.V.; Kant, L. Peas with zero shelling edible pods: A review. *Sci. Hortic.* **2021**, *288*, 110333. [CrossRef]
53. Myers, J.R.; Bagget, J.R. Improvement of snap beans. In *Common Bean Improvement in the Twenty-First Century*; Singh, S.P., Ed.; Springer: Dordrecht, The Netherlands, 1999; pp. 289–329.
54. Rucker, R.B.; Zempleni, J.; Suttie, J.W.; McCormick, D.B. *Handbook of Vitamins*, 4th ed.; CRC Press: Boca Raton, FL, USA; Taylor & Francis Group: New York, NY, USA, 2007; pp. 489–520.

55. Ntatsi, G.; Gutiérrez-Cortines, M.E.; Karapanos, I.; Barros, A.; Weiss, J.; Balliu, A.; dos Santos Rosa, E.A.; Savvas, D. The quality of leguminous vegetables as influenced by preharvest factors. *Sci. Hortic.* **2018**, *232*, 191–205. [CrossRef]
56. Favell, D.J. A comparison of the vitamin C content of fresh and frozen vegetables. *Food Chem.* **1998**, *62*, 59–64. [CrossRef]
57. Rickman, J.C.; Barrett, D.M.; Bruhn, C.M. Nutritional comparison of fresh, frozen and canned fruits and vegetables. Part 1. Vitamins C and B and phenolic compounds. *J. Sci. Food Agric.* **2007**, *87*, 930–944. [CrossRef]
58. Avilés García, M.J. & Cruz Rodríguez, J.A. Estudio Químico y Actividad Antioxidante de la Vaina de Arveja (*Pisum sativum*). Ph.D. Thesis, Universidad de Guayaquil, Facultad de Ciencias Químicas, Guayaquil, Ecuador, 2020; pp. 38–39.
59. Khah, M.E.; Arvanitoyannis, S.I. Yield, nutrient content and physico-chemical and organoleptic properties en green bean are affected by N:K ratios. *Food Agric. Environ.* **2003**, *1*, 17–26.
60. Dawo, M.; Wilkindon, J.M.; Sanders, F.; Pilbean, D.J. The yield and quality of fresh and ensiled plant material from intercropped maize (Zea mays) and beans (Phaseolus vulgaris). *J. Sci. Food Agric.* **2007**, *87*, 1391–1399. [CrossRef]
61. Fernández, N.E.P.; López, G.P.; Domínguez, C.R.; Peñuelas, V.M.L.; Izaguirre, S.C.O. Composición química, características funcionales y capacidad antioxidante de formulaciones de garbanzo (*Cicer arietinum* L.) Blanco Sinaloa 92. *Agrociencia* **2019**, *53*, 35–44.
62. Segura, M.L.; Contreras, J.I.; García, I.I.; García, M.C.; Cuadrado, I.M. Fertilización nitrogenada de judía verde bajo invernadero con criterios agroecológicos. In *VII Congreso SEAE*; Sociedad Española de Agricultura Ecológica: Catarroja, Spain, 2006.
63. Champa, W.A.H.; Weerakkody, W.A.P.; Palipane, K.B. Maturity indices for harvesting of beans (*Phaseolus vulgaris* L.) variety Kentuckey Wonder Green in dry zone cultivation in Sri Lanka. *Trop. Agric. Res.* **2008**, *20*, 123–133.
64. Cajamar. *Ficha de Transferencia: Parámetros de Calidad Interna de Hortalizas y Frutas en la Industria Agroalimentaria*; Cajamar Caja Rural: Almeria, Spain, 2014.
65. Del Río, D.; Rodríguez-Mateos, A.; Spencer, J.P.E.; Tognolini, M.; Borges, G.; Crozier, A. Dietary (poly)phenolics in human health: Structures, bioavailability, and evidence of protective effects against chronic diseases. *Antioxid. Redox Signal.* **2013**, *18*, 1818–1892. [CrossRef] [PubMed]
66. Devi, J.; Sanwal, S.K.; Koley, T.K.; Dubey, R.K.; Singh, P.M.; Singh, B. Variability and character association studies for horticultural and quality traits in garden pea (*Pisum sativum* L. var. *hortense*). *Veg. Sci.* **2018**, *45*, 161–165.
67. Arana Torres, N.M. Una Revisión de los Beneficios en la Salud Humana y Métodos de Extracción de Compuestos Bioactivos de Vaccinium Meridionale. PhD Thesis, Universidad Nacional de Frontera, Sullana, Peru, 2021.
68. Hood-Niefer, S.D.; Warkentin, T.D.; Chibbar, R.N.; Vandenberg, A.; Tyler, R.T. Effect of genotype and environment on the concentrations of starch and protein in, and the physicochemical properties of starch from, field pea and faba bean. *J. Sci. Food Agric.* **2012**, *92*, 141–150. [CrossRef]
69. Mateos-Aparicio, I.; Redondo-Cuenca, A.; Villanueva-Suárez, M.J.; Zapata-Revilla, M.A.; Tenorio-Sanz, M.D. Pea pod, broad bean pod and okara, potential sources of functional compounds. *LWT Food Sci. Technol.* **2010**, *43*, 1467–1470. [CrossRef]
70. Kumari, T.; Deka, S.C. Potential health benefits of garden pea seeds and pods: A review. *Legume Sci.* **2021**, *3*, e82. [CrossRef]
71. Mejri, F.; Khoud, H.B.; Njim, L.; Baati, T.; Selmi, S.; Martins, A.; Hosni, K. In vitro and in vivo biological properties of pea pods (*Pisum sativum* L.). *Food Biosci.* **2019**, *32*, 100482. [CrossRef]
72. Merzlyak, M.N.; Solovchenko, A.E.; Gitelson, A.A. Reflectance spectral features and non-destructive estimation of chlorophyll, carotenoid and anthocyanin content in apple fruit. *Postharvest Biol. Technol.* **2003**, *27*, 197–211. [CrossRef]
73. Murray, I.; Williams, P.C. Chemical principles of Near-infrared technolog. *Near-Infrared Technol. Agric. Ind.* **1987**, *17*, 17–34.
74. Font, R.; Velez, D.; Del Rio-Celestino, M.; De Haro-Bailon, A.; Montoro, R. Fast, clean, low-cost screening of cadmium and lead in the mussel Mytilus galloprovincialis Lmk. by visible spectroscopy and partial least squares regression. *Food Chem.* **2007**, *105*, 667–674. [CrossRef]
75. Tkachuk, R.; Kuzina, F.D. Chlorophyll analysis of whole rapeseed kernels by near infrared reflectance. *Can. J. Plant Sci.* **1982**, *62*, 875–884. [CrossRef]
76. Gomez, A.H.; He, Y.; Pereira, A.G. Non-destructive measurement of acidity, soluble solids and firmness of Satsuma mandarin using Vis/NIR-spectroscopy technique. *J. Food Eng.* **2006**, *77*, 313–319. [CrossRef]
77. Murray, I.; Williams, D. Chemical principals of near-infrared technology. In *Near Infrared Technology in the Agricultural and Food Industries*; American Association of Cereal Chemists: St Paul, MN, USA, 1987; pp. 143–167.
78. Lucas, A.; Andueza, D.; Rock, E.; Martin, B. Prediction of dry matter, fat, pH, vitamins, minerals, carotenoids, total antioxidant capacity, and color in fresh and freeze-dried cheeses by visible-near-infrared reflectance spectroscopy. *J. Agric. Food Chem.* **2008**, *56*, 6801–6808. [CrossRef]
79. Osborne, B.G.; Fearn, T.; Hindle, P.H. Introduction in Practical NIR Spectroscopy with Applications. In *Food and Beverage Analysis*; Osborne, B.G., Fearn, T., Hindle, P.H., Eds.; Longman Scientific and Technical: London, UK, 1993; pp. 1–12.
80. Hervera, M.; Castrillo, C.; Albanell, E.; Baucells, M.D. Use of near-infrared spectroscopy to predict energy content of commercial dog food. *J. Anim. Sci.* **2012**, *90*, 4401–4407. [CrossRef]
81. Hacisalihoglu, G.; Gustin, J.L.; Louisma, J.; Armstrong, P.; Peter, G.F.; Walker, A.R.; Settles, A.M. Enhanced Single Seed Trait Predictions in Soybean (*Glycine max*) and Robust Calibration Model Transfer with Near-Infrared Reflectance Spectroscopy. *J. Agric. Food Chem.* **2016**, *64*, 1079–1086. [CrossRef]
82. Hermida, M.; Rodríguez, N.; Rodríguez-Otero, J.L. Determination of moisture, starch, protein, and fat in common beans (*Phaseolus vulgaris* L.) by near infrared spectroscopy. *J. AOAC Int.* **2006**, *89*, 1039–1041. [CrossRef]

83. Carbas, B.; Machado, N.; Oppolzer, D.J.; Queiroz, M.; Brites, C.M.; Rosa, E.A.S.; Barros, A.I. Prediction of Phytochemical Composition, In Vitro Antioxidant Activity and Individual Phenolic Compounds of Common Beans Using MIR and NIR Spectroscopy. *Food Bioprocess Technol.* **2020**, *13*, 962–977. [CrossRef]
84. Hacisalihoglu, G.; Larbi, B.; Settles, A.M. Near-Infrared Reflectance Spectroscopy Predicts Protein, Starch, and Seed Weight in Intact Seeds of Common Bean (*Phaseolus vulgaris* L.). *J. Agric. Food Chem.* **2009**, *58*, 702–706. [CrossRef] [PubMed]
85. Plans, M.; Simó, J.; Casals, F.; Sabaté, J.; Rodríguez-Saona, L.E. Characterization of common beans (*Phaseolus vulgaris* L.) by infrared spectroscopy: Comparison of MIR, FT-NIR and dispersive NIR using portable and benchtop instruments. *Food Res. Int.* **2013**, *54*, 1643–1651. [CrossRef]
86. Wang, J.; Liu, H.; Ren, G. Near-infrared spectroscopy (NIRS) evaluation and regional analysis of Chinese faba bean (*Vicia faba* L.). *Crop J.* **2014**, *2*, 28–37. [CrossRef]
87. Shi, J.Y.; Zou, X.B.; Zhao, J.W.; Mel, H.; Wang, K.L.; Wang, X.; Chen, H. Determination of total flavonoids content in fresh *Ginkgo biloba* leaf with different colors using near infrared spectroscopy. *Spectrochim. Acta Part A Mol. Biomol. Spectrosc.* **2012**, *94*, 271–276. [CrossRef]
88. Xue, L.; Yang, L. Deriving leaf chlorophyll content of green-leafy vegetables from hyperspectral reflectance. *ISPRS J. Photogramm. Remote Sens.* **2009**, *64*, 97–106. [CrossRef]
89. Li, Y.; Sun, Y.; Jiang, J.; Liu, J. Spectroscopic determination of leaf chlorophyll content and color for genetic selection on *Sassafras tzumu*. *Plant Methods* **2019**, *15*, 73. [CrossRef]
90. Verardo, V.; Cevoli, C.; Pasini, F.; Gómez-Caravaca, A.M.; Marconi, E.; Fabbri, A.; Caboni, M.F. Analysis of oligomer proanthocyanidins in different barley genotypes using high-performance liquid chromatography-fluorescence detection-mass spectrometry and near-infrared methodologies. *J. Agric. Food Chem.* **2015**, *63*, 4130–4137. [CrossRef]
91. Vignolini, S.; Rudall, P.J.; Rowland, A.V.; Reed, A.; Moyroud, E.; Faden, R.B.; Baumberg, J.J.; Glover, B.J.; Steiner, U. Pointillist structural color in *Pollia* fruit. *Proc. Natl. Acad. Sci. USA* **2012**, *109*, 15712–15715. [CrossRef]

sensors

MDPI

Article

Evaluation of Minimum Preparation Sampling Strategies for Sugarcane Quality Prediction by vis-NIR Spectroscopy

Lucas de Paula Corrêdo *, Leonardo Felipe Maldaner, Helizani Couto Bazame and José Paulo Molin

Precision Agriculture Laboratory, Biosystems Engineering Department, 'Luiz de Queiroz' College of Agriculture, University of São Paulo, Av. Pádua Dias 11, 13418-900 Piracicaba, Brazil; leonardofm@usp.br (L.F.M.); helizanicouto@usp.br (H.C.B.); jpmolin@usp.br (J.P.M.)

* Correspondence: lucaspcorredo@usp.br

Abstract: Proximal sensing for assessing sugarcane quality information during harvest can be affected by various factors, including the type of sample preparation. The objective of this study was to determine the best sugarcane sample type and analyze the spectral response for the prediction of quality parameters of sugarcane from visible and near-infrared (vis-NIR) spectroscopy. The sampling and spectral data acquisition were performed during the analysis of samples by conventional methods in a sugar mill laboratory. Samples of billets were collected and four modes of scanning and sample preparation were evaluated: outer-surface ('skin') (SS), cross-sectional scanning (CSS), defibrated cane (DF), and raw juice (RJ) to analyze the parameters soluble solids content (Brix), saccharose (Pol), fibre, pol of cane and total recoverable sugars (TRS). Predictive models based on Partial Least Square Regression (PLSR) were built with the vis-NIR spectral measurements. There was no significant difference (p-value > 0.05) between the accuracy SS and CSS samples compared to DF and RJ samples for all prediction models. However, DF samples presented the best predictive performance values for the main sugarcane quality parameters, and required only minimal sample preparation. The results contribute to advancing the development of on-board quality monitoring in sugarcane, indicating better sampling strategies.

Keywords: chemometrics; proximal sensing; precision agriculture

Citation: Corrêdo, L.d.P.; Maldaner, L.F.; Bazame, H.C.; Molin, J.P. Evaluation of Minimum Preparation Sampling Strategies for Sugarcane Quality Prediction by vis-NIR Spectroscopy. *Sensors* **2021**, *21*, 2195. https://doi.org/10.3390/s21062195

Academic Editor: Lammert Kooistra

Received: 23 January 2021
Accepted: 19 March 2021
Published: 21 March 2021

Publisher's Note: MDPI stays neutral with regard to jurisdictional claims in published maps and institutional affiliations.

1. Introduction

Near-Infrared (NIR) spectroscopy is a well-established technique to monitor the quality of raw sugarcane received by sugar mills [1], and consequently, for pricing and trading with producers and growers [2]. Crop quality is estimated based on physicochemical parameters related to physiological composition, such as soluble solids content (Brix), water-insoluble solids (Fibre), and the apparent sucrose in the juice (Pol). Furthermore, all other parameters (purity, Pol of cane, reducing sugars, and total recoverable sugars) are calculated based on the former parameters [3], from which total recoverable sugars (TRS) are used for the pricing and trading of the raw material. Sugarcane quality parameters are determined by analytical methods and empirical equations described by the National Council of Sugarcane Producers (CONSECANA), which are based on the International Commission for Uniform Methods of Sugar Analysis (ICUMSA). Using calibration methods, it is now possible to obtain some crop quality parameters by NIR spectroscopy [4].

Different wavelength regions of the electromagnetic spectrum can be used in spectroscopy, such as visible (400 to 750 nm), near-infrared (NIR, 750 to 2500 nm), shortwave near-infrared (SWNIR, 750 to 1100 nm), and visible and near-infrared (vis-NIR, 400 to 2500 nm) [5–8]. However, there is no consensus on the limits between these regions. Interaction between electromagnetic radiation and matter causes molecular vibrations involving heavy atom (C, N, O, and S) attached to a hydrogen atom [8]. This basic principle has allowed substantial scientific advances to predict organic compounds of agricultural products associated with its quality using vis-NIR spectroscopy as a nondestructive and environmentally friendly analysis technique [5]. Moreover, several studies have shown promising results when using the technique to predict sugar cane quality in the sugar mill [9–12] and for breeding programs [13–15].

Despite the advances in industrial sugarcane quality monitoring, spectral methods are still a distant reality for in-field measurements in line with precision agriculture practices. Current proximal sensing technologies applied at the canopy level allow only monitoring crop yield [16,17]. However, some studies have indicated that vis-NIR could also be a viable technology for acquiring quality data of harvested products in real-time during mechanical harvesting [16]. The monitoring of crop quality parameters across the field is important to adopting precision agriculture (PA) practices, in which quality maps would show the variability of the crop and help guide site-specific management [18]. In this context, mechanical harvest opens a way to obtain a high frequency of sampling and data collection to analyze the quality variability across the field [19]. However, some requirements need to be satisfied to use vis-NIR as proximal sensing technology for this purpose: (i) the location of adaptation in the harvester, (ii) development of a sampler system, and (iii) the type of sampling required for analysis. The first two requirements are fundamentally dependent on the last one.

Nawi et al. (2014) indicated that the ideal place for implementing an on-board sugarcane quality monitor would be in the elevator of the harvester, where the sugarcane material is partially cleaned and processed in the form of billets. In this context, some studies have reported promising results on the prediction of sugarcane Brix from sensor readings made on the outer-surface ('skin') [20] or on cross-sections [21] of sugarcane billets. Furthermore, more recent studies have advanced with on-board vis-NIR spectroscopy sensor applications on the elevator of a sugarcane harvester simulator, i.e., analyzing samples at a distance and in motion [2,22]. However, Maraphum et al. [23] and Phuphaphud et al. [24] reported that the waxy material should be removed from the cane surface for maximum accuracy in the spectral data condition, even though this may be impractical for an embedded system. Associated with this fact, Phetpan et al. [2] reported on the need to evaluate the potential of the vis-NIR spectroscopy technique with data sets consisting of a larger number of sugarcane varieties. In addition, despite the advantages of nondestructive measurement, there is a lack of basic studies comparatively evaluating various forms of sampling without and with minimal processing, using extensive numbers of samples obtained over the course of a harvest. Thus, the objective of this study was to compare different sugarcane sample types, including billets, defibrated cane, and raw juice, and to analyze the spectral response of each sampling type for the prediction of quality parameters of sugarcane from vis-NIR spectroscopy.

2. Materials and Methods

2.1. Sampling

Variability of sugarcane quality parameters throughout a harvest occurs due to diverse environmental conditions, mainly temperature and precipitation, during the harvest [25]. Based on this, the data collection procedure occurred on random periods over six months of the 2019 harvest (June to November). We sought to obtain higher variability of the sugarcane quality parameters from this collection procedure throughout the harvest.

The data collection was carried out in the quality laboratory of a sugar mill. Three hundred and two samples were collected, and different levels of processing were applied. Also, all data acquisition was performed in a controlled temperature environment (20 ± 5 °C), minimizing the effects of sugar degradation by microorganisms. In addition, we collected the daily results of conventional analysis performed by sugar mill quality laboratory over the same months in which the samples were collected for spectral analysis.

The sampling procedure for vis-NIR analysis was carried out sequentially and simultaneously to the sample preparation for conventional analysis, as described:

1. An oblique probe collected a sample of sugarcane billets in each truckload to proceed with the conventional analyzes of the sugarcane transported to the sugar mill (Figure 1a). Before the laboratory processes the sample, we took a subsample composed of three random sugarcane billets (Figure 1b);
2. The remaining sample of billets was milled in a mechanical knife crusher and homogenized in a mixer (Figure 1c). Then, samples with 500 g and 50 g of the homogenized defibrated sugarcane were collected (Figure 1d) to proceed with the conventional and spectral analysis, respectively;
3. The 500 g sample was pressed in a hydraulic press under constant pressure, at 24.5 MPa (250 kgf cm^{-2}) for 60 s to obtain the raw juice for conventional analysis (Figure 1e). At the same moment, the third sample composed of extracted raw juice was collected for spectral measurements (Figure 1f).

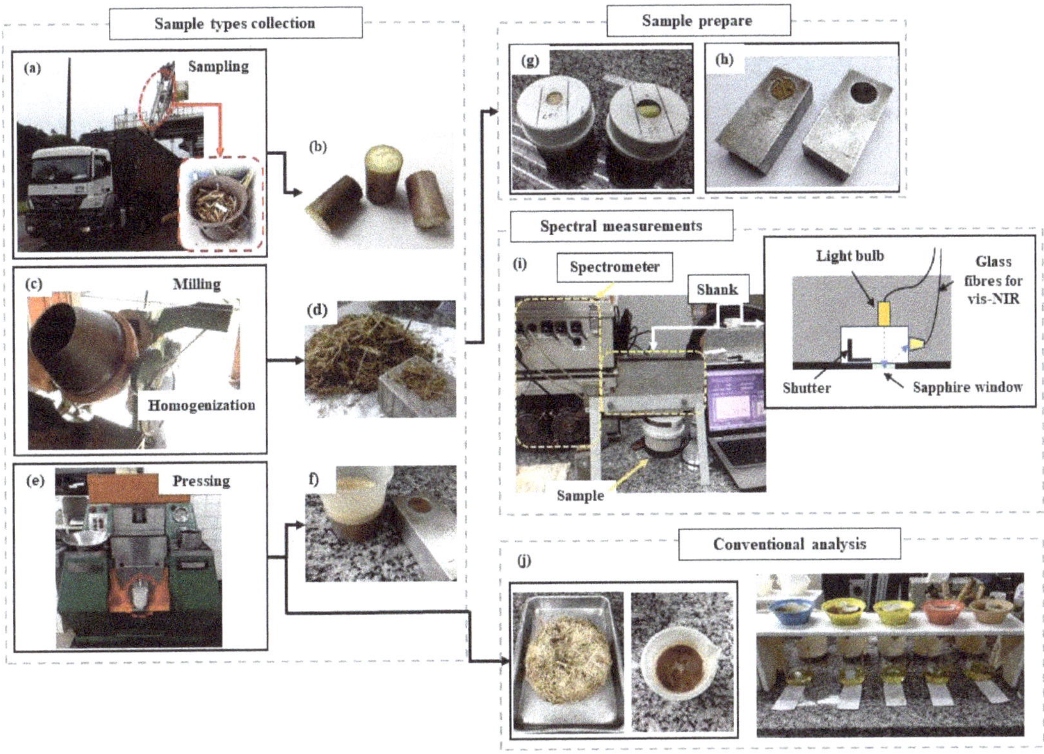

Figure 1. The sequence samples collection and spectral measurements during the preparation steps of samples for conventional analysis. (**a**) Sampling of sugarcane billets by an oblique probe in the cargo truck; (**b**) sugarcane billets for skin and cross-sectional scanning measurements; (**c**) milling and homogenization of sugarcane to defibrated sample; (**d**) defibration sample; (**e**) pressing of defibrated sample to extracting of juice; (**f**) extracted raw juice; prepared samples for vis-NIR spectral measurements: (**g**) cross-sectional and skin of billets inside pipeline chambers, (**h**) defibrated cane and raw juice; (**i**) Veris vis-NIR spectrometer and internal configuration scheme of the measurement shank; (**j**) fibrous cane residue and extracted raw juice for conventional analysis.

The sample types composed of three billets, defibrated cane, and raw juice were prepared (Figure 1g,h) and immediately used for spectral measurements at the sugar mill laboratory (Figure 1i). The remains of raw juice and the bagasse without juice (after pressed) were used for conventional analytical analysis (Figure 1j).

2.2. Sugarcane Quality Analysis

All the procedures and equations for sugarcane quality parameters determination followed the standard protocol proposed by the CONSECANA [26]. These protocols agree with the international rules from ICUMSA.

Initially, the soluble solids content (Brix) was determined by pouring raw juice into a refractometer probe (RX-5000α, ATAGO Co Ltd., Tokyo, Japan) with a maximum resolution of 0.1 Brix. Then, 14 g of a mixture composed of a 4:2:1 proportion of Celite (mineral filtering agent), aluminum chloride, and calcium hydroxide, respectively, was added to 200 mL of raw juice homogenized by a magnetic stirrer until the solution was well-mixed. The solution was filtered through filter paper to obtaining clarified juice. A volume of 70 mL was added in a digital polarimeter (Schmidt + Haensch, Polartronic NHZ 8, Berlin, Germany) to the saccharimetric reading. The result was obtained as percentage of apparent sucrose in the juice (Pol).

After the juice extraction, the remaining fibrous cane residue (Figure 1j) was transferred to metal baskets, with holes at the base. The fibrous residue was weighed on a semianalytical balance. Then, samples were maintained in a forced air circulation dryer at a temperature of 105 °C, until constant weight was achieved. The dryer samples were weighed, and the fibre content (insoluble solids) was determined.

The Pol of cane and TRS were calculated for each sample from the previous parameters as described by CONSECANA (2015). The triplicate values obtained from each sample were averaged.

It is important to mention that reducing sugars (fructose and glucose) and purity (apparent sucrose in the soluble solids content) were also determined in the laboratory. However, while these parameters are not the subject of the present study, they were used to calculate TRS by the reference method.

2.3. Acquisition of Spectral Data

The spectral measurements were performed with a Veris vis-NIR spectrometer (Veris Technologies Inc., Salina, KS, USA). This equipment was developed for on-the-go soil measurements mounted on a platform, connected to a three-point hitch, and pulled by a tractor [27]. However, it may also be used in bench mode. The equipment consists of a CCD array spectrometer (USB4000, Ocean optics, Largo, FL, USA), measuring wavelengths between 373 and 1011 nm, and an InGaAs photodiode-array spectrometer (C9914GB, Hamamatsu Photonics, Hamamatsu, Japan), with a spectral range between 1170 and 2222 nm. The system presents a resolution of around 5 nm. Spectral measurement acquisitions were performed via a sapphire window in the lower of the shank using a tungsten halogen lamp as an electromagnetic energy source (Figure 1i). Each spectrum recorded by the equipment software (Veris spectrophotometer software V1.79) corresponded to the average of 20 spectral readings. The inside shutter is operated automatically to obtain dark and reference spectra before each analysis. Four external references with different grey levels were used for spectral calibration of the spectrometer before the analysis. The spectral data were stored as absorbance units.

The three billets of each sugarcane sample were cut transversely at both extremities, and their skin was lightly cleaned with paper to remove residues from harvest (Figure 1b). An ad hoc dark chamber was constructed with PVC pipes and foam to accommodate the billets, and its inside was painted matte black. Also, magnets were placed on the cover of the chamber, next to the hole through which the spectral scans were performed, aiming to fix the chamber on the reading shank of the spectrometer. This device standardizes the distance between the sapphire window of the spectrometer and the sugarcane billets and removes interference from external lighting (Figure 1g).

The spectral scan method on billets was adapted from Nawi [20,21] and Phuphaphud [15]. The spectral scans were performed at three equidistant points (around 120°) on the skin of each billet. Furthermore, the cross-sectional scanning of billets was performed in triplicate in each cross-sectional surface of each billet, only changing the position after each reading. Therefore, each sample type measurement, skin scanning (SS), and cross-sectional scanning (CSS) of billets was represented by an average of nine successive scans.

The spectral measurements of defibrated cane and raw juice were performed in the same manner. A recipient available from the equipment itself with a volume of around 3 mL was used. The recipient was filled with sample (defibrated cane or extracted raw juice, Figure 1h), and the spectral measurements were performed in triplicate. Three replicates were performed for each sample. Thus, the average of nine spectral readings of defibrated cane (DF) and nine spectral readings of raw juice (RJ) were recorded.

2.4. Spectral Preprocessing

Data preprocessing steps were performed to remove or minimize the sources of spectral variabilities, such as noise present in the dataset, which was not related to the analytical signal [8,28].

Firstly, the spectra were preprocessed using standard normal variate (SNV) [29] to eliminate the deviations caused by particle size and scattering, which centers each spectrum on its mean and then scales it by its standard deviation. Also, the second derivative based on the Savitzky-Golay algorithm [30] was applied, with a window size of 11 points and second-order polynomial fitting to minimize hurdles such as baseline shifts drifts and to remove high-frequency noise from a spectrum and improve the signal-to-noise ratio [28]. After the preprocessing of the spectral data, Pareto scaling (PS) was applied to variables, which is the most commonly applied scaling method in infrared data [31]. The method centered all variables at their means, and then divided them by the square root of the standard deviation.

2.5. Multivariate Analysis

Firstly, the spectral data of the four sampling conditions were concatenated. Then, the data was divided into calibration (75%, 227 samples) and external validation (25%, 75 samples) data sets, based on the Kennard-Stone method [19]. This procedure allowed to obtain the same samples for calibration and external validation data set for both sample types evaluated. The spectral measurements were used to build predictive models for sugarcane quality parameters based on Partial least square regression (PLS) [20].

The models were calibrated using the venetian blinds cross-validation method with 10 splits. The optimal PLS models were determined based on the lowest number of latent variables (LV), in which the root mean square error of cross-validation (RMSECV) was not significantly higher than the minimum RMSECV [23]. The root mean square error (RMSE) was calculated as follows:

$$\text{RMSE} = \sqrt{\frac{\sum_{i=1}^{n}(y_i - \hat{y}_i)^2}{n}} \tag{1}$$

where n is the number of samples, y_i is the reference measurement of sample i, and $\hat{y}_i$ is the estimated result for sample i.

The outliers were evaluated during the calibration step for the reference lab values and spectral data. The presence of outliers in the spectral data was evaluated by the "influence plots" based on high leverage and unmodeled residuals by Hotelling T^2 and Q statistics, respectively [8]. Samples with high values in both cases, at 5% of significance level, were considered outliers and removed from the spectral data set. On the other hand, outliers in reference data were evaluated by the root mean square error in calibration (RMSEC) values. Samples that presented errors in prediction greater than $\pm 3 \times$ RMSEC were considered outliers and removed from the data set [32]. The external validation samples were considered unknown samples. In this way, the outliers were evaluated only for the spectral data set. The process was carried out at most three times in the calibration step, as recommended by ASTM E1655-7 [33].

The model accuracy was evaluated based on the RMSE for calibration, cross-validation, and prediction (RMSEC, RMSECV, RMSEP, respectively). Prediction performance was evaluated based on the determination coefficient (R^2) for calibration and prediction (R^2c and R^2p, respectively), and the ratio of performance to the interquartile range (RPIQ), which is calculated by the ratio between the interquartile difference and the RMSEP. Also, a randomization test [34] with 0.05 significance level of probability was performed. The aim was to compare the accuracy of regression models using different sugarcane sample types in the validation set. The hypothesis evaluated were:

- Null hypothesis (H_0): $RMSEP_{\text{sample type 1}} = RMSEP_{\text{sample type 2}}$ (accuracy is similar);
- Alternative hypothesis (H_1): $RMSEP_{\text{sample type 1}} \neq RMSEP_{\text{sample type 2}}$ (accuracy is not similar).

An advantage of this test is its simplicity and the fact that assumptions about normality or homoscedasticity of the data are not required (distribution-free) [35]. More details about this test, included an algorithm script, can be found in Olivieri [36].

Moreover, the variable importance for the projection (VIP) was calculated to verify the wavelengths with a more significant impact on the external validate models [27] for each sample type. The VIP was calculated as follows:

$$\mathrm{VIP_j} = \sqrt{p \sum_{k=1}^{h} \left[Z \left(\frac{w_{jk}}{||w_k||} \right)^2 \right] \cdot \left(\sum_{k=1}^{h} Z \right)^{-1}} \quad (2)$$

where VIP is the variable importance for projection (dimensionless), j is a specific wavelength (nm), p is the number of wavelengths (dimensionless), h is the number of latent variables (dimensionless), Z is the fraction of variance in the prediction explained by the latent variable (dimensionless), and w is the loading weight (dimensionless).

All models, routines, and data processing were performed in Matlab R2015a (The MathWorks, Natick, MA, USA) and PLS Toolbox 8.9 (R8.9.1; Eigenvector Research, Wenatchee, WA, USA).

3. Results and Discussion

3.1. Overview of Sugarcane Quality Reference Data and vis-NIR Spectral Measurements of Different Sample Types

From the daily results of analyses performed by conventional methods at the mill, it was possible to characterize the variation of the main parameters determined analytically (Brix, Pol, and Fibre), as well as for TRS, throughout the months in which the experiment was performed (Figure 2).

It is possible to observe an increasing trend in all parameters from June to October. Afterward, there is a tendency to decrease, except for fibre. Weather is highly influential on sucrose storage [25]. In the months corresponding to autumn (June) and winter (June to September), water stress and cooler temperatures contribute to the reduction of vegetative crop growth and favor sucrose storage [37]. With the beginning of spring (September/October) and the beginning of the rainy season, the vegetative growth of the crop is resumed, and the reserves are consumed. The sample acquisition on different periods (vertical bars in Figure 2) throughout the harvest allowed us to obtain data including different stages in this variation. The effect of this variability was reflected in the range of all samples collected during the experiment (Table 1).

On a first view, the Kennard-Stone method provides a representative calibration data set, with external validation data set between its range (Table 1). The sample acquisition method provided a satisfactory variability of data, as expected; TRS varied from 86.94 to 173.80 kg of sugar per Mg of cane.

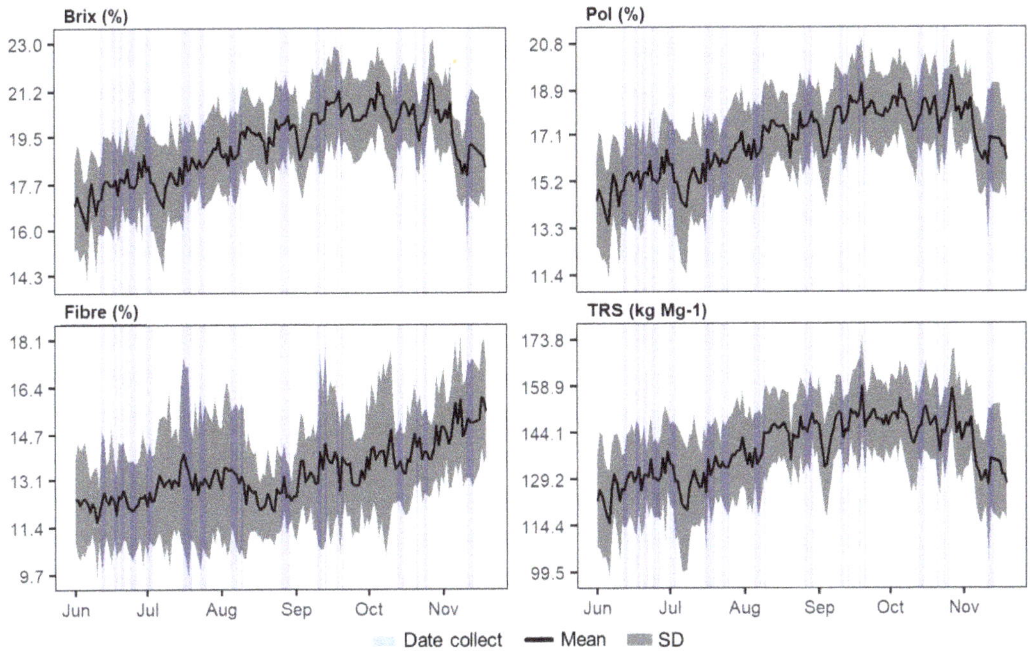

Figure 2. Mean and standard deviation (SD) of annual variation of sugarcane quality parameters, Brix, Pol, Fibre, and total recoverable sugars (TRS); and spectral data collection periods (vertical bars).

Table 1. Descriptive statistics of the reference results for the sugarcane quality attributes of all samples, calibration, and external validation data sets.

Parameter	unit	Mean ± SD	Median	Range	p_{25}	p_{75}	SEL
			All samples (n = 302)				
Brix	%	18.95 ± 1.71	18.99	13.08–23.42	17.80	20.01	0.03
Pol	%	16.67 ± 1.90	16.66	10.78–21.20	15.41	17.95	0.01
Fibre	%	13.29 ± 1.79	12.90	7.22–20.08	12.07	14.33	0.07
Pol of cane	%	13.80 ± 1.56	13.91	8.40–17.56	12.92	14.78	0.01
TRS	kg Mg^{-1}	137.66 ± 14.48	138.66	86.94–173.80	129.75	146.84	1.12
			Calibration set (n = 227)				
Brix	%	18.86 ± 1.66	18.80	13.08–23.42	17.79	19.98	-
Pol	%	16.54 ± 1.86	16.55	10.78–21.20	15.38	17.78	-
Fibre	%	13.31 ± 1.89	12.83	7.22–20.08	12.05	14.41	-
Pol of cane	%	13.69 ± 1.52	13.79	8.40–17.56	12.83	14.60	-
TRS	kg Mg^{-1}	136.65 ± 14.07	137.01	86.94–173.80	128.99	145.19	-
			Validation set (n = 75)				
Brix	%	19.24 ± 1.85	19.59	13.55–23.05	18.06	20.61	-
Pol	%	17.06 ± 1.98	17.33	11.24–20.90	15.73	18.54	-
Fibre	%	13.23 ± 1.44	13.02	10.49–17.15	12.24	14.16	-
Pol of cane	%	14.14 ± 1.64	14.34	8.96–17.14	13.27	15.47	-
TRS	kg Mg^{-1}	140.76 ± 15.36	142.59	92.16–169.02	132.35	152.44	-

SD: standard deviation; p_{25}: lower quartile; p_{75}: upper quartile; SEL: standard error of laboratory; TRS: total recoverable sugar.

The distribution of all quality parameter values had wide distribution (Figure 3). Fibre content did not positively or negatively correlate with any other parameters analyzed, with values varying from −0.16 to 0.13 ($p < 0.05$). On the other hand, the other parameters showed a positive correlation higher than 0.94 ($p < 0.05$). The highest correlation was observed between Pol of cane and TRS, close to 1.00 ($p < 0.05$). Higher correlation values are observed between the TRS with parameters analytically determined such as Brix and Pol (0.94 and 0.96, respectively, $p < 0.05$). The correlation values for these attributes are firstly explained by the composition of the soluble solids content of sugarcane, measured by Brix, in which the largest proportion corresponds to sucrose (about 15–18%), measured by Pol [38,39], and reducing sugars (fructose and glucose) in a smaller proportion (about 0.5%) [40]; note that the determination of reducing sugars was not the objective of the present study.

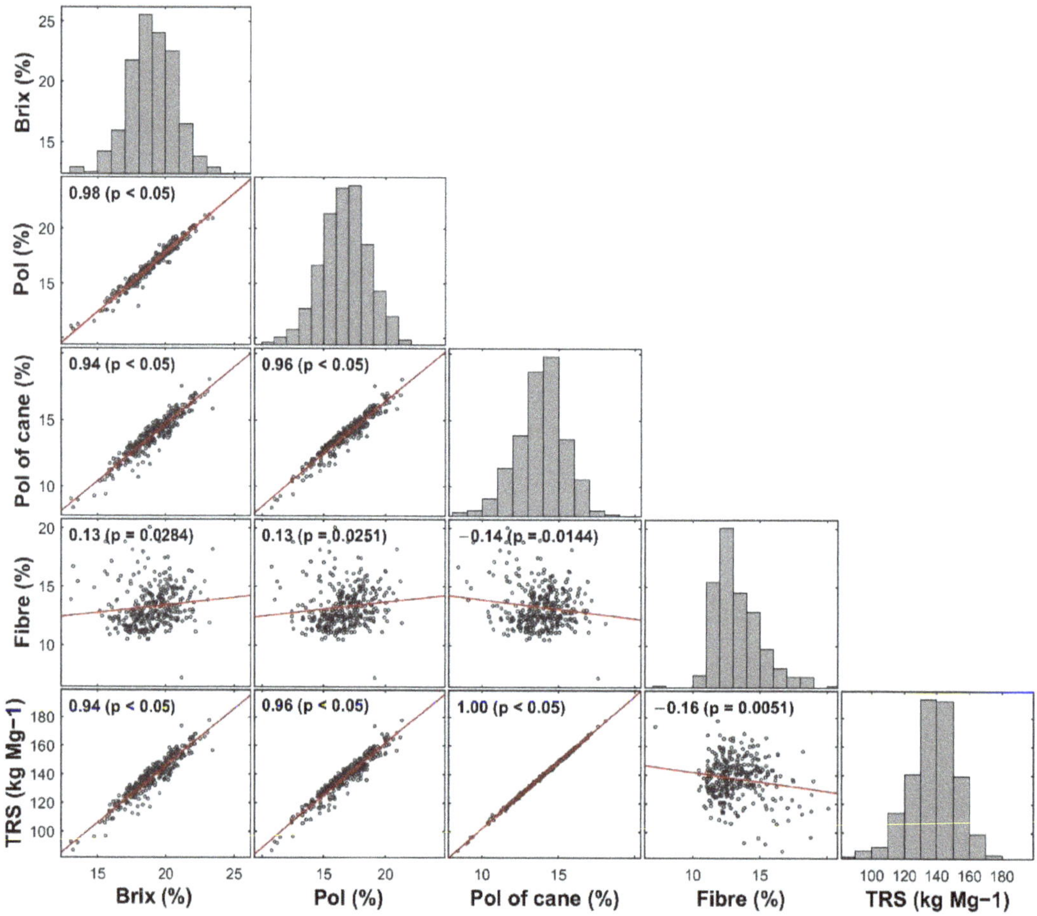

Figure 3. Correlogram of sugarcane quality parameters with frequency distributions on the principal diagonal, Pearson's correlation coefficient and respective p-values, and the correlation trend line.

The vis-NIR raw spectral data obtained for 302 samples of each sugarcane sample type are shown in Figure 4. The spectral data were evaluated to identify possible spectral errors [41]. We observed a noisy aspect in the region corresponding to the visible spectrum (400 to 698 nm), mainly for SS samples. This effect may have been attributed to the influence of skin colors of billet samples, which were obtained for several different sugarcane varieties (Table A1), or soil residues from the harvest present in the RJ samples. Therefore, this spectral region was removed from the data set. Phuphaphud [14] observed the same effect due to the skin color of sugarcane billets. Also, based on the evaluation of the coefficient of variation (CV) obtained for each spectral band, the last spectral bands showed high CV concerning their neighbors and were also removed from the dataset, similar to the procedure performed by Franceschini [27] in a study on the external effects on the spectral reading of vis-NIR of soil samples using the same equipment. Thus, only bands in the spectral range between 699 and 1010 nm and between 1070 and 2153 nm (303 spectral bands) were retained.

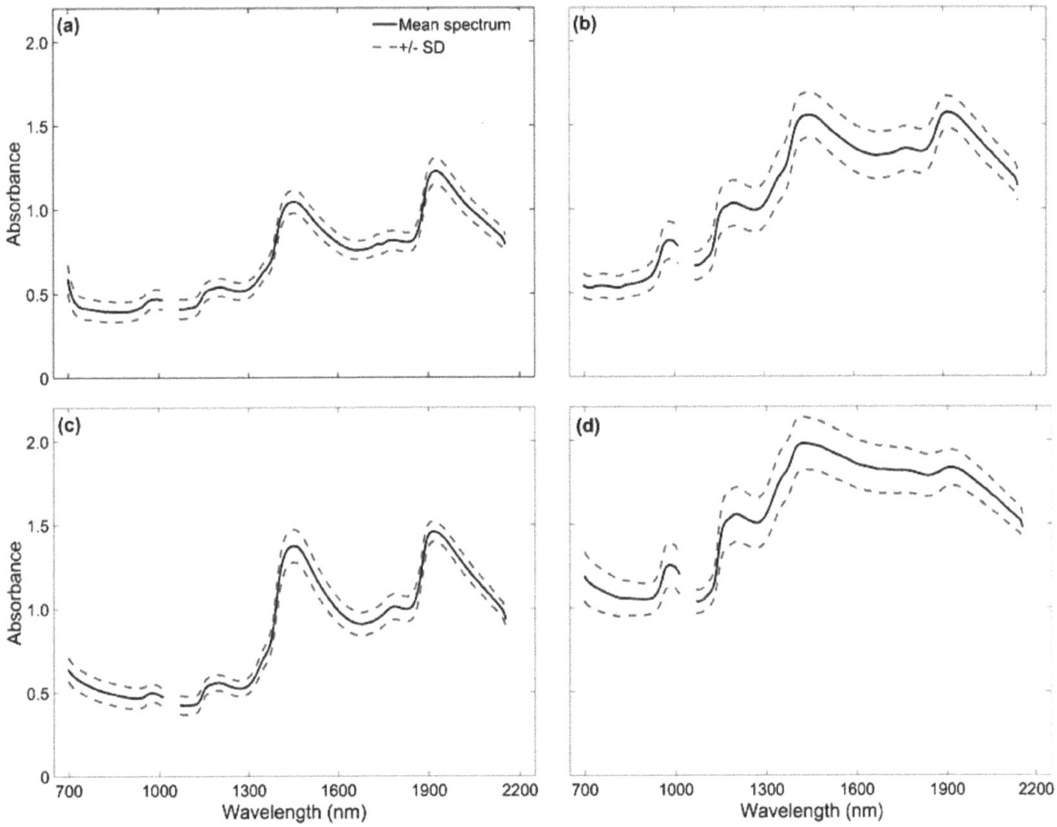

Figure 4. vis-NIR mean spectra and standard deviation (SD) of all 302 sugarcane samples for (**a**) skin and (**b**) cross-sectional scanning of billets, (**c**) defibrated, and (**d**) raw juice samples.

A PCA analysis performed an exploratory overview of the data structure. The spectral data were only mean-centered, and the classes were identified by sample type. Two principal components, PC1, and PC2, explained 98.6% of the data variance (Figure 5). The first component explained 96.8% of the data variance. The data structure was different for each sample type, as can be seen from ellipses illustrating the majority of samples (Figure 5). However, a first overview allowed us to verify the greater difficulty in explaining the variance of less processed samples, such as samples obtained by spectral readings in the skin (SS) and cross-sectional (CSS) of the billets, than processed samples, such as raw juice samples (RJ).

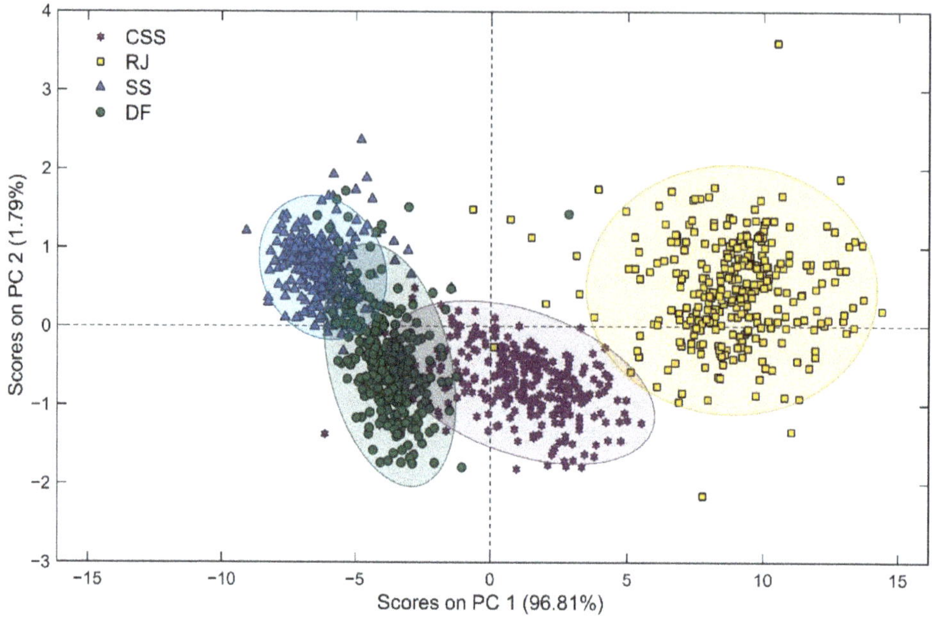

Figure 5. PCA score plot for the sugarcane sample types analyzed. SS-skin scanning of billets; CSS-cross-sectional scanning of billets; DF-defibrated samples; RJ-raw juice samples.

Vis-NIR spectroscopy may be used in a number of applications, including the classification of sugarcane varieties, with promising results [42]. The same authors showed that the spectral regions between 650 and 750 nm, corresponding to the visible spectrum, was the most suitable for sugarcane discrimination. The principal component analysis for the four sample types individually showed that the scatter plots were not categorized into groups based on sugarcane varieties (Figure A1). These results indicated that the sample set composed for many different varieties did not affect the spectral characteristics between each variety for both sample types. A similar effect was observed by Phuphaphud et al. (2020) [14] when evaluating the classification of three varieties. Therefore, the present study was conducted for all varieties combined.

The vis-NIR technique principle is based on the detection of compounds and molecules through their molecular vibration states [8]. Different varieties naturally have different concentrations of parameters such as sucrose and fibre according to genetics. Furthermore, for all of them, the plant matrix is essentially composed of water (75–82%), insoluble solids content (Fibre, 10–18%), and soluble solids (Brix, 18–25%), which are composed of nonsugars (1–2%), sucrose (14–24%), and reducing sugars (0–1.5%) [43]. However, the prediction of quality parameters related to chemical compounds of interest should be independent of sugarcane varieties.

3.2. Prediction Performance of Models Based on Different Sugarcane Sample Types

Figure 6 presents scatter plots showing reference versus predicted values of sugarcane quality parameters. There was an underestimation of high values and overestimation of lower values for all attributes and sample types evaluated. However, this effect was more intense for the less prepared sampling condition, such as SS and CSS. Also, overall, the residuals showed no trend (Figure A2).

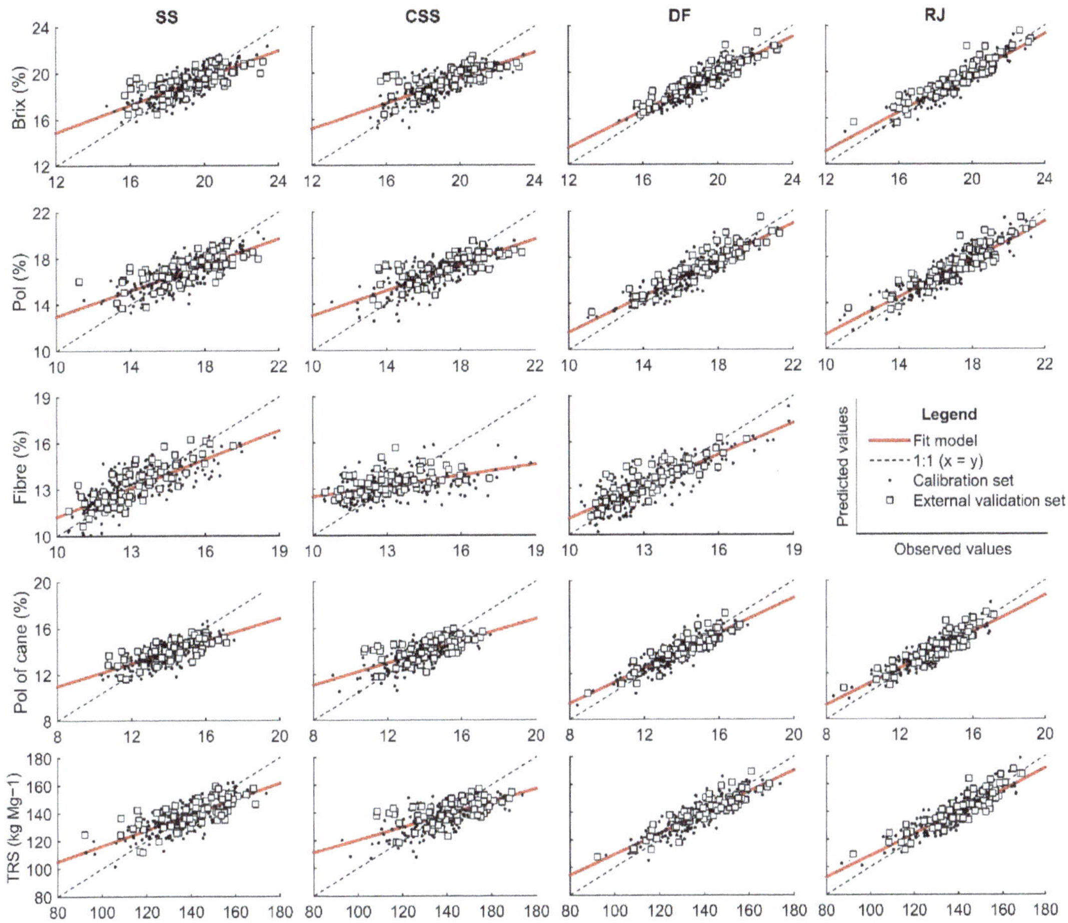

Figure 6. Plots of observed values versus predicted sugarcane quality values from vis-NIR by skin (SS) and cross-sectional (CSS) scanning of billets, defibrated (DF), and raw juice (RJ) samples. Brix, Pol, Fibre, and Pol of cane are in percentage, and TRS values are in kg Mg^{-1}.

More LVs were necessary to explain the variance of the data for models constructed from SS samples (between 7 and 10) than those obtained to predict the same parameters from other sample types (Table 2). Also, it could be observed that SS and CSS did not show similar accuracy (p-value < 0.05) for Fibre, Pol of cane, and TRS (Table 3). Moreover, the prediction performance results for these parameters by these sample types were worse than the performance results obtained by DF and RJ samples (Table 3). The RPIQ values for TRS were 40% higher on average than those for SS and CSS samples, for example. Furthermore, the model accuracy observed between sampling methods for all the other conditions was statistically nonsignificant (p-value > 0.05).

Table 2. Figures of merit for the PLSR models for all studied sugarcane quality attributes and sample types.

Attribute	Sample Type	LV	RMSEC [a]	RMSECV [a]	RMSEP [a]	R^2c	R^2p	RPIQ
Brix	SS	9	0.92	1.10	1.29	0.64	0.48	1.98
	CSS	6	0.95	1.04	1.38	0.62	0.41	1.85
	DF	7	0.67	0.75	0.84	0.81	0.80	3.05
	RJ	8	0.64	0.83	0.75	0.85	0.85	3.39
Pol	SS	8	1.09	1.26	1.42	0.60	0.48	1.98
	CSS	6	1.09	1.19	1.44	0.61	0.44	1.95
	DF	7	0.82	0.93	0.87	0.79	0.83	3.24
	RJ	7	0.80	0.97	0.90	0.82	0.81	3.12
Fibre	SS	10	1.02	1.29	0.87	0.59	0.65	2.22
	CSS	4	1.45	1.50	1.27	0.24	0.23	1.51
	DF	5	0.93	1.04	0.82	0.69	0.69	2.36
	RJ [b]	-	-	-	-	-	-	-
Pol of cane	SS	7	0.95	1.07	1.13	0.52	0.46	1.94
	CSS	5	1.01	1.09	1.27	0.52	0.31	1.73
	DF	7	0.73	0.84	0.72	0.76	0.81	3.04
	RJ	7	0.71	0.85	0.72	0.78	0.81	3.07
TRS	SS	9	8.57	10.27	10.86	0.60	0.50	1.85
	CSS	5	9.49	10.17	11.86	0.50	0.32	1.69
	DF	7	6.50	7.51	6.71	0.76	0.82	2.99
	RJ	7	6.38	7.95	6.79	0.78	0.81	2.96

SS: skin scanning of billets samples; CSS: cross-sectional scanning of billets samples; DF: defibrated samples; RJ: raw juice samples. LV: latent variable. RMSEC: root mean square error of calibration. RMSECV: Root Mean Square Error of Cross-Validation. RMSEP: Root Mean Square Error of Prediction. R^2c: calibration coefficient of determination. R^2p: prediction coefficient of determination. RPIQ: Ratio of performance to interquartile distance. [a] values for Brix, Pol, Fibre, and Pol of cane are in percentage and TRS in kg Mg^{-1}. [b] The fibre content was not determined from raw juice samples.

Table 3. *p*-Values of randomization test of external validation set for all compared sugarcane sample types.

Binary Combination (Sample Types)	Sugarcane Quality Parameters				
	Brix	Pol	Fibre	Pol of cane	TRS
SS vs. CSS	0.104	0.116	<0.001	0.036	0.008
SS vs. DF	1.00	1.00	0.667	1.00	1.00
SS vs. RJ	1.00	1.00	-	1.00	1.00
CSS vs. DF	1.00	1.00	1.00	1.00	1.00
CSS vs. RJ	1.00	1.00	-	1.00	1.00
DF vs. RJ	0.879	0.344	-	0.606	0.502

SS: skin scanning of billets; CSS: cross-sectional scanning of billets; DF: defibrated samples; RJ: raw juice samples; TRS: total recoverable sugar.

The model performance for DF and RJ samples was equivalent for practically all parameters evaluated. There was no significant difference between the model's accuracy (p-value > 0.05) and very close values of R^2p and RPIQ. Moreover, from DF samples, it was possible to obtain a satisfactory performance to predict Fibre content; this was not possible for RJ samples. On the other hand, the models performed for SS samples presented a higher number of LV than for a prepared sample. The model performance for predicting parameters related to sucrose (Brix, Pol, Pol of cane, and TRS) was not satisfactory, with R^2p and RPIQ below 0.5 and 2.0, respectively, except for Fibre prediction. The prediction results from SS samples for Fibre were close to those obtained for DF samples, as shown by the values of R^2p and RPIQ. However, the results were less promising than those obtained by Phuphaphud et al. (2019) [15], which obtained the following results: maxima of 0.81 for R^2p and 0.63 for RMSEP. Although Fibre content is an important attribute for sugarcane quality determination, it is not essential for sucrose estimation. Fibre content has no relation with some important attributes, such as Brix and Pol, and only minimally impacts TRS calculation. The prediction of this parameter is important for producing energy cane and breeding programs, as in work developed by Phuphaphud et al. (2019) [14].

Some models developed for CSS samples were similar to those developed for SS, as for Brix and Pol prediction. However, its predictive performance was lower than those obtained for Fibre, Pol of cane, and TRS predicted by SS samples, with worse results for R^2p and RPIQ. In a first investigation, Nawi et al. (2013a) obtained values of 0.87 for R^2p. The excellent performance of this index can be explained by the method of data acquisition adopted by the authors, with individualized samples according to the stem portion (lower, middle, and upper portion) and only three varieties of cane. Sucrose accumulation occurs in an ascending manner, with more accumulation in the lower portion and less in the internodes of the upper portion, close to the leaves [38,44,45]. Therefore, samples composed of different sections resulted in more variability in quality parameters. However, if we analyze the characteristics of a sugar cane harvester, after the stems pass through the chopper roll system, the distinction between portions of the sugarcane stem is not viable.

The RPIQ values for the SS method were higher than those obtained by the CSS method. The SS method on billet samples on the harvester conveyor would be the most practical method, due to the better operability of sample acquisition in that portion of the harvester [16]. However, several external factors must be considered to measure quality attributes by the SS method. A critical one is the constitution of the sugarcane skin itself, as various colors depending on the variety, black and white waxy material, and organic compounds may be present [24,46].

There are common waxy materials on the cane surface that affect vis-NIR measurements by the SS method. Maraphum et al. (2018) evaluated the effect of the waxy material on the cane surface to eliminate or avoid getting low accuracy of the models for Pol measurements. They obtained RMSEP values were around 1.20 to 1.50%, i.e., close to those found by the present study. The authors concluded that spectra acquisition by removed-wax samples was convenient for the measurement of Pol. However, other compounds could affect vis-NIR spectroscopy measurements, such as cellulose and lignin [7].

Overall, the predictive performance results of the models (based on the R^2p and RPIQ) indicate that DF and RJ samples presented similar performance and provide the best results. However, models built from DF samples require less preparation, i.e., by avoiding juice extraction, making them more attractive for an on-board system. Moreover, CSS samples presented worse performance than all other samples. SS samples presented higher values of R^2p and RPIQ than CSS samples for all quality parameters. Furthermore, SS samples showed a nonsignificant difference (p-value > 0.05) of accuracy (RMSEP) with models built from DF samples, but worse results for performance (R^2p and RPIQ), except for Fibre. Possibly DF results were satisfactory due to the exposure of the internal constituents, which overlapped concerning the waxy material that becomes visually negligible. On the other hand, the organic compounds in the sugarcane outer-surface may have interfered in the prediction models. Future studies using nonlinear processing methods [47] or advanced filtering methods, with orthogonalization of unwanted signals concerning the compounds of interest [48], may help in increasing the predictive performance of the models, which is more interesting for an on-board system.

3.3. Variable Influence on the Models

VIP scores were used to describe the importance of each wavelength to the prediction of the main sugarcane quality parameters (Figure 7).

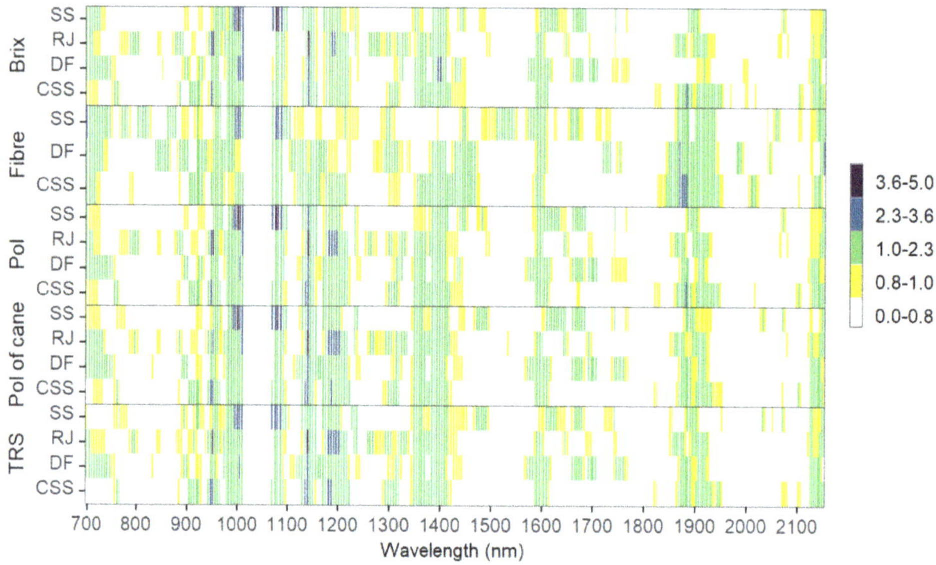

Figure 7. Heatmap of Variable Importance in Projection (VIP) for models used to predict Brix, Pol, Fibre, Pol of cane, and TRS based on different spectral sample types datasets. SS: skin scanning of billets; CSS: cross-sectional scanning of billets; DF: defibrated samples; RJ: raw juice samples; TRS–total recoverable sugars.

As a vibrational spectroscopy technique, the interaction between the vis-NIR electromagnetic radiation and the matter of the sample could be interpreted mainly by overtones and combinations of vibrational modes involving C-H, O-H, and N-H chemical bonds [8,49]. VIP values greater than 1.0 indicate variables with greater influence on the models, and VIP values between 0.8 and 1.0 indicate the moderately influential variables. All variables with VIP smaller than 0.8 are insignificant to the predictions [50]. Some substantial similarities could be observed for different samples and quality parameters on a first overview.

At around 960 nm, there is a small interval with high values of VIP (higher than 2.3), especially related to the prediction of Brix, Pol, Pol of cane, and TRS by CSS and RJ samples, corresponding to the second and third overtone of O-H and C-H stretching, respectively [51]. The region between 980 and 1030 nm can be regarded as an important contributor to quality predictions (VIP higher than 1.0). Between 972 and 1009 nm there is a characteristic signal related to saccharides and the third overtone of O-H [7]. This signal is more expressive for parameters determined for SS samples. This spectral range could be associated with cellulosic fibres, which explained the higher VIP values to SS samples. Similar observations were found by Phuphaphud et al. (2020) to predict commercial cane sugar from growing cane stalks for breeding programs using vis-NIR spectroscopy.

At 1139 nm, there is a small band with VIP values higher than 1.0 for all parameters predicted for four sample types, except for Fibre content. On the interval between 1100 and 1230 nm occurs the second vibrational frequency overtones associated with C-H stretching [51]. Also, at around 1170–1180 nm, there are VIP values higher than 1 for Fibre predicting, mainly by DF and CSS samples. In this region, the third overtone of C-H and unsaturated C=C double bonds are typically associated with fibre, such as lignin [7].

At 1360 nm, there is another expressive region with high VIP values, possibly related to C-H combinations and the O-H first overtone, respectively [49]. Then, at 1600 nm, there are highly similar VIP values possibly associated with to first vibrational frequency overtone of C-H stretching [51]. Another region shows a high contribution, with VIP values higher than 1, for Fibre predicting between 1850 and 1900 nm, mainly for CSS and DF samples. Around 1820 nm occurs the effects of O-H stretching associated with two combinations of C-O stretching commonly associated with Fibre as cellulosic [7]. This interval is lower and with lower VIP values for SS samples, possibly due to the waxy effect on the near-infrared signal [23]. Finally, in the last bands of the spectra, after 2100 nm, the intensity of VIP values is similar for all predicted attributes from any sample type due to O-H bending and C-O stretching combination [7].

The scores of the models for all measured quality parameters are displayed by their first PLS loadings (Figures A3–A7 see Appendix A), accounting for more than 95% of the data variance. Overall, the most considerable variation occurred in the spectral region between 1300 and 1500 nm and between 1800 and 1950 nm. Other authors have found similar response in these spectral regions for prediction of sugars in other products [52,53]. This effect was similar for all sample types and all parameters evaluated. Therefore, this fact proved the relationships identified by the VIP scores and the key molecular bonds related to the parameters of interest described earlier.

The gap between two spectrometers, starting at 1011 nm until 1070 nm, is not related to significative known vibrational frequency overtones associated with some bands related to sugars or fibres [7]. Therefore, the absence of information in this range would not have significantly affected the development of the models.

Processed samples allowed a more significant interaction of electromagnetic radiation corresponding to vis-NIR bands with matter constituents. This physical effect resulted in more prominent signals from specific vibrational frequency bands related to the chemical constitution of sugarcane quality parameters. Overall, defibrated samples (DF) showed performance prediction results that were close to raw juice samples (RJ). Also, the DF sample allowed us to predict Fibre content as well as other parameters, which is not possible with RJ samples. The prediction of sugarcane quality parameters from less processed samples is a desirable characteristic for mechanization of on-the-go measurements of crops, thus promoting spatial information of crops based on quality. DF samples may partially satisfy this requirement; however, this is a destructive sampling technique.

Improving the predicting performance of sugarcane quality parameters from billets for on-the-go systems may be possible [20]. Some effects, such as waxy and skin organic compounds, need to be considered and minimized [24]. Other data processing techniques such as nonlinear models [47] or advanced filtering methods such as orthogonalization [48], could be investigated to improve the performance aiming to develop reliable models for measuring sugarcane quality using billets of cane.

4. Conclusions

This study demonstrates that vis-NIR spectroscopy could be used as a quick method to assess the abundance of chemical compounds of sugarcane related to its quality. There was no significant difference (p-value > 0.05) in the accuracy (RMSEP) of prediction of whole cane samples when compared to processed samples, such as defibrated cane (DF) and extracted raw juice (RJ), for all evaluated quality parameters. Also, outer-surface measurements of sugarcane billets presented a better accuracy (RMSEP, p-value > 0.05) and performance (R^2p and RPIQ) than measurements on the cross-section.

Despite the similar accuracy (p-value > 0.05), DF and RJ sampling presented better performance than outer-surface measurements of sugarcane billets. Moreover, the performance of the models from DF and RJ samples were similar, but DF samples involve less preparation, as they do not require juice extraction of the sample.

The results showed that DF sampling could be used to predict the main sugarcane quality parameters, such as soluble solids content (Brix), saccharose (Pol), Fibre, Pol of cane, and total recoverable sugars (TRS), all of which are used for pricing and trading between mills and sugarcane producers. The DF models presented RMSEP varying between 0.72% and 0.87% for Brix, Pol, Fibre, and Pol of cane, and 6.71 kg Mg^{-1} for TRS.

The results in this study contribute to advancing the development of on-board quality monitoring in sugarcane. This information shows the spatial variability of crop quality and helps guide site-specific management of sugarcane fields.

Author Contributions: Conceptualization, L.d.P.C., and J.P.M.; methodology, L.d.P.C., and J.P.M.; software, L.d.P.C.; validation, L.d.P.C.; formal analysis, L.d.P.C.; investigation, L.d.P.C., and J.P.M.; resources, L.d.P.C. and J.P.M.; data curation, L.d.P.C.; writing—original draft preparation, L.d.P.C.; writing—review and editing, L.d.P.C., L.F.M., H.C.B., and J.P.M.; visualization, L.d.P.C., L.F.M., H.C.B., and J.P.M.; supervision, J.P.M.; project administration, L.d.P.C. and J.P.M.; funding acquisition, L.d.P.C., and J.P.M. All authors have read and agreed to the published version of the manuscript.

Funding: This research was funded by São Paulo Research Foundation (FAPESP), which provided a doctoral scholarship to the first author (grant number 2018/25008-8).

Institutional Review Board Statement: Not applicable.

Informed Consent Statement: Not applicable.

Data Availability Statement: The data presented in this study are available on request from the corresponding author.

Acknowledgments: The authors would like to thank the São Martinho group for allowed to conduct this study in one of its mills. Also, we would like to thank the researcher André Freitas Colaço, from Commonwealth Scientific and Industrial Research Organization (CSIRO), for his valuable and pertinent contributions on an earlier version of this manuscript.

Conflicts of Interest: The authors declare no conflict of interest.

Appendix A

Table A1. Number of samples of each Brazilian sugarcane variety used in the study.

Variety	# Samples	Variety	# Samples	Variety	# Samples
CT96-1007	12	CTC 9001	1	RB966928	7
CT96-3346	7	CTC 9005	2	RB975201	1
CTC 11	19	CV 6654	3	RB975952	2
CTC 14	10	IACSP95-5000	1	RB985476	1
CTC 15	12	RB855002	3	SP80-3280	9
CTC 17	7	RB855156	33	SP83-2847	31
CTC 2	19	RB855536	4	SP83-5073	2
CTC 20	26	RB867515	9	RB965621	1
CTC 22	2	RB935621	4	SP91-1049	1
CTC 4	34	RB965621	1	Various	26
CTC 7	1	RB965902	12		

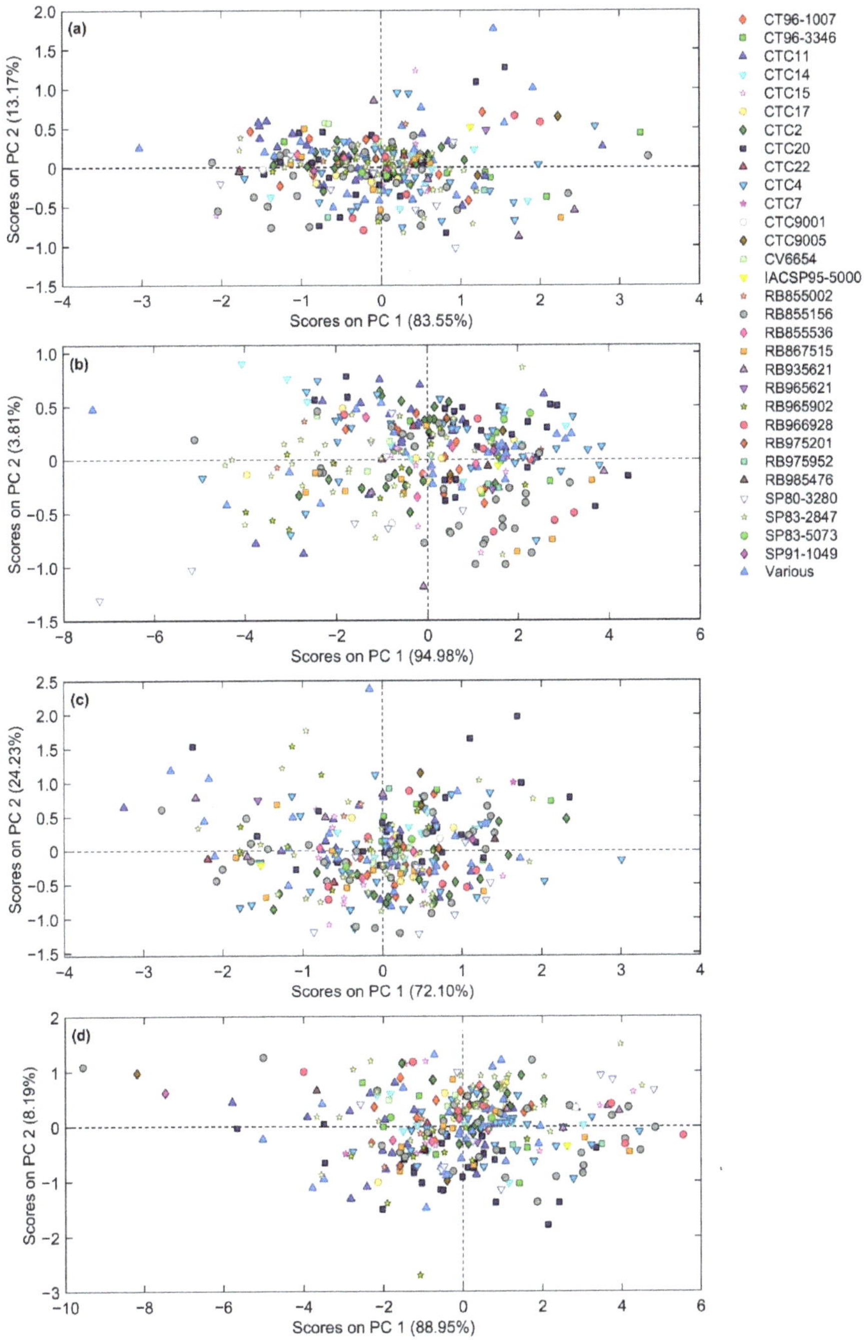

Figure A1. Principal component scores of different sugarcane varieties for spectral data of skin scanning of sugarcane billets (**a**), cross-sectional scanning of sugarcane billets (**b**), defibrated cane samples (**c**), and raw juice samples (**d**).

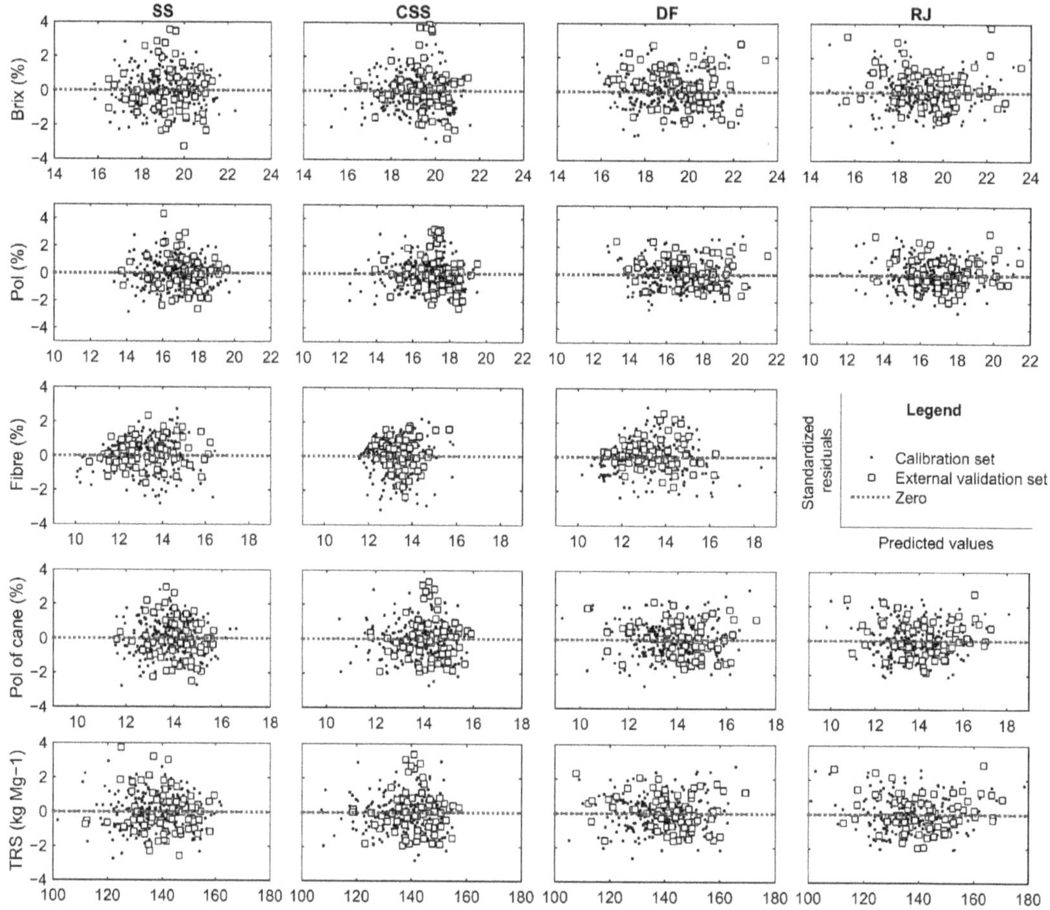

Figure A2. The plot of standardized residuals versus predicted sugarcane quality values from vis-NIR by skin (SS) and cross-sectional (CSS) scanning of billets, defibrated (DF), and raw juice (RJ) samples. Black dots are calibration samples, and red dots are external validation samples.

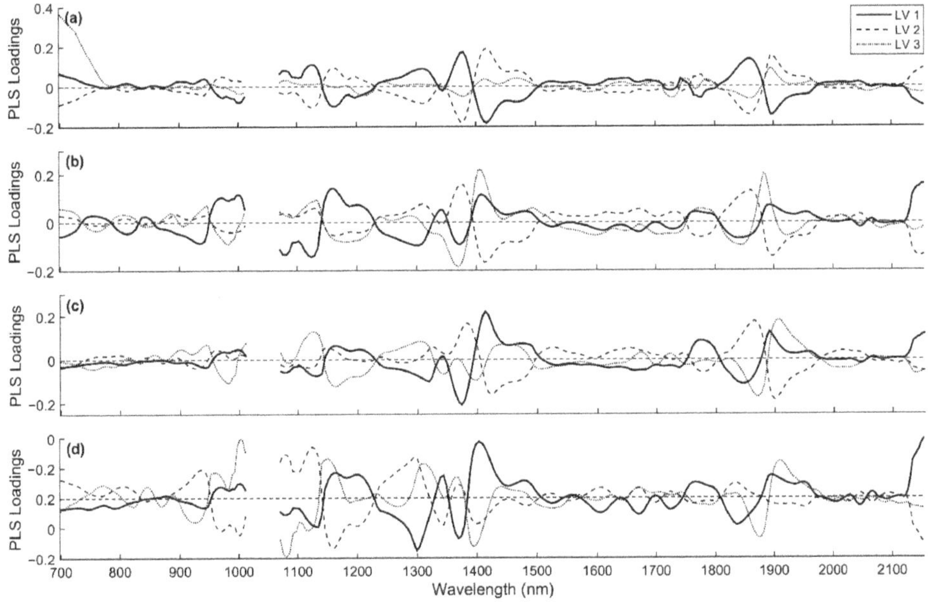

Figure A3. The three first Partial least squares loadings for Brix prediction using near-infrared reflectance spectroscopy from the skin (**a**) and cross-sectional (**b**) scanning of billets, defibrated cane (**c**), and extracted raw juice (**d**).

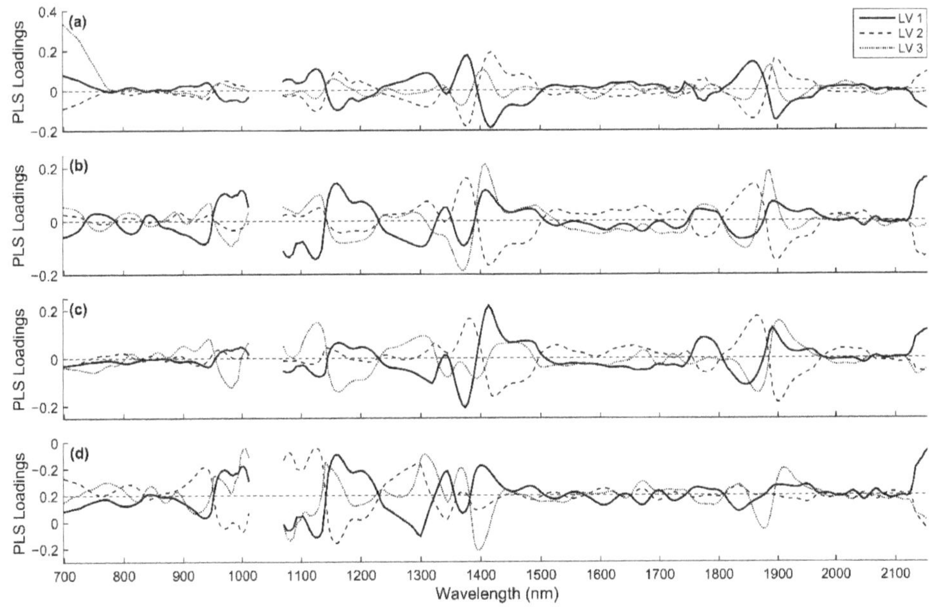

Figure A4. The three first Partial least squares loadings for Pol prediction using near-infrared reflectance spectroscopy from the skin (**a**) and cross-sectional (**b**) scanning of billets, defibrated cane (**c**), and extracted raw juice (**d**).

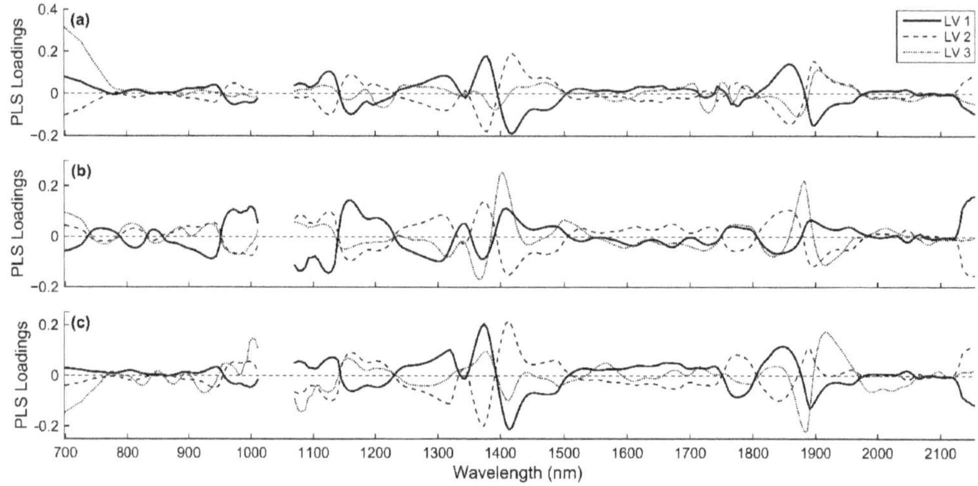

Figure A5. The three first Partial least squares loadings for Fibre prediction using near-infrared reflectance spectroscopy from the skin (**a**) and cross-sectional (**b**) scanning of billets, and defibrated cane (**c**).

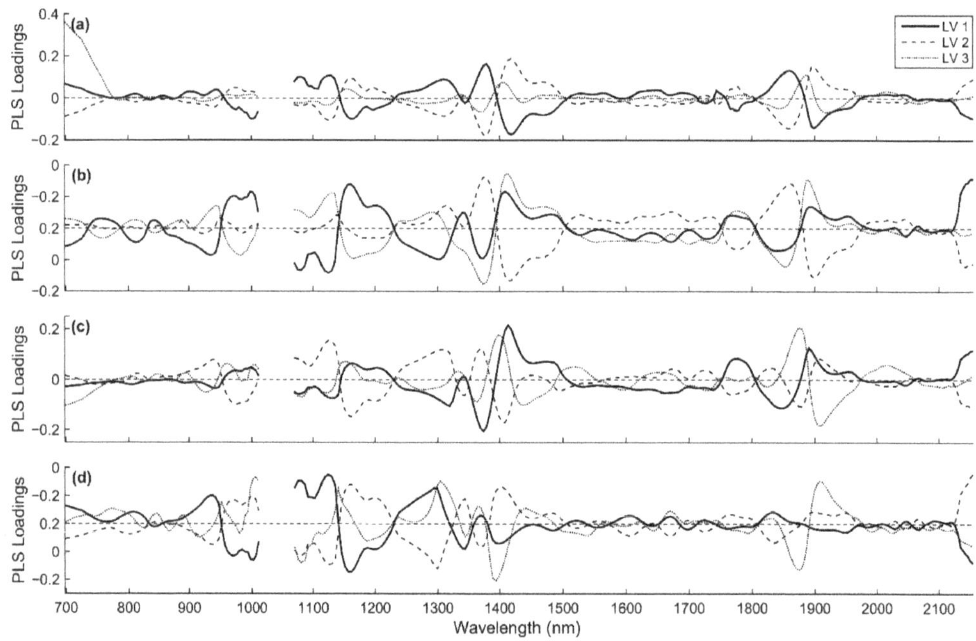

Figure A6. The three first Partial least squares loadings for Pol of cane prediction using near-infrared reflectance spectroscopy from the skin (**a**) and cross-sectional (**b**) scanning of billets, defibrated cane (**c**), and extracted raw juice (**d**).

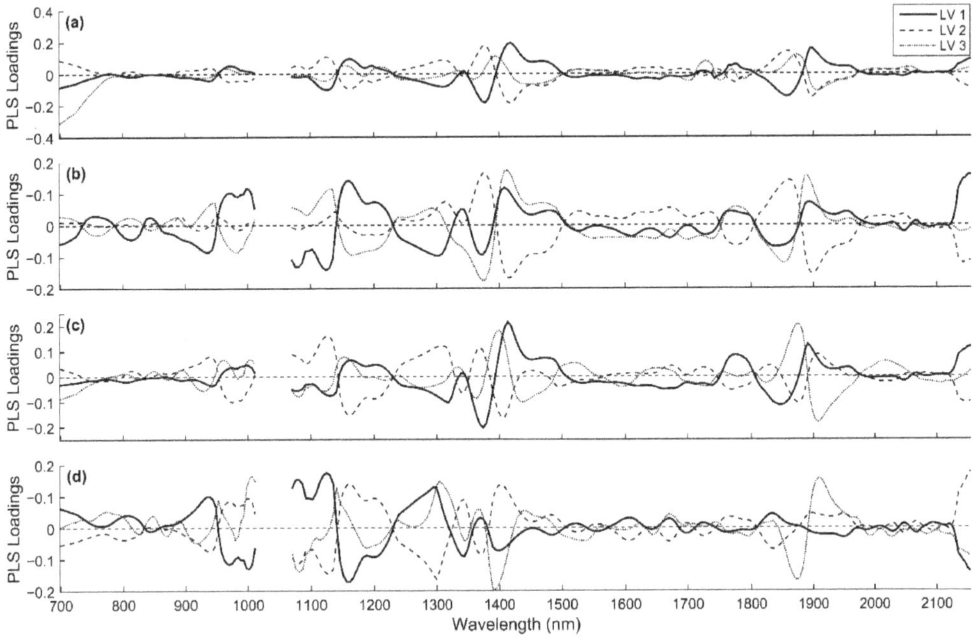

Figure A7. The three first Partial least squares loadings for total recoverable sugars (TRS) prediction using near-infrared reflectance spectroscopy from the skin (**a**) and cross-sectional (**b**) scanning of billets, defibrated cane (**c**), and extracted raw juice (**d**).

References

1. Sexton, J.; Everingham, Y.; Donald, D.; Staunton, S.; White, R. Investigating the identification of atypical sugarcane using NIR analysis of online mill data. *Comput. Electron. Agric.* **2020**, *168*, 105111. [CrossRef]
2. Phetpan, K.; Udompetaikul, V.; Sirisomboon, P. An online visible and near-infrared spectroscopic technique for the real-time evaluation of the soluble solids content of sugarcane billets on an elevator conveyor. *Comput. Electron. Agric.* **2018**, *154*, 460–466. [CrossRef]
3. Rodrigues, F.A.; Magalhães, P.S.G.; Franco, H.C.J. Soil attributes and leaf nitrogen estimating sugar cane quality parameters: Brix, pol and fibre. *Precis. Agric.* **2013**, *14*, 270–289. [CrossRef]
4. Ramírez-Morales, I.; Rivero, D.; Fernández-Blanco, E.; Pazos, A. Optimization of NIR calibration models for multiple processes in the sugar industry. *Chemom. Intell. Lab. Syst.* **2016**, *159*, 45–57. [CrossRef]
5. Cortés, V.; Blasco, J.; Aleixos, N.; Cubero, S.; Talens, P. Monitoring strategies for quality control of agricultural products using visible and near-infrared spectroscopy: A review. *Trends Food Sci. Technol.* **2019**, *85*, 138–148. [CrossRef]
6. Walsh, K.B.; Blasco, J.; Zude-Sasse, M.; Sun, X. Visible-NIR 'point' spectroscopy in postharvest fruit and vegetable assessment: The science behind three decades of commercial use. *Postharvest Biol. Technol.* **2020**, *168*, 111246. [CrossRef]
7. Workman, J., Jr.; Weyer, L. *Practical Guide and Spectral Atlas for Interpretive Near-Infrared Spectroscopy*; CRC Press: Boca Raton, FL, USA, 2012; ISBN 9781439875261.
8. Pasquini, C. Near infrared spectroscopy: A mature analytical technique with new perspectives—A review. *Anal. Chim. Acta* **2018**, *1026*, 8–36. [CrossRef] [PubMed]
9. Taira, E.; Ueno, M.; Kawamitsu, Y. Automated quality evaluation system for net and gross sugarcane samples using near infrared spectroscopy. *J. Near Infrared Spectrosc.* **2010**, *18*, 209–215. [CrossRef]
10. Sorol, N.; Arancibia, E.; Bortolato, S.A.; Olivieri, A.C. Visible/near infrared-partial least-squares analysis of Brix in sugar cane juice. A test field for variable selection methods. *Chemom. Intell. Lab. Syst.* **2010**, *102*, 100–109. [CrossRef]
11. Valderrama, P.; Braga, J.W.B.; Poppi, R.J. Validation of multivariate calibration models in the determination of sugar cane quality parameters by near infrared spectroscopy. *J. Braz. Chem. Soc.* **2007**, *18*, 259–266. [CrossRef]
12. Taira, E.; Ueno, M.; Furukawa, N.; Tasaki, A.; Komaki, Y.; Nagai, J.I.; Saengprachatanarug, K. Networking system employing near infrared spectroscopy for sugarcane payment in Japan. *J. Near Infrared Spectrosc.* **2013**, *21*, 477–483. [CrossRef]
13. Taira, E.; Ueno, M.; Saengprachatanarug, K.; Kawamitsu, Y. Direct sugar content analysis for whole stalk sugarcane using a portable near infrared instrument. *J. Near Infrared Spectrosc.* **2013**, *21*, 281–287. [CrossRef]

14. Phuphaphud, A.; Saengprachatanarug, K.; Posom, J.; Maraphum, K.; Taira, E. Non-destructive and rapid measurement of sugar content in growing cane stalks for breeding programmes using visible-near infrared spectroscopy. *Biosyst. Eng.* **2020**, *197*, 76–90. [CrossRef]
15. Phuphaphud, A.; Saengprachatanarug, K.; Posom, J.; Maraphum, K.; Taira, E. Prediction of the fibre content of sugarcane stalk by direct scanning using visible-shortwave near infrared spectroscopy. *Vib. Spectrosc.* **2019**, *101*, 71–80. [CrossRef]
16. Nawi, N.M.; Chen, G.; Jensen, T. In-field measurement and sampling technologies for monitoring quality in the sugarcane industry: A review. *Precis. Agric.* **2014**, *15*, 684–703. [CrossRef]
17. Bramley, R.G.V. Lessons from nearly 20 years of Precision Agriculture research, development, and adoption as a guide to its appropriate application. *Crop Pasture Sci.* **2009**, *60*, 197–217. [CrossRef]
18. Bramley, R.G.V.; Panitz, J.H.; Jensen, T.A.; Baillie, C.P. Within block spatial variation in CCS—Another potentially important consideration in the application of precision agriculture to sugarcane production. *Int. Sugar J.* **2013**, *115*, 1–8.
19. Corrêdo, L.d.P.; Canata, T.F.; Maldaner, L.F.; de Lima, J.J.A.; Molin, J.P. Sugarcane Harvester for In-field Data Collection: State of the Art, Its Applicability and Future Perspectives. *Sugar Tech* **2020**, *22*, 1–14. [CrossRef]
20. Nawi, N.M.; Chen, G.; Jensen, T.; Mehdizadeh, S.A. Prediction and classification of sugar content of sugarcane based on skin scanning using visible and shortwave near infrared. *Biosyst. Eng.* **2013**, *115*, 154–161. [CrossRef]
21. Nawi, N.M.; Chen, G.; Jensen, T. Visible and shortwave near infrared spectroscopy for predicting sugar content of sugarcane based on a cross-sectional scanning method. *J. Near Infrared Spectrosc.* **2013**, *21*, 289–297. [CrossRef]
22. Udompetaikul, V.; Phetpan, K.; Sirisomboon, P. Development of the partial least-squares model to determine the soluble solids content of sugarcane billets on an elevator conveyor. *Meas. J. Int. Meas. Confed.* **2021**, *167*, 107898. [CrossRef]
23. Maraphum, K.; Chuan-Udom, S.; Saengprachatanarug, K.; Wongpichet, S.; Posom, J.; Phuphaphud, A.; Taira, E. Effect of waxy material and measurement position of a sugarcane stalk on the rapid determination of Pol value using a portable near infrared instrument. *J. Near Infrared Spectrosc.* **2018**, *26*, 287–296. [CrossRef]
24. Phuphaphud, A.; Saengprachatanarug, K.; Posom, J.; Wongpichet, S.; Maraphum, K.; Taira, E. Effects of waxy types of a sugarcane stalk surface on the spectral characteristics of visible-shortwave near infrared measurement. *Eng. J.* **2019**, *23*, 13–24. [CrossRef]
25. Cardozo, N.P.; Sentelhas, P.C. Climatic effects on sugarcane ripening under the influence of cultivars and crop age. *Sci. Agric.* **2013**, *70*, 449–456. [CrossRef]
26. CONSECANA National Concill of Sugarcane Producers of São Paulo State. *Instruction Manual*, 6th ed.; CONSECANA-SP: Piracicaba, SP, Brazil, 2015; 81p. (In Portuguese)
27. Franceschini, M.H.D.; Demattê, J.A.M.; Kooistra, L.; Bartholomeus, H.; Rizzo, R.; Fongaro, C.T.; Molin, J.P. Effects of external factors on soil reflectance measured on-the-go and assessment of potential spectral correction through orthogonalisation and standardisation procedures. *Soil Tillage Res.* **2018**, *177*, 19–36. [CrossRef]
28. Oliveri, P.; Malegori, C.; Simonetti, R.; Casale, M. The impact of signal pre-processing on the final interpretation of analytical outcomes—A tutorial. *Anal. Chim. Acta* **2019**, *1058*, 9–17. [CrossRef] [PubMed]
29. Barnes, R.J.; Dhanoa, M.S.; Lister, S.J. Standard normal variate transformation and de-trending of near-infrared diffuse reflectance spectra. *Appl. Spectrosc.* **1989**, *43*, 772–777. [CrossRef]
30. Savitzky, A.; Golay, M.J.E. Smoothing and Differentiation of Data by Simplified Least Squares Procedures. *Anal. Chem.* **1964**, *36*, 1627–1639. [CrossRef]
31. Gerretzen, J.; Szymańska, E.; Jansen, J.J.; Bart, J.; Van Manen, H.J.; Van Den Heuvel, E.R.; Buydens, L.M.C. Simple and Effective Way for Data Preprocessing Selection Based on Design of Experiments. *Anal. Chem.* **2015**, *87*, 12096–12103. [CrossRef] [PubMed]
32. Valderrama, P.; Braga, J.W.B.; Poppi, R.J. Variable selection, outlier detection, and figures of merit estimation in a partial least-squares regression multivariate calibration model. A case study for the determination of quality parameters in the alcohol industry by near-infrared spectroscopy. *J. Agric. Food Chem.* **2007**, *55*, 8331–8338. [CrossRef]
33. The American Society for Testing and Materials (ASTM). *ASTM E1655-17 Standard Practices for Infrared Multivariate Quantitative Analysis*; ASTM Annual Book of Standards: West Conshohocken, PA, USA, 2017; Volume 5, 30p.
34. Van der Voet, H. Comparing the predictive accuracy of models using a simple randomization test. *Chemom. Intell. Lab. Syst.* **1994**, *25*, 313–323. [CrossRef]
35. Santana, F.B.; Giuseppe, L.O.; Souza, A.M.; Poppi, R.J. Removing the moisture effect in soil organic matter determination using NIR spectroscopy and PLSR with external parameter orthogonalization. *Microchem. J.* **2019**, *145*, 1094–1101. [CrossRef]
36. Olivieri, A.C. Practical guidelines for reporting results in single- and multi-component analytical calibration: A tutorial. *Anal. Chim. Acta* **2015**, *868*, 10–22. [CrossRef] [PubMed]
37. Pagani, V.; Stella, T.; Guarneri, T.; Finotto, G.; van den Berg, M.; Marin, F.R.; Acutis, M.; Confalonieri, R. Forecasting sugarcane yields using agro-climatic indicators and Canegro model: A case study in the main production region in Brazil. *Agric. Syst.* **2017**, *154*, 45–52. [CrossRef]
38. Wang, J.; Zhao, T.; Yang, B.; Zhang, S. Sucrose Metabolism and Regulation in Sugarcane. *J. Plant Physiol. Pathol.* **2017**, *5*, 2. [CrossRef]
39. Mancini, M.C.; Leite, D.C.; Perecin, D.; Bidóia, M.A.P.; Xavier, M.A.; Landell, M.G.A.; Pinto, L.R. Characterization of the Genetic Variability of a Sugarcane Commercial Cross Through Yield Components and Quality Parameters. *Sugar Tech* **2012**, *14*, 119–125. [CrossRef]
40. Solomon, S. Post-harvest deterioration of sugarcane. *Sugar Tech* **2009**, *11*, 109–123. [CrossRef]

41. Bazar, G.; Kovacs, Z.; Tsenkova, R. Evaluating spectral signals to identify spectral error. *PLoS ONE* **2016**, *11*, e0146249. [CrossRef] [PubMed]
42. Steidle Neto, A.J.; Lopes, D.C.; Toledo, J.V.; Zolnier, S.; Silva, T.G.F. Classification of sugarcane varieties using visible/near infrared spectral reflectance of stalks and multivariate methods. *J. Agric. Sci.* **2018**, *156*, 537–546. [CrossRef]
43. da Costa, M.V.A.; Fontes, C.H.; Carvalho, G.; de Moraes Júnior, E.C. UltraBrix: A Device for Measuring the Soluble Solids Content in Sugarcane. *Sustainability* **2021**, *13*, 1227. [CrossRef]
44. Uys, L.; Botha, F.C.; Hofmeyr, J.H.S.; Rohwer, J.M. Kinetic model of sucrose accumulation in maturing sugarcane culm tissue. *Phytochemistry* **2007**, *68*, 2375–2392. [CrossRef]
45. Wang, J.; Nayak, S.; Koch, K.; Ming, R. Carbon partitioning in sugarcane (Saccharum species). *Front. Plant Sci.* **2013**, *4*, 201. [CrossRef]
46. Inarkar, M.B.; Lele, S.S. Extraction and Characterization of Sugarcane Peel Wax. *ISRN Agron.* **2012**, *2012*, 1–6. [CrossRef]
47. Sexton, J.; Everingham, Y.; Donald, D.; Staunton, S.; White, R. A comparison of non-linear regression methods for improved on-line near infrared spectroscopic analysis of a sugarcane quality measure. *J. Near Infrared Spectrosc.* **2018**, *26*, 297–310. [CrossRef]
48. Roger, J.M.; Boulet, J.C. A review of orthogonal projections for calibration. *J. Chemom.* **2018**, *32*, 1–16.
49. Osborne, B.G. Near-Infrared Spectroscopy in Food Analysis. In *Encyclopedia of Analytical Chemistry*; John Wiley & Sons, Ltd.: Chichester, UK, 2006.
50. Steidle Neto, A.J.; Toledo, J.V.; Zolnier, S.; Lopes, D.C.; Pires, C.V.; Silva, T.G.F. Prediction of mineral contents in sugarcane cultivated under saline conditions based on stalk scanning by Vis/NIR spectral reflectance. *Biosyst. Eng.* **2017**, *156*, 17–26. [CrossRef]
51. Golic, M.; Walsh, K.; Lawson, P. Short-wavelength near-infrared spectra of sucrose, glucose, and fructose with respect to sugar concentration and temperature. *Appl. Spectrosc.* **2003**, *57*, 139–145. [CrossRef]
52. Pomares-Viciana, T.; Martínez-Valdivieso, D.; Font, R.; Gómez, P.; del Río-Celestino, M. Characterisation and prediction of carbohydrate content in zucchini fruit using near infrared spectroscopy. *J. Sci. Food Agric.* **2018**, *98*, 1703–1711. [CrossRef]
53. Bázár, G.; Romvári, R.; Szabó, A.; Somogyi, T.; Éles, V.; Tsenkova, R. NIR detection of honey adulteration reveals differences in water spectral pattern. *Food Chem.* **2016**, *194*, 873–880. [CrossRef] [PubMed]

sensors MDPI

Article

PLS-R Calibration Models for Wine Spirit Volatile Phenols Prediction by Near-Infrared Spectroscopy

Ofélia Anjos [1,2,3,*], Ilda Caldeira [4,5,*], Tiago A. Fernandes [6,7], Soraia Inês Pedro [2,3], Cláudia Vitória [8], Sheila Oliveira-Alves [4], Sofia Catarino [9,10] and Sara Canas [4,5]

1 Instituto Politécnico de Castelo Branco, Quinta da Senhora de Mércules, 6001-909 Castelo Branco, Portugal
2 Centro de Estudos Florestais, Instituto Superior de Agronomia, Universidade de Lisboa, Tapada da Ajuda, 1349-017 Lisboa, Portugal; soraia_p1@hotmail.com
3 Centro de Biotecnologia de Plantas da Beira Interior, 6001-909 Castelo Branco, Portugal
4 Instituto Nacional de Investigação Agrária e Veterinária, Quinta de Almoínha, Pólo de Dois Portos, 2565-191 Dois Portos, Portugal; sheila.alves@iniav.pt (S.O.-A.); sara.canas@iniav.pt (S.C.)
5 MED—Mediterranean Institute for Agriculture, Environment and Development, Instituto de formação avançada, Universidade de Évora, Pólo da Mitra, Ap. 94, 7006-554 Évora, Portugal
6 CQE—Centro de Química Estrutural, Associação do Instituto Superior Técnico para a Investigação e Desenvolvimento (IST-ID), Universidade de Lisboa, 1049-001 Lisboa, Portugal; tiago.a.fernandes@tecnico.ulisboa.pt
7 DCeT—Departamento de Ciências e Tecnologia, Universidade Aberta, Rua da Escola Politécnica, 141-147, 1269-001 Lisboa, Portugal
8 Faculdade de Ciências, Universidade da Beira Interior, 6201-556 Covilhã, Portugal; adriana-1998@hotmail.com
9 LEAF—Linking Landscape, Environment, Agriculture and Food Research Center, Instituto Superior de Agronomia, Universidade de Lisboa, Tapada da Ajuda, 1349-017 Lisboa, Portugal; sofiacatarino@isa.ulisboa.pt
10 CEFEMA—Center of Physics and Engineering of Advanced Materials, Instituto Superior Técnico, Universidade de Lisboa, Av. Rovisco Pais, 1, 1049-001 Lisboa, Portugal
* Correspondence: ofelia@ipcb.pt (O.A.); ilda.caldeira@iniav.pt (I.C.)

Citation: Anjos, O.; Caldeira, I.; Fernandes, T.A.; Pedro, S.I.; Vitória, C.; Oliveira-Alves, S.; Catarino, S.; Canas, S. PLS-R Calibration Models for Wine Spirit Volatile Phenols Prediction by Near-Infrared Spectroscopy. *Sensors* **2022**, *22*, 286. https://doi.org/10.3390/s22010286

Academic Editors: Mercedes Del Río Celestino and Rafael Font Villa

Received: 3 December 2021
Accepted: 29 December 2021
Published: 31 December 2021

Abstract: Near-infrared spectroscopic (NIR) technique was used, for the first time, to predict volatile phenols content, namely guaiacol, 4-methyl-guaiacol, eugenol, syringol, 4-methyl-syringol and 4-allyl-syringol, of aged wine spirits (AWS). This study aimed to develop calibration models for the volatile phenol's quantification in AWS, by NIR, faster and without sample preparation. Partial least square regression (PLS-R) models were developed with NIR spectra in the near-IR region (12,500–4000 cm^{-1}) and those obtained from GC-FID quantification after liquid-liquid extraction. In the PLS-R developed method, cross-validation with 50% of the samples along a validation test set with 50% of the remaining samples. The final calibration was performed with 100% of the data. PLS-R models with a good accuracy were obtained for guaiacol (r^2 = 96.34; RPD = 5.23), 4-methyl-guaiacol (r^2 = 96.1; RPD = 5.07), eugenol (r^2 = 96.06; RPD = 5.04), syringol (r^2 = 97.32; RPD = 6.11), 4-methyl-syringol (r^2 = 95.79; RPD = 4.88) and 4-allyl-syringol (r^2 = 95.97; RPD = 4.98). These results reveal that NIR is a valuable technique for the quality control of wine spirits and to predict the volatile phenols content, which contributes to the sensory quality of the spirit beverages.

Keywords: NIR; calibration models; PLS-R; volatile phenols; aged wine spirit

1. Introduction

Volatile phenols are low molecular weight aromatic alcohols that comprise phenol and may include substituents such as alkyl, methoxyl, vinyl and allyl. These compounds can exist in foods due to a variety of mechanisms, as summarized by Schieber and Wust [1]. Some of these compounds are responsible for characteristic odor notes of various foods [1] and alcoholic beverages such as wine [2], whisky [3], rum [4] and aged wine spirit (AWS) [5]. Like other alcoholic beverages such as rum or whisky, Wine spirits are aged in wooden barrels, and the volatile phenols are among the most important compounds, in terms of sensory impact, extracted from the wood into the beverage. The main volatile phenols identified and quantified in AWS are guaiacol, eugenol, syringol, 4-methy-lsyringol, 4-allyl-syringol, 4-methyl-guaiacol and ethyl guaiacol, which are well related to odour notes such as smoky, clove, burnt, flowery and carnation, respectively [5]. Their amounts in the AWS are usually low (from traces to 1.5 g/L), increasing over time [6,7] and influenced by the wood species and toasting level, as well as the ageing system [7,8]. Although their low concentration in alcoholic beverages, these compounds have very low detection thresholds, and for this reason, several volatile phenols have been identified as critical odorants in wooden aged alcoholic beverages [3–5].

Gas chromatography (GC), coupled to an appropriate detection system (flame ionization detection (FID) or mass spectrometry (MS)), is typically used to analyse volatile phenols in alcoholic beverages. HPLC has also been used, although less extensively than GC [9]. These analyses are commonly preceded by an extraction step, which can be made through a variety of procedures such as liquid-liquid extraction [10], solid-phase extraction [11], solid-phase microextraction [12,13] stir bar sorptive extraction [14,15], dispersive liquid-liquid microextraction [16] and ultrasound-assisted emulsification-microextraction [17].

Near-infrared spectroscopy (NIR) is an analytical technique that uses the region of the electromagnetic spectrum between 12,500 and 4000 cm^{-1}, and the collected spectrum of a sample comprises overtones and combination vibrations of molecules with different functional groups [18,19]. This analytical method has been applied in several matrices, namely foods and beverages. Compared to chemical analysis, NIR spectroscopy provides the ideal technology for quick and efficient analysis and has the advantage of being faster and requiring no sample preparation [20–23]. The most significant drawback is that the identification of small compounds is limited to a mass fraction more significant than roughly 0.1–0.5%. However, this also depends on the functional group(s) present in these compounds, which determines the magnitude of the absorption band shown in the NIR spectra. The intensity of a C–H vibration, for example, is substantially lower than that of an O–H vibration.

When paired with an appropriate chemometric methodology, NIR spectroscopy provides a rapid, non-destructive, and cost-effective method of food analysis that may be used for a wide range of products. It is used in the food sector to guarantee that the food being marketed meets the highest standards of food safety and hygiene and defend against false claims made by the food producer, processor, distributor, or retailer [20]. Its advantage is that NIR spectroscopy provides a spectrum that may be typical of a sample and may behave as a "fingerprint" by recording the response of specific chemical bonds (for example, O–H, N–H, C–H) to NIR radiation. Overtones of O-H or N-H stretching modes provide detailed data on intermolecular interactions, and NIR spectroscopy offers unique capabilities for analyzing hydrogen bonding. As a result, it is no surprise that NIR is commonly used to evaluate food compositional elements, but it can also be employed to determine more complicated attributes like texture and sensory characteristics [24].

PLS-R is a method for relating two data matrices to investigate complex problems and analyze available data more realistically. Many studies with different food products, was made using NIR data and PLS-R to perform calibration model [25,26], and in some cases with better responses than other regression techniques [27]. Additionally, the PLS-R technique is known to be affected by outliers in the data, and, in the present study, it is instrumental to eliminate possible outliers from the GC analyses. In the analyses of volatile compounds with low molecular weight, some outliers can occur and, with this technique, will be identified and eliminated more easily.

Concerning the alcoholic beverages, NIR analysis has been applied to assess the alcoholic strength of whiskies and vodkas [28] as well as other constituents of whiskies [28,29], rum and brandies [30], gin and vodka [31], and other distilled beverages [30,32–34] and to identify adulteration in distilled spirits [35]. Hanousek et al. [36] performed calibration models for major volatile compounds and phenols of wine spirits based on least squares regression. A recent study used NIR to distinguish wine spirits produced with two different wood species (oak and chestnut) and ageing technologies (barrel and alternative) with a precision of up to 90% [37].

Figure 1 shows the chemical structures and sensory properties of the most frequent volatile phenols in AWS, examined in this study.

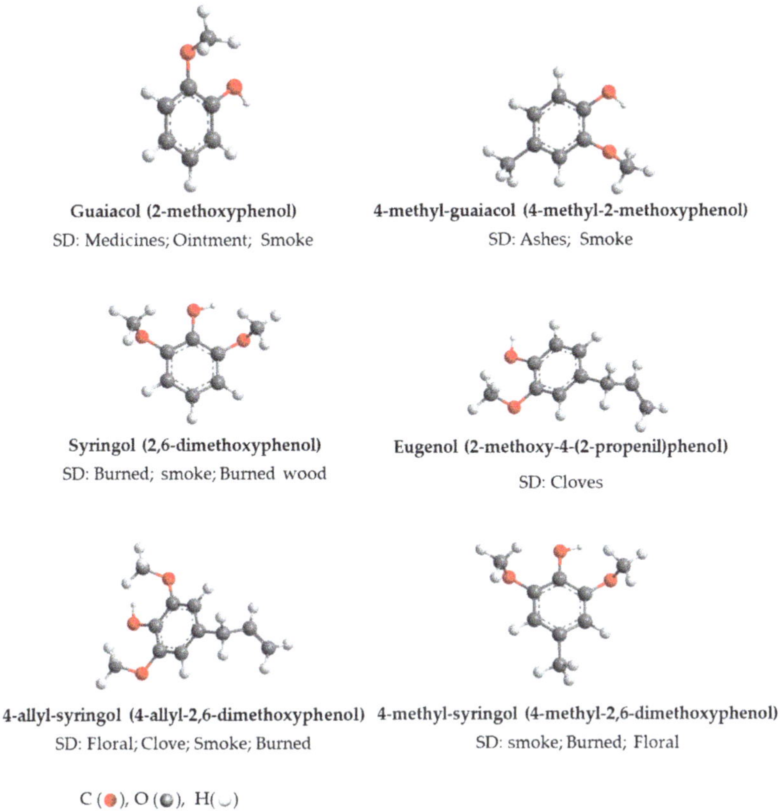

Figure 1. Chemical structure of volatile phenols studied in the AWS and their associated sensory descriptors (SD) [5].

This study aimed to assess the capability of NIR technology combined with chemometrics to perform calibration models to predict the content of volatile phenols in AWS.

2. Materials and Methods

2.1. Samples

The AWS samples used in this study were produced within the Oxyrebrand project-https://projects.iniav.pt/oxyrebrand (accessed on 14 December 2021) [6]. Briefly, samples resulting from ageing with different wood species (chestnut and oak), from traditional (250 L wooden barrel) and alternative technology (50 L glass demijohns with wood staves and micro-oxygenation-MOX) and two different periods of storage in the bottle were used. For the alternative systems, the 50 L demijohns with chestnut or oak wood staves underwent different micro-oxygenation conditions: flow rate of 2 mL/L/month during the first 15 days followed by 0.6 mL/L/month until 365 days; 2 mL/L/month during the first 30 days followed by 0.6 mL/L/month until 365 days; 2 mL/L/month during the first 60 days followed by 0.6 mL/L/month until 365 days; nitrogen application with a flow rate of 20 mL/L/month.

After the ageing process aforementioned, the AWS was bottled and stored for 2 months and analysed in the first stage of bottling (T0) and after 6 months (T6). For each modality, two essay replicates and three analytical measurements were used; a total of 120 samples were analysed, according to Table 1.

Table 1. Sample characterization and number used in the model calibration.

		Chestnut Wood (C)		Oak Wood (L)		Total
		T0	T1	T0	T1	
(B) 250 L wooden barrel		6 *	6 *	6 *	6 *	24
50 L glass demijohns with wood staves with MOX	(15) with a flow rate of 2 mL/L/month during the first 15 days followed by 0.6 mL/L/month until 365 days	6 *	6 *	6 *	6 *	24
	(30) flow rate of 2 mL/L/month during the first 30 days followed by 0.6 mL/L/month until 365 days	6 *	6 *	6 *	6 *	24
	(60) a flow rate of 2 mL/L/month during the first 60 days followed by 0.6 mL/L/month until 365 days	6 *	6 *	6 *	6 *	24
(N) nitrogen application with a flow rate of 20 mL/L/month		6 *	6 *	6 *	6 *	24
Total		30	30	30	30	120

* Two replicates of each modality were carried and the analysis was made in triplicate ($2 \times 3 = 6$).

The use of these different AWS samples is intended to ensure a high variability to have accurate models that can be applied in a broader range of this kind of beverage.

2.2. Analytical Procedures

2.2.1. Reagents

Anhydrous sodium sulfate and ethanol were acquired from Merck (Darmstadt, Germany), dichloromethane from Honeywell Riedel-de Haën (Steinheim, Germany), and silanized glass wool from Supelco (Steinheim, Germany).

The ultrapure water was achieved through the arium®comfort I equipment from Sartorius Lab Instruments, Goettingen, Germany.

GC-FID and GC-MS standards: guaiacol, eugenol, 3,4-dimethylphenol (internal standard), syringol, 5-methyl-2-hexanol (internal standard; IS) were bought from Fluka (Buchs, Switzerland); 4-methyl-syringol, 4-allyl-syringol were acquired from Aldrich (Steinheim, Germany); 4-methyl-guaiacol, were purchased from TCI (Zwijndrecht, Belgium).

2.2.2. Quantification of Volatile Phenols in AWS

Prior to GC analysis, liquid-liquid extraction with ultrasonication was performed. The wine spirits samples (100 mL), previously diluted to 20% v/v, were added with internal standards and extracted with successive additions of 30, 10 and 10 cm^3 dichloromethane and using ultrasonication according to the methodology described by Granja-Soares et al. [7]. The organic phases were collected, dried over sodium sulphate, filtered with glass wool silanized and then concentrated using a Büchi rotary concentrator (without vacuum at a temperature of 42 ± 0.5 °C) until a final volume of 0.25 mL. Each wine spirit sample was extracted in duplicate.

GC-FID analysed the concentrated extracts under the following chromatographic conditions: Agilent Technologies 6890 Series gas chromatograph (Wilmington, DE, USA) joined to a flame ionization detector (FID) and fitted out with a fused silica capillary column of polyethylene glycol (INNOWax of J&W Scientific, Folsom, CA, USA), 30 m, 0.32 mm i.d., 0.25 µm film thickness; split injection (1:25) of 0.8 µL of each extract; injector and detector temperatures (250 °C); carrier gas hydrogen (2.4 mL/min); oven temperature program: 3.5/min from 35 °C (6 min isothermal) to 55 °C, 7.5 °C/min to 130 °C, 5 °C/min to 210 °C (30 min isothermal). For each extract, three injections were done.

Hydroalcoholic solutions (20% v/v) of standards were extracted and analysed under similar conditions, and a calibration curve with five points was established for each compound. These curves were used for the quantification of volatile phenols in the AWS.

The compounds were identified by analyzing the extracts in GC-MS equipment (Magnum, Finnigan Mat, San Jose, CA, USA) under similar chromatographic conditions, with transfer line at 250 °C, working with electron impact mode at 70 eV and scanning the mass range of m/z 20–340. The compounds' identities were determined by comparing the MS fragmentation pattern with reference compounds and with mass spectra in the NIST libraries.

2.3. Spectroscopic Measurements

The spectra of the AWS samples were obtained using a NIR spectrometer (MPA Bruker) in a transmitted light mode with 1 mm quartz cells. The samples were measured at 25 °C after 2 min in the instrument before scanning; the background was air-made. The samples were measured with an 8 cm^{-1} spectral resolution and 32 scans in the wavenumber range of 12,500 to 4000 cm^{-1} [32,37]. A background scan was performed after scanning a sequence of 10 samples.

2.4. Data Analysis

To ensure that the models were produced with a significant variability for the analytical determination, two principal component analysis (PCA) was performed: the first with the analytical determination identifying the different factor variance effects, and the second one with NIR spectra of AWS. The second PCA was also useful to identify the region that best discriminated the samples and, consequently, was the best to use in the models.

The model calibration analysis was performed with the average of two replicated spectra for each AWS sample.

The vector normalization pre-processes (SNV) were applied to all spectra used in the calibration models, which first normalizes a spectrum by calculating the average intensity value and then subtracting this value from the spectrum. Following that, new pre-treatments for model construction were tested. Briefly, multiplicative scatter correction (MSC); first derivative (1stDer); second derivative (2stDer), first derivative + multiplicative scatter correction (1stDer + MSC) and first derivative + straight line elimination (1stDer + SLS).

The cross-validation process was used in model validation with the general theoretical validation criterion leave-one-out method, which is more appropriate when a small dimension data set is used. The parameters used to identify a better calibration model were: r^2—coefficient of determination (proportion of variance in the dependent variable that the independent one can explain); RPD—residual prediction deviation (by providing a metric of model validity, higher values correspond to better model's predictive capacity); RMSEP—root means square error of validation; RMSECV—root means a square error of cross-validation, and RMSEC—root mean of the standard error of calibration.

Data pre-processing methods and selection of wavenumber ranges resulted in high predictability and precise estimation of volatile phenol in AWS.

The samples were divided into two sets, one for calibration (50% of data) and the other for validation (50% of data) after the model was tested with all values (100% of data), according to a similar methodology previously used [38].

The PCA for analytical data analysis was carried out using Statistica version 7.0 software (StatSoft Inc., Tulsa, OK, USA). Calibration models were made using OPUS 8.5.29 From Bruker Optik GmbH 2019. Spectral PCA analysis was performed using the UnscramblerX 10.5 (CAMO, Oslo, Norway).

3. Results

In this study, guaiacol, 4-methyl-guaiacol, eugenol, syringol, 4-methyl-syringol and 4-allyl-syringol contents in AWS presented a wide range of values (Figure 2) and significative variability given the different ageing modalities used as variability sources, which suggests a good data scattering.

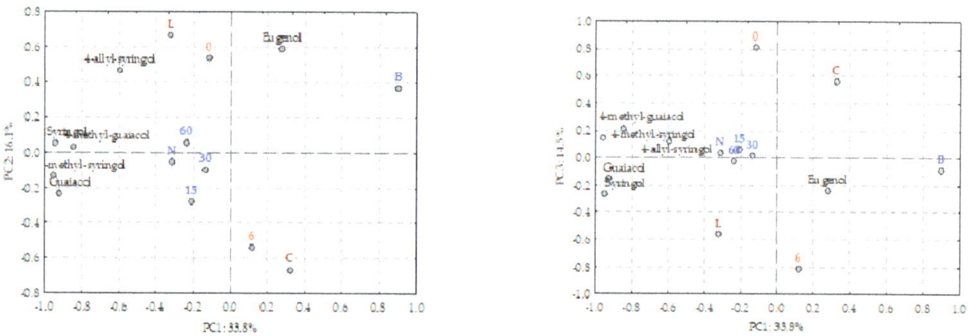

Figure 2. PCA representation of loadings and scores of all AWS samples and all volatile phenols analysed. Legend: C and L stand for the wood used in the ageing process, Chestnut and Limousin respectively; O15, 30 and 60 are the different micro-oxygenation modalities used in the alternative system; N—without micro-oxygenation; B—Barrel; 0—0 months in bottle; 6—6 months in bottle.

Regarding Figure 3, it is possible to establish that the NIR spectra followed the trend of sample differentiation, which was also observed in Figure 2. However, NIR spectra showed that other compounds present in AWS could affect their relative position along the PCA axes [8,39,40]. In Figure 3, the AWS samples aged with chestnut wood and Limousin oak wood are presented separately to understand better.

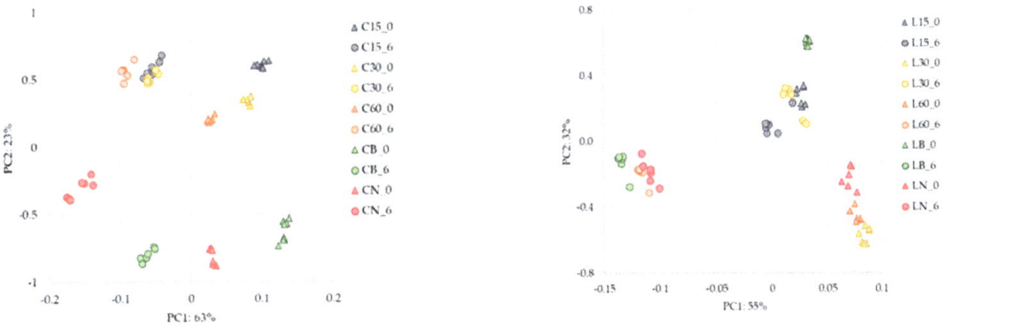

Figure 3. PCA was performed with spectral information of the AWS with chestnut (C) and with Limousin wood, acquired in NIR. Legend: C and L stand for the wood used in the ageing process, Chestnut and Limousin respectively; 15, 30 and 60 the different levels of micro-oxygenation used in the alternative system; N—no micro-oxygenation used in the alternative system; B—Barrel; 0—0 months in a bottle; 6—6 months in bottle.

Figure 4 exhibits a representative NIR spectrum of the AWS, similar to those obtained by other authors for wine spirit, grape marc spirit, fruit spirits, whisky and vodka [28,29,37,40–43].

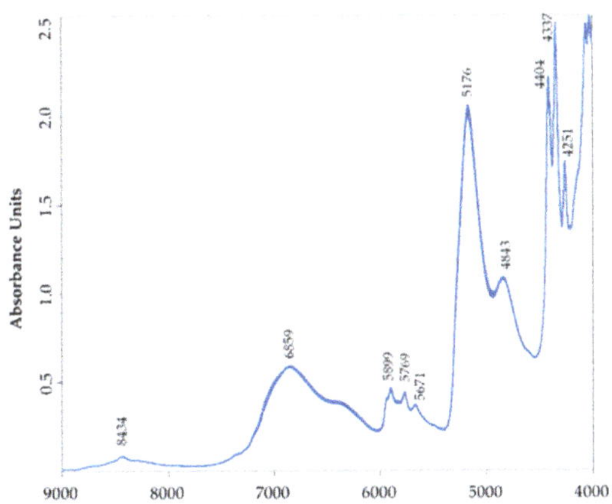

Figure 4. Representative absorption spectra of all AWS samples acquired in the NIR region measured against a background of air.

The water content in the spirits can be detected in the region around 6859 cm^{-1}, which comprises the second overtones of the stretching νO–H band and a combination of deformation and stretching vibrations of the OH group (specifically water).

The peak with lower intensity near 8434 cm^{-1} is assigned to the second overtone of the C–H stretch of ethanol, one of the main compounds in AWS. This peak is also ascribed to the combination of the bending vibration of δO–H bend and the first overtone of the stretching νO–H has given the water influence [37].

The region from 5600 to 6000 cm^{-1} presents three small peaks ascribed to the νC–H stretch of the first overtones of CH_2 and CH_3 groups [22,43] and OH from aromatic groups [44].

At 6859 cm^{-1} occurs the second overtone of the stretching vibrations of ν(O–H) of water and ethanol as well.

The strong band at 5176 cm^{-1}, characteristic of AWS [37], is related to a combination of stretching and deformation of the OH group and first overtones of water and ethanol and C–H stretch first overtones [43].

The peak at 4843 cm^{-1} can be assigned to aromatic C–H and –C=CH [44].

Volatile compounds extracted from the wooden barrel (mainly furanic and phenolic compounds) contribute to the flavour of the beverage [45,46]. Even in small amounts, soluble carbohydrates, most notably sugars, may contribute to the final flavour [46]. The ethanol, sugars and phenolic compounds have an absorption band at 4404 cm^{-1} related to the second overtone of stretching νC–H and νO-H overtones at 4338 cm^{-1} [47]. The bands at 4404 cm^{-1} and 4337 cm^{-1} are also related to the methanol content in the AWS [32,37]. The band at 4251 cm^{-1} is related to the combination of stretching and bending deformation of CH units of C–H(aromatic) and C–H(aryl) [48,49].

Table 2 presents the descriptive statistics (average, standard deviation, range, and coefficient of variation) for the volatile phenols, namely, guaiacol; 4-methyl-guaiacol, eugenol, syringol, 4-methyl-syringol, 4-allyl-syringol, content in the AWS samples used to develop the NIR calibrations. Table 3 shows the statistics of the prediction model for cross-validation of the calibration set and of the test set validation of the compounds above in the set of all samples analysed.

Table 2. Statistics of the sample sets for guaiacol, 4-methyl guaiacol, eugenol, syringol, 4-methyl-siringol and 4-allyl-syringol quantification in AWS analysed.

Volatile Phenol	Number of Samples	N	Mean ± SD	Min–Max	CV (%)	LOQ [1]
Guaiacol (mg/L)	Set1	56	0.491 ± 0.165	0.098–0.696	33.65	0.037
	Set2	56	0.489 ± 0.158	0.095–0.699	32.31	
	Set1 + Set2	112	0.487 ± 0.158	0.095–0.696	32.33	
4-methyl-guaiacol (mg/L)	Set1	56	0.279 ± 0.109	0.073–0.487	39.07	0.033
	Set2	56	0.280 ± 0.101	0.073–0.478	38.92	
	Set1 + Set2	112	0.279 ± 0.174	0.073–0.487	37.75	
Eugenol (mg/L)	Set1	54	0.291 ± 0.020	0.252–0.350	6.91	0.021
	Set2	54	0.290 ± 0.019	0.251–0.328	6.57	
	Set1 + Set2	108	0.289 ± 0.021	0.252–0.328	7.22	
Syringol (mg/L)	Set1	54	1.708 ± 0.705	0.221–3.172	41.31	0.029
	Set2	54	1.679 ± 0.683	0.244–3.106	40.66	
	Set1 + Set2	108	1.702 ± 0.695	0.221–3.172	39.65	
4-methyl-syringol (mg/L)	Set1	55	1.034 ± 0.383	0.274–1.552	37.04	0.034
	Set2	55	1.090 ± 0.395	0.259–1.536	36.21	
	Set1 + Set2	110	1.043 ± 0.393	0.259–1.552	37.66	
4-allyl-syringol (mg/L)	Set1	51	0.414 ± 0.076	0.273–0.55	18.32	0.043
	Set2	51	0.416 ± 0.078	0.255–0.578	18.80	
	Set1 + Set2	102	0.417 ± 0.075	0.255–0.578	17.87	

[1] LOQ—limit of quantification; CV—coefficient of variation (CV = SD/mean); SD—standard deviation; min—minimum value observed in the corresponding set; max—maximum value observed in the corresponding set.

Table 3. Cross-validation and validation set results of the calculated models obtained for different determinations.

Volatile Phenol	Spectral Range (cm^{-1})	Pre-Process		Rk	r^2	RMSEP	RMSECV	RMSEC	RPD	Bias
Guaiacol	9118.1–5415.3	1stDer + MSC	Set 1	10	96.80	0.0296			5.90	−0.0095
			Set 2	5	96.84		0.0270		5.63	0.0004
			Set 1 + 2	8	96.34			0.0298	5.23	
4-methyl-guaiacol	8304.2–7347.7 6869.4–5434.6 4956.3–4478	1stDer + SLS	Set 1	10	96.34	0.0233			5.36	−0.0052
			Set 2	10	92.70		0.0204		3.7	0.0006
			Set 1 + 2	10	96.10			0.0218	5.07	
Eugenol	9337.9–5446.2	1stDer + SLS	Set 1	7	95.30	0.0049			4.92	−0.0017
			Set 2	10	92.30		0.0053		3.59	0.0001
			Set 1 + 2	10	96.06			0.0044	5.04	
Syringol	6101.9–5446.2	1stDer + SLS	Set 1	9	97.81	0.1170			6.76	−0.0028
			Set 2	8	93.74		0.1560		4.50	−0.0028
			Set 1 + 2	10	97.32			0.1170	6.11	
4-methyl-syringol	9160.5–4512.7	1stDer + SLS	Set 1	10	94.88	0.0874			4.45	−0.0108
			Set 2	10	90.42		0.0653		3.23	−0.0024
			Set 1 + 2	10	95.79			0.0772	4.88	
4-allyl-syringol	9353.3–7498.1 6101.9–5446.2	1stDer + MSC	Set 1	8	90.05	0.0176			3.19	−0.0018
			Set 2	10	92.44		0.0243		3.64	−0.0011
			Set 1 + 2	10	95.97			0.0159	4.98	

MSC—multiplicative scatter correction; SLS—straight line elimination; 1stDer—first derivative; 2ndDer—second derivative; r^2—coefficient of determination; RMSECV—root mean square error of cross-validation; RMSEP—root mean square error of prediction; RMSEC: root mean square error of calibration; RPD—ratios of performance to deviation; Bias—mean value of deviation, also called systematic error; Rk—rank.

For the calibration models development, the entire infrared spectral region (12,000–4000 cm^{-1}) was considered for spectral acquisition after eliminating the redundant spectra based on the spectral PCA analyses.

As shown in Table 2, a wide-ranging concentration value was found in the AWS for each volatile phenol, indicating a good scattering for such model development.

The more accurate model, for each analysed compound, obtained with NIR raw spectral data regressed against their GC–FID determination is summarised in Table 3 for validation set (50% of the samples), cross-validation (50% of the samples) and calibration (100% of the samples). Figure 5 represents the deviation observed with the final calibration model.

The model selection was based on the analyses of all error parameters. Only the model with higher RPD, lower standard error of prediction of the test-set and calibration model (given by the root mean square error of cross-validation (RMSECV) and root mean square error of prediction (RMSEP)) and lower rank used in the prediction, were selected and presented. Bias analysis was also performed to confirm the adjustment of the model, and the value must be as nearer as possible to zero.

PLS was used to perform the calibration model with the more appropriate pre-treatments to increase the performance of the predictive models in the selected spectral range. Regarding Table 3, different spectral ranges were identified for each volatile phenol comprising wave number values from 9300 to 4500 cm^{-1}.

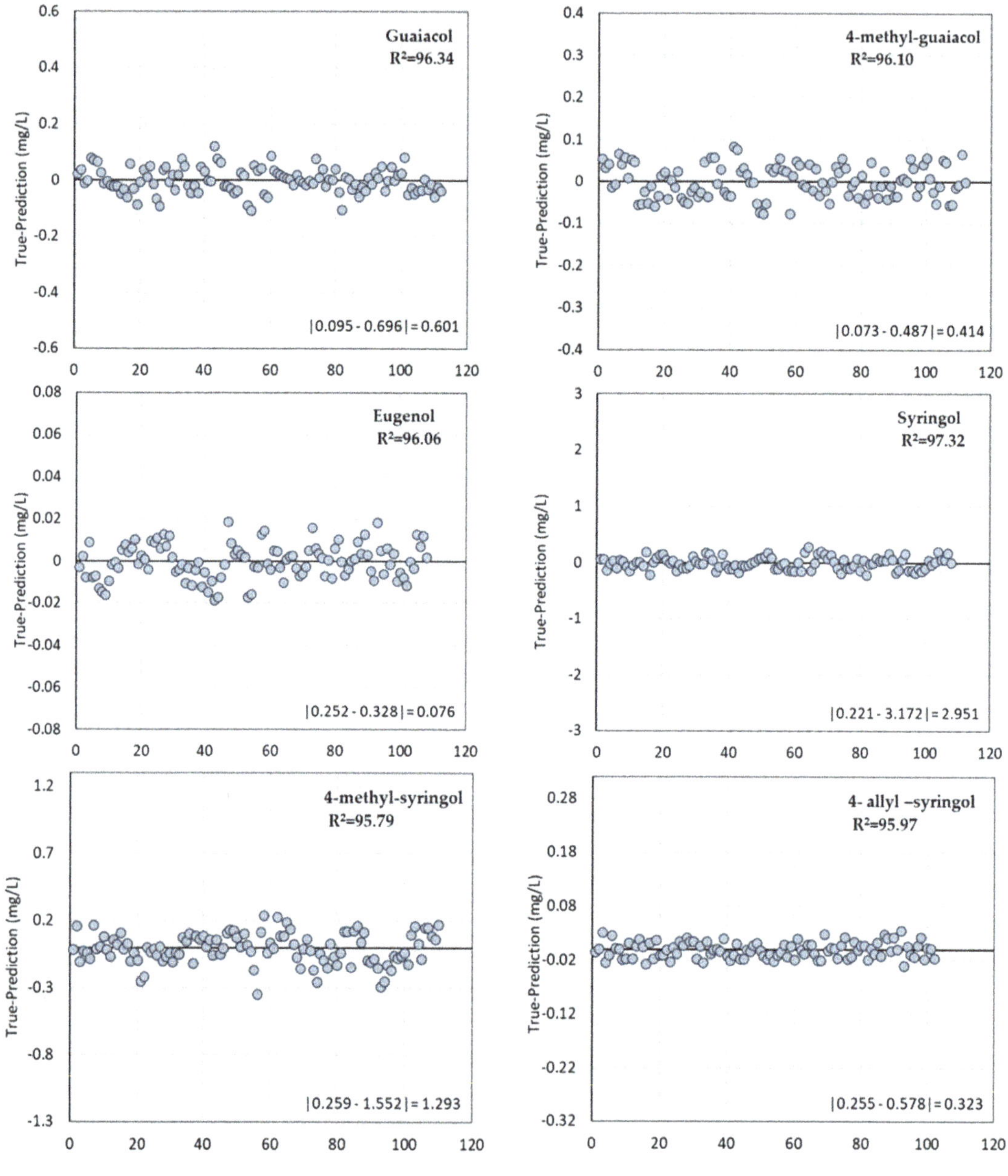

Figure 5. True value−Prediction value of each volatile compound analysed compared to the difference between the minimum and maximum a value.

Thus, for guaiacol quantification, the spectral range from 9118.1 to 5415.3 cm^{-1} was selected; for 4-methyl-guaiacol, three spectral ranges 8304.2 to 7347.7 cm^{-1} + 6869.4 to 5434.6 cm^{-1} + 4956.3 to 4478 cm^{-1}, were selected; for eugenol, the spectral range was between 9337.9 and 5446.2 cm^{-1}; for syringol, the spectral range between 6101.9 and 5446.2 cm^{-1} was selected; for 4-methyl-syringol, the spectral range was between 9160.5 and 4512.7 cm^{-1}; for 4-allyl-syringol, the spectral range from 9353.3 to 7498.1 cm^{-1} +

from 6101.9 to 5446.2 cm^{-1} were selected. Each chemical structure influences the analyte's absorption bands' position, shape, and size. Concerning the results mentioned above, the wavelength range selected in all calibration models was the region from 6000 to 5500 cm^{-1} ascribed to the νC–H stretch of the first overtones of CH_3 and CH_2 groups [42,49], and OH from aromatic groups [44]. According to the ageing time, these regions were also identified as good discriminants of wine spirits aged with different kinds of wood and ageing systems [37]. All these groups are presented in the volatile phenols studied, as shown in Figure 1, and some of them can even be differentiators when thoroughly examined. The hydroxyl groups arrangement (or even other substituent groups) at the aromatic phenolic skeleton has a significant impact on the absorption band shown in the NIR spectra, such as some of their chemical properties: dipole moment, bond dissociation enthalpy for the O–H bond, ionization potential or the antioxidant activity, among others. As a result, various skeleton and structural parameters, including the number and position of hydroxyl groups, the presence of other functional groups, their position in relation to hydroxyl groups, and stereochemical impediment, may affect the distinctive bands of each compound [50,51].

According to Jakubíková et al. [40], which used NIR spectroscopy to distinguish fruit spirits, the spectral region of 6050–5500 cm^{-1} is the more accurate to discriminate the different beverages analysed using PCA with linear discriminant analysis and general discriminant analysis models that giving 100% classification of spirits.

Concerning the pre-process selected, the one identified as better in the calibration model was the first derivative with 17 smoothing points combined with the multiplicative scatter correction or straight-line elimination (Table 3).

Regarding Table 3, all values of r^2 are higher than 90.05%, which can be classified as excellent precision [52]. The values of r^2 ranged between 90.05% for 4-allyl-syringol and 97.81 for syringol.

Several authors defined different threshold values for the accuracy of the model given by RPD that report the ratio between the standard deviation of the reference data of the validation set and the standard error of cross-validation prediction or the test set validation. According to Workman and Weyer [47], RPD must be higher than 2.5 to have good calibration. Conzen [53] states that a good calibration model must have an RPD higher than 3.0. In the present study, all models have values of RPD higher than 3.19.

The RPD values obtained for the analysed compounds ranged between 3.19 and 6.76 to predict 4-allyl-syringol and syringol, respectively. As far as we know, no studies were published about calibration models for volatile phenols. Therefore, it is only possible to compare with other volatile compounds of the AWS, but even these are scarce in the bibliography.

As aforementioned, the RMSEs (root mean square errors) of the validation set, cross-validation and calibration was also used to evaluate the ability of the PLS-R models developed to predict these parameters. All obtained values are low, denoting an accurate calibration model.

The NIR spectroscopy ability to monitor the distillation process of ethanol and methanol (two compounds that have legal limits for this beverage) from wine has been demonstrated by Dambergs et al. [54]. In this case, the more relevant regions studied for methanol and ethanol were 4401 cm^{-1} (related to CH combinations from the CH_3 group) and 4337 cm^{-1} (associated with the CH_2 group), respectively, which were also visible in the spectra obtained in the present study (Figure 4). At 5176 cm^{-1}, the most significant peak is related to OH vibration combinations found in WS compounds and the volatile compounds that rise with the ageing process. These compounds are major volatiles of the WS, so they are easier to identify by NIR, and consequently, with more accurate models than those obtained for volatile phenols in this work. PLS and multiple linear regression (MLR) methods were tested for NIR calibrations using gas chromatography as the reference method in the study mentioned above. The PLS calibrations show better results with r^2 of

0.96, a calibration error of 0.08% *v/v* for ethanol, and r^2 of 0.99 and a calibration error of 0.06 g/L for methanol [54].

Yang et al. [32] proposed using two-dimensional NIR to determine the concentration of methanol in the white spirit combined with multivariate analysis, obtaining values of relative error of 2.97 and root mean square error of 0.079%.

In another research work [55], NIR was used to discriminate sugarcane spirits according to their origin using PLS-R, PLS combined with linear discriminant analysis, successive projection algorithm and genetic algorithm, which allowed identifying the authenticity of the studied beverages. Among the statistical approaches performed, the PLS-R model exhibited accurate values to predict the ethanol content of sugarcane spirits in the quality control process.

Figure 5 exhibits that the concentration value measured by GC (assumed as actual value) subtracted from the prediction value given by the corresponding proposed model for each volatile compound. Each graphic is represented in the spaces of the higher possible variance given by the minimum and maximum value difference observed in each analytical parameter. The results show the excellent performance of the models and the low deviation of the predicted value to the actual value one.

For the first time, this research shows the applicability of NIR spectroscopy to assess the volatile phenol's contents, namely guaiacol, 4-methyl-guaiacol, eugenol, syringol 4-methyl-syringol and 4-allyl-syringol and confirms the ability of this technique to quantify those compounds in AWS.

4. Conclusions

The results attained in this study demonstrate that NIR spectroscopy can be used as an easy and quick method, without sample preparation and good reproducibility, to assess the content of volatile phenols in AWS. The performance of the models, given by the values of RPD, which are higher than 3.19 with a coefficient of determination higher than 90% and low root mean square error, are promising results for the use of this methodology at an industrial scale. However, further studies are needed to compare the ability of NIR with other methodologies, namely FTIR and RAMAN, using samples from other aged spirits, such as grape marc spirits, to increase the accuracy of the models and to extend this prediction analytical approach to other volatile compounds.

Author Contributions: Conceptualization, O.A. and I.C.; methodology, O.A.; technological assay, O.A., I.C., S.O.-A., T.A.F., S.C. (Sofia Catarino) and S.C. (Sara Canas); formal analysis, O.A., I.C., S.I.P. and C.V.; writing—original draft preparation, O.A., T.A.F. and I.C.; writing—review and editing, all team; project administrations, S.C. (Sara Canas). All authors have read and agreed to the published version of the manuscript.

Funding: This research was funded by National Funds through FCT—Foundation for Science and Technology under the Project POCI-01-0145-FEDER-027819 (PTDC/OCE-ETA/27819/2017). This work is also funded by National Funds through FCT—Foundation for Science and Technology under the Projects UIDB/00239/2020 [CEF], UIDB/05183/2020 [MED]; UIDB/00100/2020, UIDP/00100/2020 [CQE]; UID/AGR/04129/2020, DL 57/2016/CP1382/CT0025 [LEAF].

Institutional Review Board Statement: Not applicable.

Informed Consent Statement: Not applicable.

Data Availability Statement: Not applicable.

Acknowledgments: The authors thank "Centro de Biotecnologia de plantas" for the equipment availability and Vitor de Freitas as the Scientific Consultant of the Project PO-CI-01-0145-FEDER-027819. The authors also thank João Pedro Catela, Nádia Santos, Manuela Gomes, Eugénia Gomes and Inês Antunes from Adega Cooperativa da Lourinhã, José Abílio Gonçalves and Sérgio Gonçalves from Tanoaria J.M. Gonçalves, Ana Partidário from INIAV—Oeiras, Pedro Rodrigues and Diogo Rodrigues from AZ3Oeno Portugal, João Amaral from INIAV—Pólo de Dois Portos, A. Pedro Belchior, and Laurent Fargeton and Benjamin Boissier from Vivelys for the technical support.

Conflicts of Interest: The authors declare no conflict of interest.

References

1. Schieber, A.; Wüst, M. Volatile Phenols—Important Contributors to the Aroma of Plant-Derived Foods. *Molecules* **2020**, *25*, 4529. [CrossRef]
2. Waterhouse, A.L.; Sacks, G.L.; Jeffery, D.W. Volatile Phenols. In *Understanding Wine Chemistry*; John Wiley & Sons, Ltd.: Chichester, UK, 2016; pp. 105–111.
3. Jeleń, H.H.; Majcher, M.; Szwengiel, A. Key odorants in peated malt whisky and its differentiation from other whisky types using profiling of flavor and volatile compounds. *LWT* **2019**, *107*, 56–63. [CrossRef]
4. Franitza, L.; Granvogl, M.; Schieberle, P. Characterization of the Key Aroma Compounds in Two Commercial Rums by Means of the Sensomics Approach. *J. Agric. Food Chem.* **2016**, *64*, 637–645. [CrossRef]
5. Caldeira, I.; de Sousa, R.B.; Belchior, A.P.; Climaco, M.C. A sensory and chemical approach to the aroma of wooden agend Lourinha wine brandy. *Cienc. Tec. Vitiv.* **2008**, *23*, 97–110.
6. Caldeira, I.; Vitória, C.; Anjos, O.; Fernandes, T.A.; Gallardo, E.; Fargeton, L.; Boissier, B.; Catarino, S.; Canas, S. Wine Spirit Ageing with Chestnut Staves under Different Micro-Oxygenation Strategies: Effects on the Volatile Compounds and Sensory Profile. *Appl. Sci.* **2021**, *11*, 3991. [CrossRef]
7. Granja-Soares, J.; Roque, R.; Cabrita, M.J.; Anjos, O.; Belchior, A.P. Effect of innovative technology using staves and micro-oxygenation on the odorant and sensory pro fi le of aged wine spirit. *Food Chem.* **2020**, *333*, 127450. [CrossRef]
8. Caldeira, I.; Santos, R.; Ricardo-da-Silva, J.M.; Anjos, O.; Mira, H.; Belchior, A.P.; Canas, S. Kinetics of odorant compounds in wine brandies aged in different systems. *Food Chem.* **2016**, *211*, 937–946. [CrossRef]
9. Vanbeneden, N.; Delvaux, F.; Delvaux, F.R. Determination of hydroxycinnamic acids and volatile phenols in wort and beer by isocratic high-performance liquid chromatography using electrochemical detection. *J. Chromatogr. A* **2006**, *1136*, 237–242. [CrossRef]
10. Caldeira, I.; Pereira, R.; Clímaco, M.C.; Belchior, A.P.; de Sousa, R.B. Improved method for extraction of aroma compounds in aged brandies and aqueous alcoholic wood extracts using ultrasound. *Anal. Chim. Acta* **2004**, *513*, 125–134. [CrossRef]
11. Valente, I.M.; Santos, C.M.; Moreira, M.M.; Rodrigues, J.A. New application of the QuEChERS methodology for the determination of volatile phenols in beverages by liquid chromatography. *J. Chromatogr. A* **2013**, *1271*, 27–32. [CrossRef]
12. Castro Mejías, R.; Natera Marín, R.; de Valme García Moreno, M.; García Barroso, C. Optimisation of headspace solid-phase microextraction for the analysis of volatile phenols in wine. *J. Chromatogr. A* **2003**, *995*, 11–20. [CrossRef]
13. Pizarro, C.; Pérez-del-Notario, N.; González-Sáiz, J.M. Optimisation of a simple and reliable method based on headspace solid-phase microextraction for the determination of volatile phenols in beer. *J. Chromatogr. A* **2010**, *1217*, 6013–6021. [CrossRef]
14. Díez, J. Optimisation of stir bar sorptive extraction for the analysis of volatile phenols in wines. *J. Chromatogr. A* **2004**, *1025*, 263–267. [CrossRef]
15. Zhou, Q.; Qian, Y.; Qian, M.C. Analysis of volatile phenols in alcoholic beverage by ethylene glycol-polydimethylsiloxane based stir bar sorptive extraction and gas chromatography–mass spectrometry. *J. Chromatogr. A* **2015**, *1390*, 22–27. [CrossRef]
16. Fariña, L.; Boido, E.; Carrau, F.; Dellacassa, E. Determination of volatile phenols in red wines by dispersive liquid–liquid microextraction and gas chromatography-mass spectrometry detection. *J. Chromatogr. A* **2007**, *1157*, 46–50. [CrossRef]
17. Pizarro, C.; Sáenz-González, C.; Pérez-del-Notario, N.; González-Sáiz, J.M. Ultrasound-assisted emulsification-microextraction for the sensitive determination of Brett character responsible compounds in wines. *J. Chromatogr. A* **2011**, *1218*, 8975–8981. [CrossRef]
18. Downey, G. Tutorial review. Qualitative analysis in the near-infrared region. *Analyst* **1994**, *119*, 2367. [CrossRef]
19. Wang, X. Near-infrared spectroscopy for food quality evaluation. In *Evaluation Technologies for Food Quality*; Elsevier: Amsterdam, The Netherlands, 2019; pp. 105–118.
20. Woodcock, T.; Downey, G.; O'Donnell, C.P. Better Quality Food and Beverages: The Role of near Infrared Spectroscopy. *J. Near Infrared Spectrosc.* **2008**, *16*, 1–29. [CrossRef]
21. Lohumi, S.; Lee, S.; Lee, H.; Cho, B.-K. A review of vibrational spectroscopic techniques for the detection of food authenticity and adulteration. *Trends Food Sci. Technol.* **2015**, *46*, 85–98. [CrossRef]
22. Cozzolino, D.; Smyth, H.E.; Gishen, M. Feasibility Study on the Use of Visible and Near-Infrared Spectroscopy Together with Chemometrics to Discriminate between Commercial White Wines of Different Varietal Origins. *J. Agric. Food Chem.* **2003**, *51*, 7703–7708. [CrossRef]
23. Da Paixão Teixeira, J.L.; Dos Santos Caramês, E.T.; Baptista, D.P.; Gigante, M.L.; Pallone, J.A.L. Vibrational spectroscopy and chemometrics tools for authenticity and improvement the safety control in goat milk. *Food Control* **2020**, *112*, 107105. [CrossRef]
24. Cozzolino, D.; Smyth, H.E.; Lattey, K.A.; Cynkar, W.; Janik, L.; Dambergs, R.G.; Francis, I.L.; Gishen, M. Relationship between sensory analysis and near infrared spectroscopy in Australian Riesling and Chardonnay wines. *Anal. Chim. Acta* **2005**, *539*, 341–348. [CrossRef]
25. Özdemir, İ.S.; Karaoğlu, Ö.; Dağ, Ç.; Bekiroğlu, S. Assessment of sesame oil fatty acid and sterol composition with FT-NIR spectroscopy and chemometrics. *TURKISH J. Agric. For.* **2018**, *42*, 444–452. [CrossRef]
26. Macedo, L.; Araújo, C.; Vimercati, W.; Hein, P.R.; Pimenta, C.J.; Saraiva, S. Evaluation of chemical properties of intact green coffee beans using near-infrared spectroscopy. *J. Sci. Food Agric.* **2021**, *101*, 3500–3507. [CrossRef]

27. Dhaulaniya, A.S.; Balan, B.; Sodhi, K.K.; Kelly, S.; Cannavan, A.; Singh, D.K. Qualitative and quantitative evaluation of corn syrup as a potential added sweetener in apple fruit juices using mid-infrared spectroscopy assisted chemometric modeling. *LWT* **2020**, *131*, 109749. [CrossRef]
28. Nordon, A.; Mills, A.; Burn, R.T.; Cusick, F.M.; Littlejohn, D. Comparison of non-invasive NIR and Raman spectrometries for determination of alcohol content of spirits. *Anal. Chim. Acta* **2005**, *548*, 148–158. [CrossRef]
29. Mendes, L.S.; Oliveira, F.C.C.; Suarez, P.A.Z.; Rubim, J.C. Determination of ethanol in fuel ethanol and beverages by Fourier transform (FT)-near infrared and FT-Raman spectrometries. *Anal. Chim. Acta* **2003**, *493*, 219–231. [CrossRef]
30. Palma, M. Application of FT-IR spectroscopy to the characterisation and classification of wines, brandies and other distilled drinks. *Talanta* **2002**, *58*, 265–271. [CrossRef]
31. Kolomiets, O.A.; Lachenmeier, D.W.; Hoffmann, U.; Siesler, H.W. Quantitative Determination of Quality Parameters and Authentication of Vodka Using near Infrared Spectroscopy. *J. Near Infrared Spectrosc.* **2010**, *18*, 59–67. [CrossRef]
32. Yang, Y.R.; Ren, Y.F.; Dong, G.M.; Yang, R.J.; Liu, H.X.; Du, Y.H.; Zhang, W.Y. Determination of Methanol in Alcoholic Beverages by Two-Dimensional Near-Infrared Correlation Spectroscopy. *Anal. Lett.* **2016**, *49*, 2279–2289. [CrossRef]
33. Mosedale, J.; Puech, J. Wood maturation of distilled beverages. *Trends Food Sci. Technol.* **1998**, *9*, 95–101. [CrossRef]
34. França, L.; Grassi, S.; Pimentel, M.F.; Amigo, J.M. A single model to monitor multistep craft beer manufacturing using near infrared spectroscopy and chemometrics. *Food Bioprod. Process.* **2021**, *126*, 95–103. [CrossRef]
35. Power, A.C.; Jones, J.; NiNeil, C.; Geoghegan, S.; Warren, S.; Currivan, S.; Cozzolino, D. What's in this drink? Classification and adulterant detection in Irish Whiskey samples using near infrared spectroscopy combined with chemometrics. *J. Sci. Food Agric.* **2021**, *101*, 5256–5263. [CrossRef]
36. Hanousek-Cica, K.; Pezer, M.; Mrvcic, J.; Stanzer, D.; Cacic, J.; Jurak, V.; Krajnovic, M.; Gajdos-Kljusuric, J. Identification of phenolic and alcoholic compounds in wine spirits and their classification by use of multivariate analysis. *J. Serb. Chem. Soc.* **2019**, *84*, 663–677. [CrossRef]
37. Anjos, O.; Caldeira, I.; Roque, R.; Pedro, S.I.; Lourenço, S. Screening of Different Ageing Technologies of Wine Spectroscopy and Volatile Quantification. *Processes* **2020**, *8*, 8–18.
38. Anjos, O.; Santos, A.J.A.; Estevinho, L.M.; Caldeira, I. FTIR–ATR spectroscopy applied to quality control of grape-derived spirits. *Food Chem.* **2016**, *205*, 28–35. [CrossRef]
39. Canas, S.; Caldeira, I.; Anjos, O.; Belchior, A.P. Phenolic profile and colour acquired by the wine spirit in the beginning of ageing: Alternative technology using micro-oxygenation vs traditional technology. *LWT* **2019**, *111*, 260–269. [CrossRef]
40. Jakubíková, M.; Sádecká, J.; Kleinová, A.; Májek, P. Near-infrared spectroscopy for rapid classification of fruit spirits. *J. Food Sci. Technol.* **2016**, *53*, 2797–2803. [CrossRef]
41. Schiavone, S.; Marchionni, B.; Bucci, R.; Marini, F.; Biancolillo, A. Authentication of Grappa (Italian grape marc spirit) by Mid and Near Infrared spectroscopies coupled with chemometrics. *Vib. Spectrosc.* **2020**, *107*, 103040. [CrossRef]
42. Giannetti, V.; Mariani, M.B.; Marini, F.; Torrelli, P.; Biancolillo, A. Grappa and Italian spirits: Multi-platform investigation based on GC–MS, MIR and NIR spectroscopies for the authentication of the Geographical Indication. *Microchem. J.* **2020**, *157*, 104896. [CrossRef]
43. Li, S.; Shan, Y.; Zhu, X.; Zhang, X.; Ling, G. Detection of honey adulteration by high fructose corn syrup and maltose syrup using Raman spectroscopy. *J. Food Compos. Anal.* **2012**, *28*, 69–74. [CrossRef]
44. Chen, H.; Tan, C.; Wu, T.; Wang, L.; Zhu, W. Discrimination between authentic and adulterated liquors by near-infrared spectroscopy and ensemble classification. *Spectrochim. Acta Part A Mol. Biomol. Spectrosc.* **2014**, *130*, 245–249. [CrossRef]
45. Schwarz, M.; Rodríguez, M.; Martínez, C.; Bosquet, V.; Guillén, D.; Barroso, C.G. Antioxidant activity of Brandy de Jerez and other aged distillates, and correlation with their polyphenolic content. *Food Chem.* **2009**, *116*, 29–33. [CrossRef]
46. Tsakiris, A.; Kallithraka, S.; Kourkoutas, Y. Grape brandy production, composition and sensory evaluation. *J. Sci. Food Agric.* **2014**, *94*, 404–414. [CrossRef]
47. Workman, J., Jr.; Weyer, L. *Practical Guide to Interpretive Near-Infrared Spectroscopy*; Taylor & Francis Group: Boca Raton, FL, USA, 2007; ISBN 978-1-57444-784-2.
48. Yu, H.; Zhou, Y.; Fu, X.; Xie, L.; Ying, Y. Discrimination between Chinese rice wines of different geographical origins by NIRS and AAS. *Eur. Food Res. Technol.* **2007**, *225*, 313–320. [CrossRef]
49. Cozzolino, D.; Corbella, E. Determination of honey quality components by near infrared reflectance spectroscopy. *J. Apic. Res.* **2003**, *42*, 16–20. [CrossRef]
50. Pardeshi, S.; Dhodapkar, R.; Kumar, A. Quantum chemical density functional theory studies on the molecular structure and vibrational spectra of Gallic acid imprinted polymers. *Spectrochim. Acta Part A Mol. Biomol. Spectrosc.* **2013**, *116*, 562–573. [CrossRef]
51. Badhani, B.; Sharma, N.; Kakkar, R. Gallic acid: A versatile antioxidant with promising therapeutic and industrial applications. *RSC Adv.* **2015**, *5*, 27540–27557. [CrossRef]
52. Williams, P.; Dardenne, P.; Flinn, P. Tutorial: Items to be included in a report on a near infrared spectroscopy project. *J. Near Infrared Spectrosc.* **2017**, *25*, 85–90. [CrossRef]
53. Conzen, J.P. *Multivariate Calibration. A Practical Guide for the Method Development in the Analytical Chemistry*, 2nd ed.; Bruker Optick: Ettlingen, Germany, 2006.

54. Dambergs, R.G.; Kambouris, A.; Schumacher, N.; Francis, I.L.; Esler, M.B.; Gishen, M. Wine quality grading by near infrared spectroscopy. In *Near Infrared Spectroscopy: Proceedings of the 10th International Conference*; IMPublications Open: Chichester, UK, 2001; pp. 187–190. ISBN 978-1-906715-22-9.
55. De Carvalho, L.C.; de Lelis Medeiros de Morais, C.; de Lima, K.M.G.; Cunha, L.C., Jr.; Martins Nascimento, P.A.; Bosco de Faria, J.; de Almeida Teixeira, G.H. Determination of the geographical origin and ethanol content of Brazilian sugarcane spirit using near-infrared spectroscopy coupled with discriminant analysis. *Anal. Methods* **2016**, *8*, 5658–5666. [CrossRef]

sensors

MDPI

Article

NIR Spectroscopy for Discriminating and Predicting the Sensory Profile of Dry-Cured Beef "Cecina"

Isabel Revilla [1,*], Ana M. Vivar-Quintana [1], María Inmaculada González-Martín [2], Miriam Hernández-Jiménez [1], Iván Martínez-Martín [1] and Pedro Hernández-Ramos [3]

1 Food Technology, University of Salamanca Escuela Politécnica Superior de Zamora, Avenida Requejo 33, 49022 Zamora, Spain; avivar@usal.es (A.M.V.-Q.); miriamhj@usal.es (M.H.-J.); ivanm@usal.es (I.M.-M.)
2 Analytical Chemistry, Nutrition and Bromatology, University of Salamanca Calle Plaza de los Caídos s/n, 37008 Salamanca, Spain; inmaglez@usal.es
3 Graphic Expression in Engineering, University of Salamanca Escuela Politécnica Superior de Zamora, Avenida Requejo 33, 49022 Zamora, Spain; pedrohde@usal.es
* Correspondence: irevilla@usal.es

Received: 2 November 2020; Accepted: 1 December 2020; Published: 2 December 2020

Abstract: For Protected Geographical Indication (PGI)-labeled products, such as the dry-cured beef meat "cecina de León", a sensory analysis is compulsory. However, this is a complex and time-consuming process. This study explores the viability of using near infrared spectroscopy (NIRS) together with artificial neural networks (ANN) for predicting sensory attributes. Spectra of 50 samples of cecina were recorded and 451 reflectance data were obtained. A feedforward multilayer perceptron ANN with 451 neurons in the input layer, a number of neurons varying between 1 and 30 in the hidden layer, and a single neuron in the output layer were optimized for each sensory parameter. The regression coefficient R squared (RSQ > 0.8 except for odor intensity) and mean squared error of prediction (MSEP) values obtained when comparing predicted and reference values showed that it is possible to predict accurately 23 out of 24 sensory parameters. Although only 3 sensory parameters showed significant differences between PGI and non-PGI samples, the optimized ANN architecture applied to NIR spectra achieved the correct classification of the 100% of the samples while the residual mean squares method (RMS-X) allowed 100% of non-PGI samples to be distinguished.

Keywords: near infrared spectra; chemometry; dry meat; artificial neural networks; organoleptic parameters; prediction; protected geographical indication distinguishing

1. Introduction

The dry-cured beef meat called Cecina is a meat of intermediate moisture from different anatomic parts which undergoes a process of profiling, salting, washing, settling, smoking and drying; the whole procedure takes between 7 and 12 months after salting. In Spain, cecinas produced exclusively in the province of León from hind leg pieces of beef cattle (at least 5 years old and 400 kg of weight) and produced following the processing scheme established by the Supervisory Council of cecina de León may be awarded the quality label "Protected Geographical Indication" (PGI) [1]. The most remarkable characteristics of this product include its intense red color, smoked odor, slightly salty taste, and soft texture owing to which it is highly appreciated by consumers. These sensory characteristics constitute the distinguishing marks of this product. One of the activities of the Supervisory Council of PGI cecina de León is, therefore, to perform a sensory analysis to determine the existence of defects and to certify its typicality in order to differentiate it from that of non-PGI products [2].

However, a descriptive sensory analysis is a very complex and time-consuming process which involves the recruitment, selection, training and qualification of assessors following the ISO 8586 and

ISO 5492 standards [3,4]. Although it is possible to find literature on the creation of sensory panels to certificate Protected Denomination of Origin (PDO) or PGI products [5,6], the information related to meat products is scarce [7]. This illustrates the difficulty of performing a sensory analysis for products under quality labels and means that it would be desirable to replace it with a fast, reliable, and cheap instrumental analysis.

Near infrared spectroscopy (NIRS) is a fast, accurate, multiparametric and non destructive technique which, due to its agility and can be implemented on-line. This technique has been shown to be a powerful tool for discriminating products according to the geographical origin of the samples. Therefore, it has been used for rice wines [8], honey [9], tea [10], and lentils [11] among other products. These results indicate that this technique can be a simple way of distinguishing PGI from non-PGI meat products. However, there are numerous discriminating methods which could be applied to NIR spectra which have been assayed for discriminating between food products belonging to PDO or PGI quality labels. Examples include principal component analysis for almonds [12], factorial discriminant analysis (FDA) for Swiss cheeses [13], partial least squares discriminant analysis (PLS-DA) for vinegars [14], K-nearest neighbors (KNN) for hazelnuts [15], the residual mean squares method (RMS-X residuals) for dry sausages [16] or artificial neural networks (ANN) for cheeses [17]. However, all this research has also shown that the most suitable distinguishing method is closely linked to the product; a detailed study should be carried out for each specific PDO or PGI label or food matrix.

Moreover, NIRS has also shown promising results for predicting the sensory characteristics of meat [18,19] and meat products such as sausages [20] or dry-cured ham [21–23]. However, a determinant step is the selection of the chemometric tool for multivariate analysis of data. In this sense, previous studies indicate that ANNs are more suitable for predicting sensory parameters than multiple regression tools. In general, higher regression coefficient R squared (RSQ) and lower mean squared error of prediction (MSEP) are observed when ANN are used for predictions [24–26] and it is possible to predict a higher number of parameters [16].

This study is part of a larger project which aims to develop a fast, objective, and reliable methodology for classifying and predicting the sensory parameters of meat products with quality labels [20,23]. Taking this into account, the aim of this study was to assess the feasibility of the prediction of sensory parameters of dry-cured beef meat cecina using NIR spectroscopy, as a fast and non destructive method, together with Levenberg–Marquardt feedforward ANN of the multi-layer perceptron type, which have shown the best results for other meat products. Moreover, the study also aims to determine the most suitable multivariate classification tool for distinguishing whether a sample belongs to a quality label such as the PGI Cecina de León only by using its NIR spectra.

2. Materials and Methods

2.1. Samples

Fifty samples of dry-cured beef Cecina were used for this study. Of these, 25 samples belonged to the "cecina de León" Protected Geographical Indication and were produced according to the specifications of the Regulatory Board of the cecina de León PGI published in the Official Bulletin of Castilla y León [27]. To this end, the hind leg was selected and after trimming the external connective tissue the pieces were covered in salt, placed in piles alternating meat and salt, and kept at 3–4 °C and 85–90% relative humidity (RH) for 3 days. The pieces were subsequently rinsed to remove the salt and moved again to the cold-storage room under controlled temperature and humidity for 50 days' post-salting. The smoking process was carried out using oak wood in a smoking chamber (5 days at 12–15 °C and 65–75% RH). Finally, the pieces were dried and aged by reducing the RH from 80–85% to 60–65% at 12–16 °C for a variable period of time. The total period of salting, rinsing, post-salting, smoking, drying and aging ranged between 7 and 12 months.

The remaining 25 samples did not belong to the PGI and were acquired in shops in the same province (León) as those of the PGI and their productive process was similar to that of "cecina de León".

2.2. Sensory Analysis

The panel that carried out the sensory analysis was formed by 10 assessors with a wide previous experience of dry-cured meat analysis as previously described [16,23]. The training for the specific sensory profiling of this product involved 8 sessions lasting 1–1.5 h each. The choice, definition and consensus of the evaluation methodology of the parameters were established during four sessions.

The selected parameters for appearance were veined, fat color, color intensity, exudate and white dots. For flavor, the evaluated parameters were odor intensity, cured odor, rancid odor, flavor intensity, cured flavor, saltiness, sweetness, rancidity and aftertaste. Finally, for texture the parameters were hardness, juiciness, fibrousness, chewiness and gumminess. The description and score criteria for these parameters were the same as previously selected for dry-cured ham by this sensory panel [23]. Together with these parameters, the following which were characteristic of dry-cured beef meat were also selected by the assessors: smoked odor and smoked flavor described as intensity of odor and flavor respectively produced by the presence of smoke, moldy odor described as the presence of a characteristic odor recalling that of mushrooms, and pungency described as the intensity of a pungent sensation. Other parameters such as color homogeneity, atypical aroma, fat flavor intensity, sourness, atypical flavor, heterogeneity of the texture, or chewing residue which were assessed in dry-cured ham were not included by the panel for dry-cured beef meat characterization.

During the next four qualification sessions, the same sample was assessed three times each session. The results were used to calculate the reproducibility of the panel whose maximum uncertainty has to be lower than 0.5 and repeatability of the panel whose maximum uncertainty has to be lower than 0.8 [5]. Finally, 14 quantification sessions were held with four samples being tested in each of them. Some selected samples were tasted three times to check the continuous accuracy of the panel. A slice of approximately 1.5 mm thick from each sample was presented to the assessors. The samples were coded with three digits codes and kept at room temperature. A structured 9-point scale was used for each of the attributes, ranged from low intensity (1) to high intensity (9) of the parameter.

A statistical analysis of the sensory data was carried out in the form of an analysis of variance (ANOVA) using the SPSS Package 25 (IBM, Chicago, IL, USA).

2.3. Near Infrared (NIR) Spectroscopy

All samples were analyzed by using Foss NIR System 5000 equipment (Foss, Hillerod, Denmark) to obtain the NIR spectra. The NIR measurement was recorded by applying directly the quartz window of 5 cm × 5 cm to a slice of the product cut with a slicing machine transversally to the direction of the muscle fiber (Figure 1a).

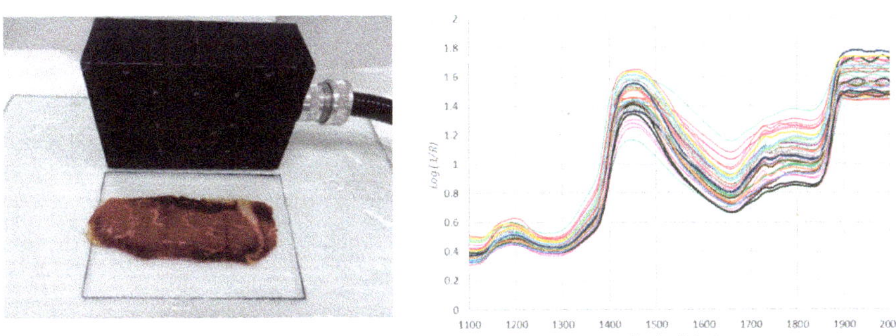

(a) (b)

Figure 1. Near infrared (NIR) measurements: (**a**) a record of NIR Spectra; (**b**) spectra of the cecina samples.

The NIR System is coupled with a remote reflectance fiber-optic probe of 1.5 m 210/210 (Ref no. R6539-A) and uses a remote reflectance system and a ceramic plate as a reference. The spectra of the samples were recorded between 1100 and 2000 nm (Figure 1b), at intervals of 2 nm, i.e., 451 data for each spectrum; 32 scans for both the reference and the samples were recorded for each sample. Three NIR records were made for each slice of cecina, one in the centre of the slice and the other two at the ends of the slice. The average representative spectrum of each slice of cecina was correlated with the sensory data provided by the sensory panel, as the assessors were given the whole slice and consumed it completely for evaluation. Spectra were stored as the logarithm reciprocal of reflectance log (1/R) (R = reflectance). The software used for spectral collection and data handling was Win ISI 1.50 (Foss, Hillerod, Denmark) installed on a Hewlett-Packard Pentium III computer.

2.4. Discriminant Analysis

Samples of cecina were analyzed according to whether they belonged to the "cecina de León" PGI or not. The significance of the effect of belonging to the PGI as determined by sensory attributes was assessed by the analysis of variance (ANOVA). Indeed, different discriminant analysis was carried out using the whole NIR spectrum from 1100 to 2000 nm measuring each 2 nm, i.e., 451 data.

Soft independent modelling of class analogy (SIMCA) using principal component analysis (PCA) was used to group the samples. The data for SIMCA modeling were normalized, scaled, and mean-centered. Subsequently the original variables were linearly transformed into a new set of variables (principal components, PCs) which preserve the information of the original set. The number of PCs for classifying purposes was determined by selecting those with an eigenvalue of > 1. The projection of the samples in the space determined by the principal components allowed the detection of groups present in the samples. An orthogonal projection latent structure discriminant analysis (OPLS-DA) was then carried out. To do so, the total of the samples were randomly divided into a training set (80% of the data) and a validation set (20% of the samples) in order to test the robustness of the discriminant model. Moreover, it was possible to calculate the discriminating plot which identifies the wavelengths or bands which may have a higher impact on the classification ability of the model. The software used was SIMCA-P software version 14.1 (Umetrics, Sartorius Stedim Biotech AS, Umeå, Sweden).

The RMS-X residuals analysis was carried out with the Win ISI 1.50 (Infrasoft International, State College, PA, USA) software using the whole NIR spectrum. Different combinations of the following mathematical treatments (none, multiplicative scatter correction (MSC), standard normal variate (SNV), detrend (DT) or SNV-DT), first or second derivatives, and several gaps over the derivative were calculated, and different numbers of data points in a running average and one or two smoothing were assayed and coded as follows (None 2,10,10,1) as previously described by González-Martín et al. [28]. The best mathematical treatment for distinguishing between the samples was selected taking into account the highest percentage of correctly classified samples.

The ANN selected for product classification was a multi-layer perceptron feedforward of the backpropagation type. This ANN type uses the tangent sigmoid function in the hidden layer and the softmax transfer function in the output layer. The following ANN learning or training algorithms were tested in order to minimize the error process: gradient descent, gradient descent with adaptive learning rate backpropagation, gradient descent with momentum, gradient descent with momentum and adaptive learning rate, scaled conjugate gradient, conjugate gradient with Powell–Beale restarts, conjugate gradient with Fletcher–Reeves restarts, conjugate gradient with Polak–Ribiere restarts and Levenberg–Marquardt.

For each of the learning algorithms the ANN architecture includes an input layer with 451 data, a hidden layer with a number of neurons between 1 and 30 and one output layer with two nominal variables that was PGI or non-PGI. If the classification is in the correct class output the target output value is 1 and 0 for the other nominal variable. The data (NIR spectra) set was randomly divided into three sets: the training set with 70% of the data, the validation set with 15%, and the test set with 15%. For all the ANN structures 500 trainings with different initial seed values were held in order to select the ANN with the best performance, which was established by the highest percentage of correctly classified samples. The software used was the Deep Learning Toolbox of MatLab (MathWorks®) in its R2020a version.

2.5. *Artificial Neural Network for Predicting Sensory Parameters*

In this case, a feedforward artificial neural network of the multi-layer perceptron type was used for processing the data. The input layer had 451 neurons (i.e., 451 values of log (1/R) recorded by NIR spectroscopy), a varying number of neurons between 1 and 30 were also tested in the hidden layer, and the output layer had only one neuron that showed the predicted value of the sensory parameter. An ANN was constructed for each of the 24 sensory parameters. As previously reported by Hernández-Jiménez et al. [16], the best training algorithm for predicting sensory parameters is the Levenberg-Marquardt backpropagation. The hyperbolic tangent sigmoid function was selected for the hidden layer and the pure linear transfer function was used for the output layer. The weight and bias matrix were randomly initialized and a known seed value number was used and stored. As previously reported, this will allow the reproducibility of the ANN data [17]. For all the ANN, the data (NIR-expected sensory parameter) were randomly divided into three sets: the training set with 70% of the data, the validation set with 15%, and the test set with 15%. In order to achieve the best prediction capability, 100 trainings of architectures were tested for each sensory parameter. The best ANN architecture was selected according to the highest value of the RSQ and the lowest value of the MSEP. The software used was the Deep Learning Toolbox of MatLab (MathWorks®) in its R2020a version.

3. Results and Discussion

3.1. *Sensory Data*

The mean values together with the minimum, maximum and standard deviation for the 50 samples analyzed are shown in Table 1. These values show that the range of variation was wide enough to guarantee an adequate margin for calibration purposes. The observed variation is a consequence of using different producers, ripening times, and pieces of muscle [2,29].

The sensory parameters were divided into three groups: appearance parameters, flavor parameters including odor, flavor and taste characteristics and texture characteristics to coincide with previous studies which used these three modalities for the attributes of cecina [2]. Regarding appearance parameters, veined showed a mean between the mean values previously reported for beef cecina [2,29]. Mean values for fat color and color intensity were nearer to those described by Rodríguez-Lázaro et al. [2], while the mean value for exudate, equivalent to brightness of lean described by these authors, was slightly lower. The amount of white spots was not previously quantified for cecina but is a parameter that is usually applied in dry-cured ham descriptions [7,22,30]. These white crystals appear during the meat curing process and are composed mainly of tyrosine [31]. The mean value was low as is usually reported for dry-cured ham [22,30] but some of the samples showed high values of this parameter.

The flavor characteristics included 13 odor, flavor and taste parameters, i.e., a higher number than that usually reported for dry-cured meat [2,7,29,32]. Odor and flavor intensity, cured odor, and flavor and after taste showed mean values close to those reported by Rodriguez-Lazaro et al. [2] and higher than those reported by Molinero Sastre et al. [29]. The mean value for rancid odor was similar to that reported by Lorenzo and Carballo [32], and the rancidity that was perceived during consumption was slightly higher than the rancid odor. Moldy odor has been used for describing other dry meat products [33] and has been pointed out as one of the typical odor notes of this product owing to the presence of *Penicillium* and *Aspergillus* on the surface of the product [34]. Cecina is characterized by its smoked flavor and slightly salty taste [35] which justify the values close to 5 observed for these two parameters (4.90 and 4.35, respectively). These values were slightly higher than those previously found for dry-cured beef cecina [2]. Sweetness or a sweet taste, which is related to the amount of amino acid released during the maturation process [36], is a frequent descriptor of dry-cured ham [7,30] which gives in general higher values than those observed for cecina. Finally, the pungent or burnt taste related to the presence of aldehydes and ketones released from lipid oxidation [36] showed low values.

Table 1. Mean, minimum, maximum and standard deviation for 50 samples of cecina.

	Mean	Minimum	Maximum	SD
Appearance				
Veined	4.11	1.86	8.71	1.68
Fat color	5.97	4.43	7.50	0.59
Color intensity	5.78	3.57	8.00	1.09
Exudate	3.11	1.29	7.71	1.15
White spots	1.43	1.00	7.57	1.14
Flavor				
Odor intensity	5.83	4.14	7.00	0.56
Cured odor	5.39	3.67	6.71	0.67
Smoked odor	4.90	3.14	7.00	0.72
Rancid odor	1.36	1.00	3.00	0.40
Moldy odor	1.11	1.00	2.33	0.23
Flavor intensity	6.14	4.00	7.14	0.62
Cured flavor	5.56	3.67	6.86	0.74
Saltiness	4.35	3.33	5.14	0.42
Sweetness	1.54	1.00	2.29	0.30
Smoked flavor	4.49	2.33	5.86	0.71
Rancidity	1.61	1.00	3.83	0.57
Pungency	1.38	1.00	2.00	0.25
Aftertaste	5.42	3.29	6.71	0.58
Texture				
Hardness	3.88	2.33	6.43	0.94
Juiciness	4.49	2.33	6.00	0.81
Fatness	3.17	1.50	6.29	0.96
Fibrousness	3.11	1.71	5.29	0.78
Chewiness	3.60	2.29	5.43	0.84
Gumminess	2.63	1.67	4.50	0.70

The texture parameters of hardness, juiciness and chewiness showed a range of variation within the margins reported by Rodríguez-Lázaro et al. and Molinero Sastre et al. [2,29], while fibrousness was lower and juiciness was higher than the mean values reported by these authors and within the margin found by Lorenzo and Carballo [32]. The fatness sensation which has been previously reported as an important descriptor of dry-cured meat products [33] showed a higher mean value and a wider range of variation than juiciness.

3.2. Discrimination of the Samples According to Protected Geographical Indication (PGI) Quality Label

"Cecina de León" has the PGI quality label; only the products manufactured in León (Castilla y León, Spain) which follow the scheme approved by the PGI Supervisory Council can bear the PGI label. However, cecina is also produced outside of the PGI but in the same province (León) and following similar manufacturing techniques; therefore, it is necessary to have tools that can help to distinguish whether a product belongs to the PGI or not.

As far as we know, no studies exist comparing the sensory characteristics of cecina depending on whether the product bears the PGI label. The sensory profile of both groups of cecina, PGI cecina de León and non-PGI, is shown in Figure 2.

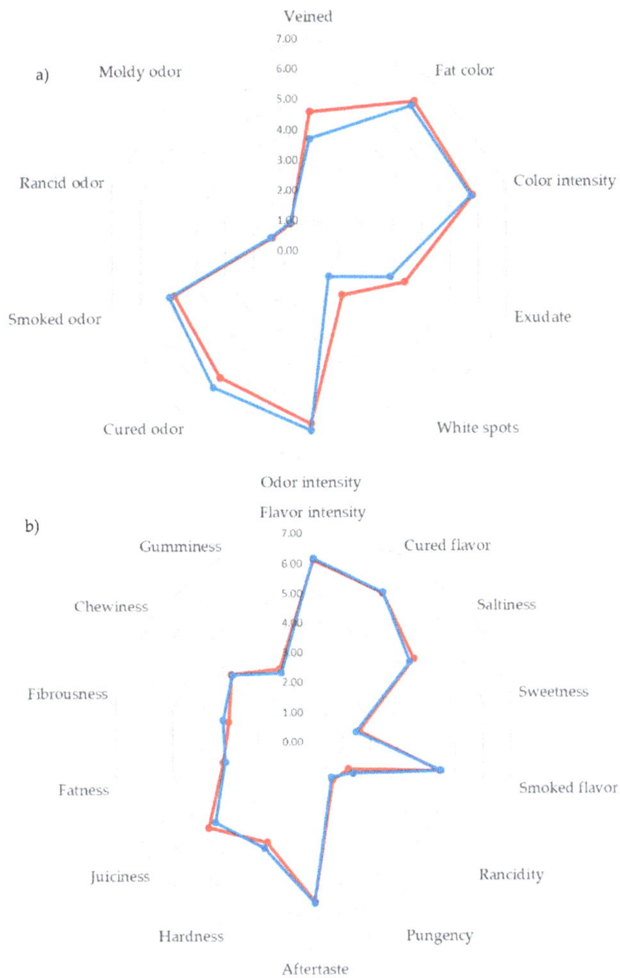

Figure 2. The mean values of (**a**) the appearance and odor attributes and (**b**) flavor and texture attributes of the Protected Geographical Indication (PGI) (red) and non-PGI (blue) samples.

The statistical analysis of the sensory parameters showed that there were significant differences between PGI and non-PGI cecina in only two parameters: white spots ($p = 0.019$), with the PGI cecina de León samples giving a slightly higher value than non-PGI cecina (1.83 vs. 1.06), and cured odor ($p = 0.044$). In this case, however, the samples from the cecina de León PGI showed a lower value (5.18 vs. 5.58). Veined tended ($p = 0.067$) to be higher in cecina de León PGI samples (4.57 vs. 3.68) but the differences were statistically significant at 90% level. These results show that differences between the sensory characteristics of both groups are small. This is due to the fact that both "cecina de León" and not-PGI cecina are produced from four different pieces of hind leg: thick flank, rump, silver side and topside, that show significant differences in sensory properties [2] and physico-chemical composition [29]. Moreover, the production area is very small and sometimes the same producer manufactures cecina with and without the PGI label.

Therefore, it is very important to have fast and reliable tools that allow the classification of the samples according to their origin. Previous studies have shown that a combination of NIR spectroscopy and chemometric tools is very useful for this purpose [9–11,16].

The analysis of the whole NIR spectra using SIMCA shows that after the Principal Component Analysis 4 PCs with an eigenvalue > 1 were obtained which explained the 98.8% of the total variance. The projection plot of the samples on the space defined by the three first PCs, which accounts for the 95.5% of the variance shown in Figure 3. It can be observed that both groups are not well separated. In fact, the classification obtained by OPSL-DA (orthogonal projection latent structure discriminant analysis) showed that the 68.4% of the PGI samples and the 70% of non-IGP samples were correctly classified in the calibration and only a 25% of the IGP samples and the 40% of the not-IGP samples were correctly classified in the external validation.

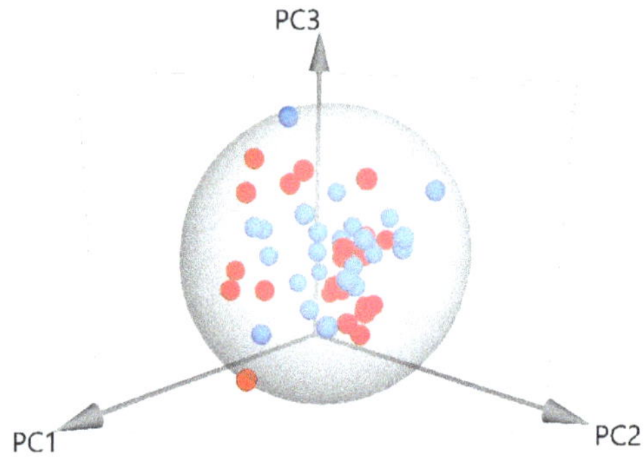

Figure 3. Projection plot of the samples in the space defined by the firsts three components.

An analysis of the whole NIR spectra using the RMS-X residuals was also carried out. This analysis implies the pre-treatment of the spectra with different combinations of mathematical treatments (MSC, SNV, DT or SNV-DT), derivatives, and smoothing procedures. The optimal treatment is that giving the highest percentage of correctly classified samples. In this case two treatments, None 2,10,10,1 and detrend 2,10,10,1 (Table 2), classify correctly 100% of the samples of non-PGI and 84% of the samples of the cecina de León PGI. The average success rate of the procedure is 92%, which indicates that it is a promising technique as the samples analyzed are very similar to each other.

Table 2. Discrimination results (number of samples and percentage of samples correctly classified) of residual mean squares (RMS-X) residuals method for some of the mathematical treatments assayed.

	None 2,4,4,1		None 2,10,10,1		SNV 1,4,4,1		Detrend 1,4,4,1		Detrend 2,10,10,1	
	PGI	**Not PGI**	**PGI**	**Not PGI**	**PGI**	**Not PGI**	**PGI**	**Non-PGI**	**PGI**	**Not PGI**
PGI	18	7	21	4	22	3	18	7	21	4
Non-PGI	0	25	0	25	3	22	2	23	0	25
Hit rate	72%	100%	84%	100%	88%	88%	72%	92%	84%	100%

PGI: Samples from Protected Geographical Indication, Non-PGI: Samples not belonging to Protected Geographical Indication, SNV: Standard Normal Variate.

The differences in the average spectra of PGI and non PGI samples in the absorption bands are shown in Figure 4a. These differences are due to physical and chemical changes that occur during the maturation process of the cecina. The different treatments that have been applied to correct the scattering of the spectra (None, SNV, DT, SNV-DT) and to achieve the optimization of the discrimination among samples using the RMS-X residual method give rise to different results, as can be seen in Table 2. The differences are due to the fact that the scattering treatments are conditioned by the moisture content of the samples, both of the two groups PGI and non-PGI, moisture that influences the size of the particles and the variations in homogeneity.

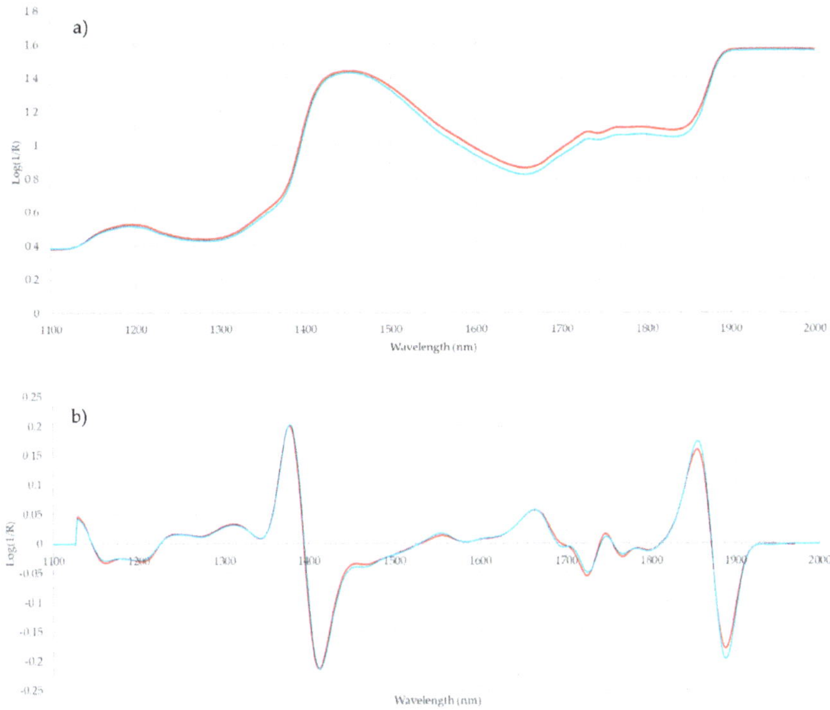

Figure 4. Plot of average near infrared (NIR) spectra of Protected Geographical Indication (PGI) (red) and non-PGI (blue) samples (**a**) without processing and (**b**) processed with detrend 2,10,10,1 mathematical treatment.

Figure 4b plots the average NIR spectra of PGI and non-PGI samples processed with detrend 2,10,10,1 mathematical pre-treatment in which differences between the two groups are observed at certain wavelengths.

Thus, the bands at 1450 nm are correlated with the first overtone of the O-H bond of the water, i.e., with the moisture of the samples. Previous works show that in meat products the 1450 nm band has been mainly related to moisture content; as well as to the third overtone of the C=O bond (1450 nm); and to the first overtone of the NH bond (urea at 1460 nm and $CONH_2$ at 1463 [37,38].

On the other hand, a correlation between NIR spectra and C–H–oil and C–O–oil groups at the wavelengths 1720 and 1760 nm, in agreement with the differences observed between PGI and non-PGI samples for veined. Furthermore, the odor intensity which also showed significant differences between the two groups of samples, and which depend on the amount of volatiles such as ketones, aldehydes and other compounds that are produced from lipolysis and fat oxidation [36] are related to the C=O bonds of the ketones at 1896 nm and to the C–H bond of aromatic structures at 1686, 1690 and 1696 nm. Finally, the salt content is reflected in the C–Cl bond at 1856 nm.

Applying ANNs for discriminating the samples, it was possible to find several ANNs with a classifying capability of over 90% for all the learning algorithms assayed. However, it is noteworthy that the number of successful ANNs was low for all of them. The results for the best ANN architecture for the nine learning algorithms, the number of the neurons in the hidden layer and the percentage of the samples correctly classified for the training, validation and test set, together with the average percentage for the total of the samples are shown in the Table 3.

Table 3. Architecture and discrimination results of the best artificial neural network (ANN) find for each of the assayed learning methods.

	Neurons	Percentage of Samples Correctly Classified			
		Training Set	Validation Set	Test Set	Total
Gradient Descent	27	95.6	85.7	85.7	92.7
Gradient Descent with Adaptive Learning Rate	30	98.5	85.7	85.7	94.8
Gradient Descent with Momentum	9	89.7	100	85.7	90.6
Gradient Descent with Momentum and Adaptive Learning Rate	19	98.5	85.7	100	96.9
Scaled Conjugate Gradient	29	98.5	100	100	98.9
Conjugate Gradient with Powell-Beale	10	100	100	92.8	98.9
Conjugate Gradient with Fletcher-Reeves	18	98.5	100	85.7	96.9
Conjugate Gradient with Polak-Ribiere	7	98.5	100	100	98.9
Levenberg-Marquardt	13	100	100	100	100

The results show that the gradient descent, gradient descent with adaptive learning rate, gradient descent with momentum, gradient descent with momentum and adaptive learning rate algorithms showed the lowest values of the samples correctly classified ranging from 90.6% to 96.9%. The group of the learning algorithms which uses variations of the conjugate gradient (scaled conjugate gradient, conjugate gradient with Powell–Beale restarts, conjugate gradient with Fletcher–Reeves restarts and conjugate gradient with Polak–Ribiere restarts) obtained higher values of correctly classified samples of between 96.9% and 98.9%. However, Levenberg-Marquardt was the method giving the best results because it was possible to find a significantly higher number of ANN architectures which correctly classified over 90% of the samples and it was also possible to find ANN architectures which correctly classified 100% of the samples.

These results showed the feasibility of the fast and accurate classification of unknown samples according to their origin using NIR spectroscopy.

3.3. *Prediction of the Sensory Parameters of Cecina*

The best ANN architecture, which is obtained by the higher RSQ and the lower MSEP, used for the prediction of cecina the sensory parameters is shown in Table 4. As previously reported, the Levenberg–Marquardt algorithm using the hyperbolic tangent sigmoid for the hidden layer and the linear functions for the output layer is the most suitable for this purpose. With this algorithm, it is possible to find an ANN with a higher RSQ and a lower MSEP than with other algorithms such as gradient scalar. The network was optimized as follows: the data were divided into a training set with 34 samples which was used to obtain the predicting neural network. The accuracy of the ANN (a comparison of the reference with the predicted value) is given by the RSQ. The validation set, which included 8 new samples, was subsequently used to avoid the overfitting of the network. The test set constituted by 8 new different samples was used to check the performance of the network so that new RSQ and MSEP values were obtained. These parameters were used to select the most suitable ANN., so this process was repeated using from 1 to 30 neurons in the hidden layer and for each architecture up to 100 training sessions with different and known seed values were carried out because previous works [23] have shown that the higher the number of trainings the better the prediction capacity of the ANN. The number of neurons for the best ANN architecture, together with the RSQ and the MSEP for the prediction of all the sensory parameters analyzed and for the total of the samples analyzed, is shown in Table 4.

Table 4. The number of neurons in the hidden layer, correlation coefficient R squared (RSQ), and mean square error or prediction (MSEP) of the best ANN for each sensory parameter.

	Neurons	RSQ	MSEP
Appearance			
Veined	15	0.90	0.293
Fat color	18	0.84	0.054
Color intensity	8	0.89	0.135
Exudate	13	0.87	0.190
White dots	1	0.99	0.008
Flavor			
Odor intensity	9	0.65	0.133
Cured odor	14	0.87	0.066
Smoked odor	25	0.73	0.183
Rancid odor	9	0.84	0.025
Moldy odor	6	0.91	0.005
Flavor intensity	22	0.80	0.097
Cured flavor	14	0.81	0.108
Saltiness	7	0.83	0.037
Sweetness	6	0.83	0.014
Smoked flavor	8	0.81	0.101
Rancidity	25	0.87	0.044
Pungency	25	0.79	0.013
Aftertaste	12	0.88	0.042
Texture			
Hardness	13	0.90	0.090
Juiciness	24	0.95	0.036
Fatness	19	0.90	0.101
Fibrousness	9	0.88	0.067
Chewiness	18	0.92	0.050
Gumminess	15	0.93	0.033

The results show that it was possible to predict all the sensory parameters with very high RSQ values (>0.80) with the exception of odor intensity. These values were higher than those previously reported for the prediction of sensory parameters in meat [39–41] and meat products [16,23]. The highest RSQ were observed for white dots (0.99) while odor intensity showed the lowest (0.68), which coincides with that reported by Hernández-Jiménez et al. [16] for dry sausages. In general, the highest RSQ values were observed in texture parameters as previously observed in cheeses by Curto et al. [42], with values close to or higher than 0.9, followed by the appearance parameters with RSQ values close to or higher than 0.85.

The MSEPs varied between 0.005 for moldy odor and 0.293 for veined and showed lower values than those previously found for dry-cured ham [23]. Furthermore, as in this study the lowest values of MSEP were observed in general for flavor parameters while the highest values were those of appearance attributes. The small differences between the reference and the predicted values can be seen in Figure 5 in which both sets of data are compared for some of the sensory parameters predicted.

These results point out the t feasibility of the prediction of sensory parameters of dry-cured beef meat cecina using NIR spectroscopy, as a fast and non destructive method, together with ANN using the Levenberg–Marquardt algorithm after the correct optimization process of the ANN.

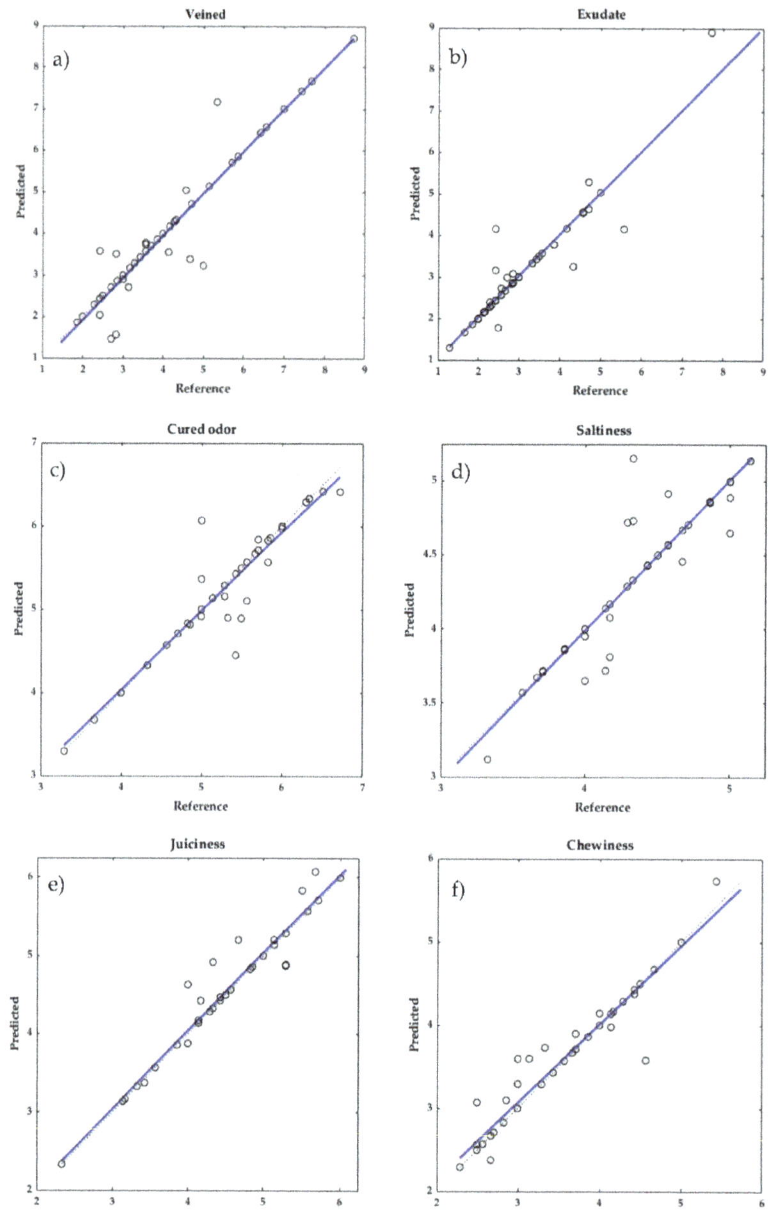

Figure 5. Reference vs. predicted values for (**a**) veined, (**b**) exudate, (**c**) cured odor, (**d**) saltiness, (**e**) juiciness, and (**f**) chewiness.

4. Conclusions

From the results obtained it can be concluded that near infrared spectroscopy together with artificial neural networks allow the accurate prediction of almost all (23 out of 24) the sensory parameters selected for an exhaustive characterization of dry-cured beef meat cecina quality with RSQ values higher than 0.8. Taking into account that a sensory analysis is compulsory for cecina de León PGI products, this result stresses the fact that it would possible to substitute the sensory panel by a faster, reliable, non destructive and cheaper instrumental technique that may be implemented on site. Moreover, after an optimization procedure, the ANNs also allow differentiation between the PGI and non-PGI samples produced in the same geographical area with 100% of samples being correctly classified, while the average percentage of correct classification when RMS-X residual was applied was 92%. This reveals that NIRS technology can be a powerful tool to ensure the quality of the product and to prevent fraud.

Author Contributions: Conceptualization, I.R. and A.M.V.-Q.; methodology M.I.G.-M.; validation, M.H.-J. and P.H.-R., formal analysis, I.M.-M., writing—reviewing and editing, I.R., A.M.V.-Q. and M.I.G.-M.; supervision and project administration, I.R. All authors have read and agreed to the published version of the manuscript.

Funding: This research was funded by Junta de Castilla y León project reference SA039P17.

Conflicts of Interest: The authors declare no conflict of interest.

References

1. European Commission. *Commission Regulation (EC) N° 1107/96 of 12 June 1996 on the Registration of Geographical Indications and Designations of Origin under the Procedure Laid down in Article 17 of Council Regulation (EEC) No 2081/92*; European Commission: Brussels, Belgium, 1996; L148:1.
2. Rodríguez-Lázaro, D.; Hernández-Pérez, M.; Capita, R.; Alonso-Calleja, C. Descriptive sensory properties of Cecina de León. *Ital. J. Food Sci.* **2018**, *30*, 249–255. [CrossRef]
3. The International Organization for Standardization. *ISO 5492. Sensory Analysis–Vocabulary*; ISO: Geneva, Switzerland, 2008.
4. The International Organization for Standardization. *ISO 8586. Sensory Analysis–General Guidelines for the Selection, Training and Monitoring of Selected Assessors and Expert Sensory Assessors*; ISO: Geneva, Switzerland, 2012.
5. Pérez Elortondo, F.J.; Ojeda, M.; Albisu, M.; Salmerón, J.; Etayo, I.; Molina, M. Food quality certification: An approach for the development of accredited sensory evaluation methods. *Food Qual. Prefer.* **2007**, *18*, 425–439. [CrossRef]
6. Etaio, I.; Albisu, M.; Ojeda, M.; Gil, P.F.; Salmerón, J.; Pérez Elortondo, F.J. Sensory quality control for food certification: A case study on wine. Method development. *Food Control* **2010**, *21*, 533–541. [CrossRef]
7. González-Casado, A.; Jiménez-Carvelo, A.M.; Cuadros-Rodríguez, L. Sensory quality control of dry-cured ham: A comprehensive methodology for sensory panel qualification and method validation. *Meat Sci.* **2019**, *149*, 149–155. [CrossRef]
8. Yu, H.; Zhou, Y.; Fu, X.; Xie, L.; Ying, Y. Discrimination between Chinese rice wines of different geographical origins by NIRS and AAS. *Eur. Food Res. Technol.* **2007**, *225*, 313–320. [CrossRef]
9. Herrero Latorre, C.; Peña Crecente, R.M.; García Martín, S.; Barciela García, J. A fast chemometric procedure based on NIR data for authentication of honey with protected geographical indication. *Food Chem.* **2013**, *141*, 3559–3565. [CrossRef]
10. Firmani, P.; De Luca, S.; Bucci, R.; Marini, F.; Biancolillo, A. Near infrared (NIR) spectroscopy-based classification for the authentication of Darjeeling black tea. *Food Control* **2019**, *100*, 292–299. [CrossRef]
11. Revilla, I.; Lastras, C.; González-Martín, M.I.; Vivar-Quintana, A.M.; Morales-Corts, R.; Gómez-Sánchez, M.A.; Pérez-Sánchez, R. Predicting the physicochemical properties and geographical ORIGIN of lentils using near infrared spectroscopy. *J. Food Compos. Anal.* **2019**, *77*, 84–90. [CrossRef]
12. Barreira, J.C.M.; Casal, S.; Ferreira, I.C.F.R.; Peres, A.M.; Pereira, J.A.; Oliveira, M.B.P.P. Supervised Chemical Pattern Recognition in Almond (*Prunus dulcis*) Portuguese PDO Cultivars: PCA- and LDA-Based Triennial Study. *J. Agric. Food Chem.* **2012**, *60*, 9697–9704. [CrossRef]
13. Karoui, R.; Mazerolles, G.; Bosset, J.-O.; de Baerdemaeker, J.; Dufour, E. Utilisation of mid-infrared spectroscopy for determination of the geographic origin of Gruyère PDO and L'Etivaz PDO Swiss cheeses. *Food Chem.* **2007**, *105*, 847–854. [CrossRef]
14. Ríos-Reina, R.; García-González, D.L.; Callejón, R.M.; Amigo, J.M. NIR spectroscopy and chemometrics for the typification of Spanish wine vinegars with a protected designation of origin. *Food Control* **2018**, *89*, 108–116. [CrossRef]

15. Moscetti, R.; Radicetti, E.; Monarca, D.; Cecchini, M.; Massantini, R. Near infrared spectroscopy is suitable for the classification of hazelnuts according to Protected Designation of Origin. *J. Sci. Food Agric.* **2015**, *95*, 2619–2625. [CrossRef]
16. Hernández-Jiménez, M.; Hernández-Ramos, P.; Martínez-Martín, I.; Vivar-Quintana, A.M.; González-Martín, I.; Revilla, I. Comparison of artificial neural networks and multiple regression tools applied to near infrared spectroscopy for predicting sensory properties of products from quality labels. *Microchem. J.* **2020**, *159*, 105459. [CrossRef]
17. Pillonel, L.; Bütikofer, U.; Schlichtherle-Cerny, H.; Tabacchi, R.; Bosset, J.O. Geographic origin of European Emmental. Use of discriminant analysis and artificial neural network for classification purposes. *Int. Dairy J.* **2005**, *15*, 557–562. [CrossRef]
18. Ripoll, G.; Albertí, P.; Panea, B.; Olleta, J.L.; Sañudo, C. Near-infrared reflectance spectroscopy for predicting chemical, instrumental and sensory quality of beef. *Meat Sci.* **2008**, *80*, 697–702. [CrossRef]
19. Prieto, N.; Andrés, S.; Giráldez, F.J.; Mantecón, A.R.; Lavín, P. Potential use of near infrared reflectance spectroscopy (NIRS) for the estimation of chemical composition of oxen meat samples. *Meat Sci.* **2006**, *74*, 487–496. [CrossRef]
20. Ellekjaer, M.R.; Isaksson, T.; Solheim, R. Assessment of Sensory Quality of Meat Sausages Using Near Infrared Spectroscopy. *J. Food Sci.* **1994**, *59*, 456–464. [CrossRef]
21. Ortiz, M.C.; Sarabia, L.; García-Rey, R.; Luque de Castro, M.D. Sensitivity and specificity of PLS-class modelling for five sensory characteristics of dry-cured ham using visible and near infrared spectroscopy. Anal. *Chim. Acta* **2006**, *558*, 125–131. [CrossRef]
22. García-Rey, R.M.; García-Olmo, J.; De Pedro, E.; Quiles-Zafra, R.; Luque de Castro, M.D. Prediction of texture and colour of dry-cured ham by visible and near infrared spectroscopy using a fiber optic probe. *Meat Sci.* **2005**, *70*, 357–363. [CrossRef]
23. Hernández-Ramos, P.; Vivar-Quintana, A.M.; Revilla, I.; González-Martín, M.I.; Hernández-Jiménez, M.; Martínez-Martín, I. Prediction of Sensory Parameters of Cured Ham: A Study of the Viability of the Use of NIR Spectroscopy and Artificial Neural Networks. *Sensors* **2020**, *20*, 5624. [CrossRef]
24. Dongre, V.B.; Gandhi, R.S.; Singh, A.; Ruhil, A.P. Comparative efficiency of artificial neural networks and multiple linear regression analysis for prediction of first lactation 305-day milk yield in Sahiwal cattle. *Livest. Sci.* **2012**, *147*, 192–197. [CrossRef]
25. Niazian, M.; Sadat-Noori, S.A.; Abdipour, M. Modeling the seed yield of Ajowan (*Trachyspermum ammi* L.) using artificial neural network and multiple linear regression models. *Ind. Crops Prod.* **2018**, *117*, 224–234. [CrossRef]
26. Torkashvand, A.M.; Ahmadi, A.; Nikravesh, N.L. Prediction of kiwifruit firmness using fruit mineral nutrient concentration by artificial neural network (ANN) and multiple linear regressions (MLR). *J. Integr. Agric.* **2017**, *16*, 1634–1644. [CrossRef]
27. Boletín Oficial del Estado. *Orden de 27 de Junio por la que se Ratifica el Reglamento de la Denominación Específica "Cecina de León" y su Consejo Regulador*; BOE: Madrid, Spain, 1994.
28. González-Martín, I.; Hernández-Hierro, J.M.; Revilla, I.; Vivar-Quintana, A.; Lobos Ortega, I. The mineral composition (Ca, P, Mg, K, Na) in cheeses (cow's, ewe's and goat's) with different ripening times using near infrared spectroscopy with a fibre-optic probe. *Food Chem.* **2011**, *127*, 147–152. [CrossRef]
29. Molinero, C.; Martínez, B.; Rubio, B.; González-Fernández, C.; García-Cachán, M.D.; Jaime, I. The Anatomical Origin of Raw Meat Affects the Sensory and Chemical Characteristics of a Dry-Cured Beef Product: Cecina de León. *Food Sci. Technol. Int.* **2008**, *14*, 225–232. [CrossRef]
30. Marušić, N.; Vidaček, S.; Janči, T.; Petrak, T.; Medić, H. Determination of volatile compounds and quality parameters of traditional Istrian dry-cured ham. *Meat Sci.* **2014**, *96*, 1409–1416. [CrossRef]
31. Arnau, J.; Guerrero, L.; Hortós, M.; García-Regueiro, J.A. The Composition of White Film and White Crystals Found in Dry-Cured Hams. *J. Sci. Food Agric.* **1996**, *70*, 449–452. [CrossRef]
32. Lorenzo, J.M.; Carballo, J. Influence of Anatomical Retail Cut on Physicochemical and Sensory Characteristics of Foal "Cecina". *Int. J. Food Prop.* **2016**, *19*, 802–813. [CrossRef]
33. Ruiz Pérez-Cacho, M.P.; Galán-Soldevilla, H.; León Crespo, F.; Molina Recio, G. Determination of the sensory attributes of a Spanish dry-cured sausage. *Meat Sci.* **2005**, *71*, 620–633. [CrossRef]
34. Molinero Sastre, C. Caracterización y Optimización del Proceso Tecnológico de Elaboración de la Cecina de León. Ph.D. Thesis, University of Burgos, Burgos, Spain, 2009.

35. Rubio, B.; Martínez, B.; González-Fernández, C.; García-Cachán, M.D.; Rovira, J.; Jaime, I. Influence of storage period and packaging method on sliced dry cured beef "Cecina de León": Effects on microbiological, physicochemical and sensory quality. *Meat Sci.* **2006**, *74*, 710–717. [CrossRef]
36. Aristoy, M.C.; Toldrá, F. Isolation of flavor peptides from raw pork meat and dry-cured ham. *Dev. Food Sci.* **1995**, *37*, 1323–1344. [CrossRef]
37. Shenk, J.S.; Workman, J.J.; Westerhaus, M.O. Application of NIR spectroscopy to agricultural products. In *Handbook of Near-Infrared Analysis*, 3rd ed.; Burns, D.A., Ciurczak, E.W., Eds.; Marcel Dekker, Inc.: New York, NY, USA, 2001; pp. 419–474.
38. Cozzolino, D.; Murray, I. Identification of animal meat muscles by visible and near infrared reflectance spectroscopy. *LWT- Food Sci. Technol.* **2004**, *37*, 447–452. [CrossRef]
39. Tian, Y.Q.; McCall, D.G.; Dripps, W.; Yu, Q.; Gong, P. Using computer vision technology to evaluate the meat tenderness of grazing beef. *Food Aust.* **2005**, *57*, 322–326.
40. Chandraratne, M.R.; Samarasinghe, S.; Kulasiri, D.; Bickerstaffe, R. Prediction of lamb tenderness using image surface texture features. *J. Food Eng.* **2006**, *77*, 492–499. [CrossRef]
41. Balejko, J.A.; Nowak, Z.; Balejko, E. Artificial neural network as the tool in prediction rheological features of raw minced meat. *Acta Sci. Pol. Technol. Aliment.* **2012**, *11*, 273–281.
42. Curto, B.; Moreno, V.; García-Esteban, J.A.; Blanco, F.J.; González, I.; Vivar, A.; Revilla, I. Accurate Prediction of Sensory Attributes of Cheese Using Near-Infrared Spectroscopy Based on Artificial Neural Network. *Sensors* **2020**, *20*, 3566. [CrossRef]

sensors

MDPI

Article

Near-Infrared Reflectance Spectroscopy for Predicting the Phospholipid Fraction and the Total Fatty Acid Composition of Freeze-Dried Beef

Guillermo Ripoll [1,2,*], Sebastiana Failla [3], Begoña Panea [1,2], Jean-François Hocquette [4], Susana Dunner [5], Jose Luis Olleta [2,6], Mette Christensen [7], Per Ertbjerg [8], Ian Richardson [9], Michela Contò [3], Pere Albertí [1], Carlos Sañudo [6] and John L. Williams [10,11]

1 Centro de Investigación y Tecnología Agroalimentaria de Aragón (CITA), Avda. Montañana, 930, 50059 Zaragoza, Spain; bpanea@cita-aragon.es (B.P.); perealbertilasalle@gmail.com (P.A.)
2 Instituto Agroalimentario de Aragón—IA2 (CITA-Universidad de Zaragoza), Avda. Miguel Servet, 177, 50013 Zaragoza, Spain; olleta@unizar.es
3 CREA, Consiglio per la Ricerca in Agricoltura e L'analisi dell'Economia Agraria, 00015 Monterotondo, Italy; sebastiana.failla@crea.gov.it (S.F.); michela.conto@creagov.onmicrosoft.com (M.C.)
4 INRAE, Institut National de Recherche pour l'Agriculture, l'alimentation et l'Environnement-VetAgro Sup, UMR1213, Unité de Recherches sur les Herbivores, Theix, F-63122 Saint-Genès Champanelle, France; jean-francois.hocquette@inrae.fr
5 Departamento de Producción Animal, Facultad de Veterinaria, Universidad Complutense, 28040 Madrid, Spain; dunner@vet.ucm.es
6 Departamento de Producción Animal, Universidad de Zaragoza, Avda. Miguel Servet, 177, 50013 Zaragoza, Spain; csanudoastiz@gmail.com
7 Frontmatec Smoerum A/S, Hassellunden 9, DK-2765 Smoerum, Denmark; mch@frontmatec.com
8 Department of Food and Nutrition, University of Helsinki, 00014 Helsinki, Finland; per.ertbjerg@helsinki.fi
9 Bristol Veterinary School, Faculty of Health Sciences, Langford, Bristol BS40 5DU, UK; isnandimelda@gmail.com
10 Davies Research Centre, School of Animal and Veterinary Sciences, University of Adelaide, Roseworthy, SA 5371, Australia; john.williams01@adelaide.edu.au
11 Department of Animal Science, Food and Nutrition, Università Cattolica del Sacro Cuore, via Emilia Parmense, 84, 29122 Piacenza, Italy
* Correspondence: gripoll@aragon.es

Citation: Ripoll, G.; Failla, S.; Panea, B.; Hocquette, J.-F.; Dunner, S.; Olleta, J.L.; Christensen, M.; Ertbjerg, P.; Richardson, I.; Contò, M.; et al. Near-Infrared Reflectance Spectroscopy for Predicting the Phospholipid Fraction and the Total Fatty Acid Composition of Freeze-Dried Beef. *Sensors* **2021**, *21*, 4230. https://doi.org/10.3390/s21124230

Academic Editors: Mercedes Del Río Celestino and Rafael Font Villa

Received: 31 May 2021
Accepted: 16 June 2021
Published: 20 June 2021

Abstract: Research on fatty acids (FA) is important because their intake is related to human health. NIRS can be a useful tool to estimate the FA of beef but due to the high moisture and the high absorbance of water makes it difficult to calibrate the analyses. This work evaluated near-infrared reflectance spectroscopy as a tool to assess the total fatty acid composition and the phospholipid fraction of fatty acids of beef using freeze-dried meat. An average of 22 unrelated pure breed young bulls from 15 European breeds were reared on a common concentrate-based diet. A total of 332 longissimus thoracis steaks were analysed for fatty acid composition and a freeze-dried sample was subjected to near-infrared spectral analysis. 220 samples (67%) were used as a calibration set with the remaining 110 (33%) being used for validation of the models obtained. There was a large variation in the total FA concentration across the animals giving a good data set for the analysis and whilst the coefficient of variation was nearly 68% for the monounsaturated FA it was only 27% for the polyunsaturated fatty acids (PUFA). PLS method was used to develop the prediction models. The models for the phospholipid fraction had a low R^2p and high standard error, while models for neutral lipid had the best performance, in general. It was not possible to obtain a good prediction of many individual PUFA concentrations being present at low concentrations and less variable than other FA. The best models were developed for Total FA, saturated FA, 9c18:1 and 16:1 with R^2_p greater than 0.76. This study indicates that NIRS is a feasible and useful tool for screening purposes and it has the potential to predict most of the FA of freeze-dried beef.

Keywords: NIRS; muscle; bovine; chemometrics; MUFA; PUFA; SFA

1. Introduction

Near-infrared spectroscopy has been used for many years to measure the chemical composition of raw materials in the agri-food industry because it is a rapid, clean and accurate tool. The meat industry routinely uses infrared spectroscopy to analyse the chemical composition of meats [1–6].

Research on the fatty acid (FA) composition of meats has been growing in response to consumer concerns about their healthiness. It is widely known that the intake of monounsaturated FA (MUFA) and polyunsaturated FA (PUFA), mainly n-3 FA, reduces the prevalence of coronary heart disease and cholesterol levels [7] and other inflammatory and immune disorders [8]. Conversely, high intakes of saturated fatty acids (SFA) is associated with increased susceptibility to heart attacks due to the formation of blood clots [9], although that relationship remains unclear [10]. However, whilst high concentrations of PUFA in meat may be nutritionally desirable [11], in the absence of adequate concentrations of antioxidant, it can increase meat colour intensity (saturation) and fat oxidation resulting in poor sensory quality [12,13]. The FA composition of meat is usually determined by gas chromatography. When this method is optimized and long columns are used it allows many fatty acids to be quantified. However, gas chromatography is expensive, slow and uses dangerous chemicals. NIR spectroscopy is fast, cheap and clean and is useful in estimating multiple characteristics at the same time. It is not used widely because not all minor FA are accurately determined [14–17], partly because of the high water content of meat that absorbs more infrared light than the solutes [18,19], and as a result, calibration often fails. The absorbance of materials with high moisture content is temperature dependent [20] which also makes it difficult to calibrate the analyses. Consequently, removing water from materials before NIRS analysis is likely to improve the quantification of certain substances [19,21–23]. Freeze drying meat prior to analysis has been shown to improve the determination of minor fatty acids such as individual FA [17]. In addition to improving performance compared with raw meat because water absorption bands are reduced, freeze-drying also concentrates substances in beef around fourfold [24]. NIRS can determine fatty acid profiles of phospholipid and neutral lipid fractions independently. Using this technique SFA and MUFA concentrations in the neutral fraction have been shown to be higher than in the phospholipid fraction while PUFA is higher in the phospholipid fraction than the neutral lipid fraction [25]. The ability to study the fatty acids of neutral and phospholipid fractions is of interest because phospholipids are the building blocks of the cell membrane. The fatty acid composition of the membrane is mainly controlled by the genes involved in fatty acid metabolism, whilst the fatty acid composition of neutral lipid is influenced by the diet [26,27]. The positive impact of dietary phospholipids on human health is well established [28] and in addition, phospholipids contribute to the flavour of meat together with the Maillard reactions [26,29].

This work evaluated near-infrared reflectance spectroscopy as a tool to assess the total fatty acid composition and the phospholipid fraction of fatty acids of beef using freeze-dried meat.

2. Materials and Methods

2.1. Animal and Rearing Conditions

The care and use of animals were in accordance with the European Union Directive 2010/63 on the protection of animals used for experimental and other scientific purposes [30] because, at the time of the experiment, the member states of the European Union had no obligation to have an Ethics Committee for Animal Experiments.

A total of 332 unrelated pure breed young bulls from 15 European breeds were reared on commercial farms or in experimental research centres, depending on the experimental facilities of each country in France, Denmark, Italy, Spain and the United Kingdom. A uniform beef management system, representative of those used in European Union countries, was used for all breeds to standardise, as far as possible, the influence of diet, management and rearing systems on meat quality. The breeds and number of animals were: Aberdeen

Angus (27), Highland (24), Jersey (25), South Devon (20), Danish Red (24), Holstein (25), Simmental (20), Asturiana de la Montaña (22), Asturiana de los Valles (20), Avileña-Negra Ibérica (22), Pirenaica (20), Marchigiana (22), Piemontese (20), Charolais (21) and Limousin (20). At 9 months of age, all the animals were transferred to the experimental farms, where they were divided into groups of 7 to 8 animals, and fed the standardised diet, which consisted of a concentrate compounded from barley flakes (80 to 84%), soya bean meal (7.5 to 11%) sodium bicarbonate (0.6%) with vitamin supplements (1.5%) and barley straw, fed ad libitum. The energy density ratio ranged from 12.9-13.5 ME/kg DM. The protein content was 160 g Crude Protein/kg DM up to 10 months of age and then decreased to 150 g CP/kg DM to slaughter. Performance, body size and carcass characteristics of the fifteen breeds have been previously reported by Albertí, et al. [31].

2.2. Sampling and Measurements

At 75% mature bull weight, which was at about 15 months of age, animals were slaughtered as reported in Albertí, et al. [31]. The carcasses were chilled at 4 °C for 24 h. Then, the longissimus thoracis (LT) muscle was excised from the left side of the carcass between the 6th and the 13th ribs and was stored at 2 °C ± 1 °C until 48 h post-mortem. Then, a steak per animal was taken from around the position of the 8th vertebra and split into two pieces. Both pieces were vacuum packed and frozen at −18 °C. One-piece was transported in polystyrene boxes filled with dry ice to the CREA-ZA (Monterotondo, Italy) for NIR analysis, while the other was transported to the University of Bristol (Bristol, United Kingdom) for fatty acid analysis.

2.3. Collection of NIR Reflectance Spectra

The intermuscular fat covering of the LT was discarded and the remaining sample was freeze-dried and stored at −70 °C until spectra collection. The sample was homogenized using a meat mincer Moulinex D-56 (Groupe SEB, Écully, France) and kept at room temperature for 1 h before recording the spectra. The minced freeze-dried sample was inserted into a cylindrical quartz glass cup with an internal diameter of 35 mm and a depth of 10 mm. Reflectance spectra were scanned and collected twice per sample with a FOSS NIRSystems 5000 (FOSS NIRSystems Inc., Silver Spring, MD, USA). Spectra were recorded from 10,000 to 4000 cm^{-1} each 2 cm^{-1} interval (1000 to 2500 nm each 0.5 nm interval) and recorded as $\log\left(\frac{1}{R}\right)$.

2.4. Fatty Acid Composition Analysis

The samples for fatty acid determination were stored at –70 °C until analysis. After thawing in tap water, the muscle was blended and the lipids were extracted from 10 g samples using chloroform:methanol (2:1, *v/v*) [32], separated into neutral lipid (NL) and phospholipid (PL) fractions, using silicic acid chromatography (Isolute Si, Jones Chromatography, Hengoed, Glamorgan, UK) and methylated as described in Scollan, et al. [33] using a solution of diazomethane in diethyl ether. Total lipid content was taken as the sum of the neutral lipid and phospholipid fractions. Samples were analysed by gas chromatography by injection in the split mode, 70:1, onto a CP Sil 88, 50 m × 0.25 mm fatty acid methyl esters (FAME) column (Chrompack UK Ltd., London, UK) with helium as the carrier gas. The individual peaks of each FA were identified and quantified as described in detail by Scollan, et al. [33]. Only the major fatty acids were reported, which represented over 90% of the total FA present.

2.5. Chemometrics

The samples from each breed were assigned to a Calibration or a Validation set randomly. Random numbers between 0 and 1 were generated using MS-Excel and they were assigned to each sample. Therefore, samples with a number lower than 0.67 were assigned to the Calibration set and the remaining samples were assigned to Validation set. Hence, the Calibration set comprised 222 samples (67%), while the Validation set had

110 samples (33%). Mathematical treatments and pre-treatments such as scatter correction and derivatives, as well as gap and smooth segments, were investigated. Forty-eight 48 optimal models were developed including FA of the total and phospholipid fraction and groups of FA. The number of factors for each model were selected to optimize the R^2. Spectral ranges and individual wavelengths were selected according to the loadings and regression coefficients of the models, and then tested to obtain the best calibration model. Because many of the data points in the spectrum were highly co-linear, they were compressed using few factors [34] to derive the calibration equation. Compression was carried out using Partial Least Squares. The performance of the different predictive models obtained were determined from calibration and validation. The standard errors of calibration (SEC) and validation (SEP) sets, the coefficients of determination ($R^2{}_c$ and $R^2{}_p$) of calibration and validation, respectively the residual predictive value (RPD) and Consistency were used to test the accuracy of the calibration models [35] and to choose the best model. RPD was calculated as ($RPD = \frac{SD}{SEP}$), where SD is the standard deviation of the laboratory (SD). Consistency was calculated as ($C = \frac{SEC}{SEP} \cdot 100$) and expressed as a percentage [36]. The Hotelling statistic (H statistics) was calculated and samples with an H statistic greater than 10 were defined as outliers. When outliers were eliminated from the calibration set the model improved. Chemometrics and spectral data management were carried out with Unscrambler X (Camo Software AS, Norway). Calibration equations were derived for the phospholipid fraction of FA and total FA but not the neutral FA as these can be calculated by the difference between total and phospholipid fraction.

3. Results and Discussion

3.1. Sample Composition

Twenty fatty acids were detected, and 4 FA groups were calculated. Tables 1 and 2 shows the descriptive statistics of the phospholipid fraction of FA and total FA, respectively. Means of total FA were higher than means of the FA of phospholipid fraction but differences were much higher for SFA and MUFA than PUFA. In general, total FA was more variable than the phospholipid fraction, as expected, because the different carcass fatness of the animals used in the study is related to the neutral lipid fraction, while the phospholipid fraction is more constant and less susceptible to differences in bodycomposition [37]. Therefore, the coefficient of variation (CV) of total FA from 11 FA was higher than 60% while only 4 FA had a CV of the phospholipid fraction higher than this value. Moreover, total SFA, total MUFA and total FA also had a CV above 65% while the phospholipid fraction had a CV below 30%. The total FA (Table 2) ranged from 2.7% to 10.9%.

In the calibration set, neutral and phospholipid fractions were 81.7% and 18.3%, respectively, in agreement with MacKintosh, et al. [37]. The phospholipid fraction contained much more PUFA than the neutral lipid (13.5% of SFA, 9.6% of MUFA and 74.1% of PUFA) while 18:1c9, 18:1t9 and CLA had lower percentages in phospholipid fraction (9.9%, 9.0% and 9.1% respectively) than in neutral lipid, which is in agreement with the findings of Wood, et al. [38] and MacKintosh, et al. [37]. Conversely, the percentages of FA with 20 or more carbons were greater in the phospholipid fraction than in neutral lipid. Therefore, the percentages of 20:3n-6, 20:4n-6, 20:5n-3, 22:4n-6, 22:5n-3 and 22:6n-3 in the phospholipid fraction were 92.3%, 98.1%, 96.2%, 93.5%, 95.4% and 95.6%, respectively [9].

Table 1. Descriptive statistics of the phospholipid fraction of fatty acids (mg/100 g of meat) of calibration and validation sets.

	Calibration Set					Validation Set				
Fatty Acid	**Mean**	**Min**	**Max**	**S.D.**	**C.V.**	**Mean**	**Min**	**Max**	**S.D.**	**C.V.**
12:0	0.06	0.00	0.53	0.08	133.33	0.09	0.57	1.32	0.16	177.78
14:0	1.64	0.34	6.34	1.13	68.90	1.62	2.49	4.96	1.05	64.81
16:0	64.42	34.05	109.85	14.34	22.26	65.92	64.20	103.23	14.93	22.65
16:0 ald	23.42	3.91	43.46	6.42	27.41	24.00	4.650	40.99	7.27	30.29
16:1	7.36	3.07	19.35	2.33	31.66	7.47	6.17	13.79	2.30	30.79
18:0	52.38	29.16	83.93	8.85	16.90	53.44	69.77	73.38	8.81	16.49
18:0 ald	15.87	3.47	26.64	4.48	28.23	16.64	3.36	31.87	5.47	32.87
18:1 t9	4.39	0.93	14.20	2.75	62.64	4.47	6.95	16.85	2.91	65.10
18:1 c9	76.72	26.06	182.51	24.35	31.74	77.08	53.77	137.56	24.19	31.38
18:1 c11	14.58	7.87	25.92	3.43	23.53	14.79	11.29	26.03	3.69	24.95
18:2 n-6	124.47	62.45	210.43	29.61	23.79	127.88	67.43	199.28	29.67	23.20
20:1	0.57	0.00	1.20	0.22	38.60	0.59	0.30	1.26	0.23	38.98
18:3 n-3	6.37	1.51	22.07	4.35	68.29	6.57	3.58	19.57	4.55	69.25
18:2 9c11tCLA	0.75	0.18	2.49	0.41	54.67	0.77	0.70	2.71	0.45	58.44
20:3 n-6	8.43	4.94	15.00	2.00	23.72	8.44	3.60	13.66	2.22	26.30
20:4 n-6	37.84	20.29	69.55	9.39	24.82	38.63	16.27	69.77	9.69	25.08
20:5 n-3	4.02	1.00	12.07	2.01	50.00	4.20	1.23	10.97	2.27	54.05
22:4 n-6	4.59	1.34	10.40	1.82	39.65	4.67	1.10	10.21	1.94	41.54
22:5 n-3	9.17	4.26	21.13	2.94	32.06	9.35	3.43	20.38	3.30	35.29
22:6 n-3	0.87	0.00	4.85	0.48	55.17	0.85	0.00	2.28	0.40	47.06
Total FA	493.81	270.43	813.78	85.49	17.31	504.13	467.25	726.78	89.74	17.80
SFA	157.80	89.26	252.16	28.679	18.17	161.71	101.33	228.035	31.09	19.23
MUFA	103.61	46.35	228.31	28.18	27.20	104.400	47.82	169.93	28.26	27.07
PUFA	196.51	107.55	319.65	39.61	20.16	201.36	103.90	297.84	40.64	20.18

SD, standard deviation; FA, fatty acid; SFA, saturated fatty acid; MUFA, monounsaturated fatty acid; PUFA, polyunsaturated fatty acid; C.V., coefficient of variation.

Table 2. Descriptive statistics of the total fatty acid composition (mg/100 g of meat) of calibration and validation sets.

	Calibration Set					Validation Set				
Fatty Acid	**Mean**	**Min**	**Max**	**S.D.**	**C.V.**	**Mean**	**Min**	**Max**	**S.D.**	**C.V.**
12:0	2.10	0.00	8.86	1.87	89.05	2.06	0.06	8.69	1.80	87.38
14:0	73.63	0.87	313.13	58.97	80.09	71.57	2.49	256.53	55.15	77.06
16:0	651.20	51.20	2878.25	480.58	73.80	639.85	64.20	2420.91	442.30	69.13
16:0 ald	23.42	3.91	43.46	6.42	27.41	24.00	4.65	40.99	7.27	30.29
16:1	92.50	5.55	414.79	73.77	79.75	89.04	6.17	266.67	64.18	72.08
18:0	401.08	50.04	1741.09	254.19	63.38	398.22	69.77	1497.58	237.68	59.69
18:0 ald	15.87	3.47	26.64	4.48	28.23	16.64	3.36	31.87	5.47	32.87
18:1 t9	77.48	3.47	625.86	76.14	98.27	75.20	6.95	363.93	68.72	91.38
18:1 c9	854.17	31.86	4125.90	658.56	77.10	831.04	53.77	2866.55	573.11	68.96
18:1 c11	49.02	11.24	191.83	29.18	59.53	47.77	11.29	119.64	25.02	52.38
18:2 n-6	183.30	83.15	500.18	62.86	34.29	183.30	67.43	314.36	54.76	29.87
20:1	4.03	0.27	22.16	3.28	81.39	3.77	0.30	13.58	2.58	68.44
18:3 n-3	15.16	2.70	69.20	12.27	80.94	14.99	3.58	48.81	11.81	78.79
18:2 9c11t CLA	8.14	0.45	43.18	6.39	78.50	7.84	0.70	27.41	5.26	67.09
20:3 n-6	9.14	5.15	16.12	2.48	27.13	9.11	3.60	16.29	2.67	29.31
20:4 n-6	38.56	20.29	71.23	9.60	24.90	39.32	16.27	70.47	10.01	25.46
20:5 n-3	4.18	1.00	15.56	2.18	52.15	4.35	1.23	11.41	2.37	54.48
22:4 n-6	4.91	1.34	11.33	2.15	43.79	4.96	1.10	11.70	2.26	45.56
22:5 n-3	9.61	4.26	21.85	3.11	32.36	9.82	3.43	20.65	3.47	35.34
22:6 n-3	0.91	0.00	4.95	0.61	67.03	0.89	0.00	5.56	0.60	67.42
Total FA	2701.26	452.75	10922.01	1780.47	65.91	2652.12	467.25	8701.06	1596.84	60.21
SFA	1167.30	133.90	4981.92	790.24	67.70	1152.34	157.41	4182.02	734.39	63.73
MUFA	1077.19	54.47	4916.56	818.18	75.96	1046.83	78.48	3570.11	713.46	68.15
PUFA	265.76	139.73	607.29	79.13	29.77	266.73	114.32	439.52	72.54	27.20

SD, standard deviation; FA, fatty acid; SFA, saturated fatty acid; MUFA, monounsaturated fatty acid; PUFA, polyunsaturated fatty acid.

Variations in the FA profiles from the 15 breeds used in the current study are representative of the variation present in European beef FA. The 15 breeds included different cattle types, milk, meat and dual purpose. The individual profiles and different lipid ratios of the 15 European cattle breeds included in the study are given in Sevane, et al. [39] and their correlations with sensory traits, such as flavour, texture or juiciness in Sevane, et al. [40]. However, as the animals were fed a standardized diet the variations were lower than if they had been fed different diets, especially in PUFA [27]. Andueza, et al. [17] noted that the variation in FA in cattle is limited because of the biohydrogenation of ruminants during the digestion process. The goodness of calibration models relies on the variability of samples in the data set used to develop the prediction models [14,24], but increasing the variability in the data does not always increase the accuracy of the calibration. Khan, et al. [41] used meat from four different species to increase the variability and showed that the calibration for the chemical composition could not be improved. The total amount of fat influences the fatty acid composition. Therefore, increasing the age of animals and fattening could increase the variation in FA, but this does not help for the phospholipid fraction which remains constant even though the total lipids increase [42].

Most of the recent studies on beef used absolute values to develop the calibration models, and have achieved good statistical results [14,43] although other authors still express the results as a percentage of total fatty acids [24]. The absorbance varies linearly with the parameter concentration, not the relative ratio, therefore using absolute content gives a better calibration than the use of percentages or relative amounts [44].

The use of freeze-dried samples increases the relative concentration of fatty acids and so improves the results [24]. In addition, freeze-drying reduces the water absorption, improving the resolution of the absorption spectra for muscle. Freeze-drying also has the advantage of fine grinding, which is expected to improve the calibration [45], because it has been demonstrated that mincing meat prior to analysis gives better results than intact meat [46]. As intact muscle fibres and myofibrils tend to conduct NIR light by absorbing more energy [47]. Expressing the results as an absolute value indicates the nutritional value of meat, but differences in the treatment of the data and the way results are expressed make it difficult to compare results among studies.

3.2. Spectral Characteristics

The mean spectrum for the 15 breeds started with values of 0.4 (minimum value above 0.2) which was sustained until 1600 nm and finished at 2500 nm with values of 1.0 and a maximum of almost 1.4 (Figure 1).

Values of absorbance and the shape of the spectrum are similar to the spectrum of freeze-dried beef reported by Andueza, et al. [17]. The absorbance for freeze-dried beef was lower than the absorbance of fresh beef [14,17,48], broiler breast [48] and pork [49] but the shape of the spectrum is similar among all the meats. The low absorbance values in the ranges 1440–1470 nm and 1920–1960 nm are due to the absence of water [50]. Giaretta, et al. [24] compared spectra from fresh and freeze-dried beef and reported similar values to ours at 1000 nm but much higher values at 2500 nm. In addition, these authors found more sharp peaks around 1700 nm and also in the C–H resonance region (2200–2500 nm). Within that latter band, Zhou, et al. [48] reported absorbance peaks, specifically 2310 nm and 2348 nm, which are related to lipids. In the present study, this region had subtle peaks which were highlighted with the use of scattering corrections and derivatives (Figure 1a,b). In that region, we also found the highest regression coefficients for the prediction of total FA and many individual FA (data not shown), which is in agreement with the findings of Prieto, et al. [51].

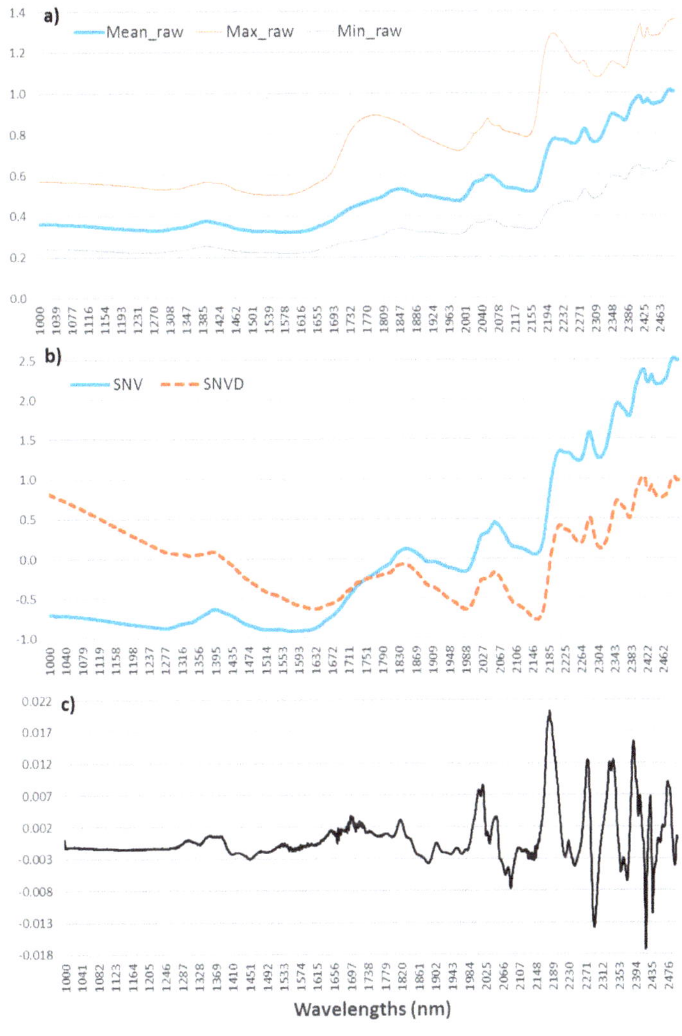

Figure 1. (**a**) Average (bold line), maximum and minimum (thin lines) of raw NIR spectra. (**b**) Standard normal variate (SNV) and standard normal variate plus detrending (SNVD) pre-treatments. (**c**) SNVD and first-order derivative pre-treatments. Spectra were recorded as log(1/R).

3.3. Prediction Models

The spectral pre-treatments and factors used to develop calibrations of phospholipids and total fatty acids used in this study are shown in Tables 3 and 4.

The offset baseline correction was used in 62.5% of the prediction models for the phospholipid fraction, while it was only used in the 8.3% of the prediction models for the total fraction of fatty acids. Area normalization was used in 50% of the prediction models for the phospholipid fraction and 12.5% of the prediction models for the total fraction of FA. The extended multiplicative scatter correction, SNV and SNVD were used in most of the calibrations of phospholipids (58.3%) but it was only required for 33.3% of calibrations of total fat fraction, mainly SNVD. The Savitzky–Golay derivative of first order with the second polynomial order and a smoothing gap of 3 or 5 were used in 15 calibrations while a Norris gap first-order derivative with large gap sizes (from 11 to 27) was useful to develop 6 calibrations. The most useful mathematical treatments were SNVD with or without first-order derivatives although many optimal models were developed without mathematical treatment. Indeed, the best prediction models of total FA, SFA, 9c18:1 and 16:1 used offset baseline correction, non-treatment, first-order Norris-Gap derivative and non-treatment, respectively. Figure 2 shows the regression coefficients of the model of total fatty acids.

Table 3. Spectral treatments and factors included in the prediction models of phospholipid fraction of fatty acids of beef.

Fatty Acid	Baseline Correction	Spectra Normalization	Scatter Correction [a]	Smooth	Mathematical Treatment [b]	F [c]
12:0	Offset	None	EMSC	None	None	6
14:0	Offset	None	EMSC	None	SG-1-2-5	5
16:0	Offset	Area	SNV+D	None	SG-1-2-3	4
16:0 ald	None	None	None	None	SG-1-2-3	3
16:1	Offset	None	None	None	None	4
18:0	Offset	Area	EMSC	None	None	4
18:0 ald	Offset	Area	SNV+D	None	None	6
18:1 t9	None	None	EMSC	None	None	3
18:1 c9	None	None	None	None	SG-1-2-5	2
18:1 c11	Offset	Area	SNV+D	None	None	3
18:2 n-6	None	None	None	None	None	8
20:1	Offset	Area	SNV	None	None	6
18:3 n-3	Offset	None	None	None	None	10
18:2 9c11t CLA	Offset	Area	SNV+D	None	None	5
20:3 n-6	None	None	EMSC	None	None	3
20:4 n-6	None	Area	SNV	None	None	12
20:5 n-3	Offset	Area	EMSC+D	None	None	8
22:4 n-6	None	None	None	None	SG-1-2-3	5
22:5 n-3	Offset	Area	None	None	None	10
22:6 n-3	None	Area	None	None	None	1
Total FA	Offset	Area	None	None	SG-1-2-5	5
SFA	Offset	None	SNV	None	None	6
MUFA	None	None	SNV	None	None	5
PUFA	Offset	Area	None	None	SG-1-2-3	5

[a] EMSC, extended multiplicative scatter correction; MSC, multiplicative scatter correction; SNV, standard normal variate; D, detrending. [b] SG, Savitzky–Golay derivative—derivative order—polynomial order—smoothing points; NG, Norris Gap derivative—derivative order—gap size. [c] F, number of factors.

Table 4. Spectral treatments and factors included in the prediction models of total fatty acids of beef.

Fatty Acid	Baseline Correction	Spectra Normalization	Scatter Correction [a]	Smooth	Mathematical Treatment [b]	F [c]
12:0	None	None	SNV+D	None	SG-1-2-3	4
14:0	None	Area	SNV+D	None	SG-1-2-3	3
16:0	None	Area	SNV+D	None	SG-1-2-3	3
16ald	None	None	None	None	None	2
16:1	None	None	None	SG1-1-1	None	8
18:0	None	None	None	None	None	5
18ald	None	None	SNV+D	None	None	8
18:1t9	None	None	SNV+D	None	None	2
9c18:1	None	None	None	None	NG-1-13	7
11c18:1	None	Area	SNV+D	None	None	11
18:2n-6	None	None	None	SG1-2-2	SG-1-2-3	5
20:1	None	None	None	None	SG-1-2-3	4
18:3n-3	Offset	None	None	None	None	9
9c11tCLA	None	None	MSC	None	SG-1-2-3	4
20:3n-6	None	None	None	None	NG-1-7	7
20:4n-6	None	None	None	None	SG-1-2-3	7
20:5n-3	None	None	None	None	NG-1-15	7
22:4n-6	None	None	None	None	SG-1-2-3	6
22:5n-3	None	None	SNV	None	NG-1-7	5
22:6n-3	None	None	None	None	None	1
TotalFA	Offset	None	None	None	None	9
SFA	None	None	None	None	None	9
MUFA	None	None	None	None	NG-1-27	7
PUFA	None	None	None	None	NG-1-11	6

[a] EMSC, extended multiplicative scatter correction; MSC, multiplicative scatter correction; SNV, standard normal variate; D, detrending. [b] SG, Savitzky–Golay derivative—derivative order—polynomial order—smoothing points; NG, Norris Gap derivative—derivative order—gap size. [c] Number of factors included in the calibrations.

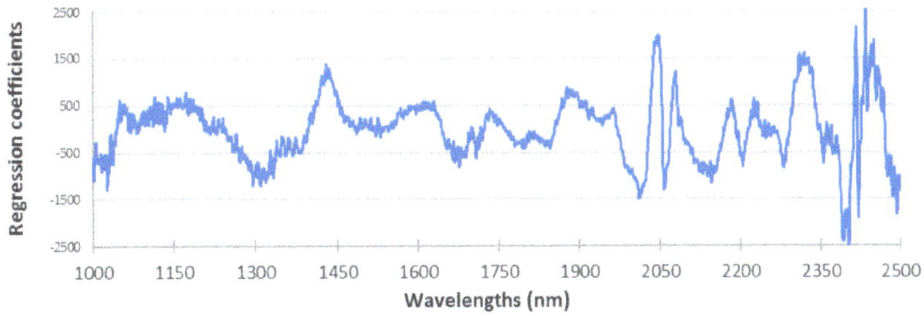

Figure 2. Regression coefficients of each wavelength for the model of total fatty acids.

There are no studies focused on the prediction of FA phospholipids, so the total FA fraction will be discussed here. The best statistical results have been found by applying SNV (with or without detrend) with the second-order derivative for most of the FA [49,52,53]. SNVD treatment reduced multicollinearity and the deleterious effects of a baseline shift and curvature while derivatives increase the resolution of peaks and reduce scattering [52]. However, the use of finely ground freeze-dried samples, means that for many calibrations pre-treatments are not needed, which has also been reported by Andueza, et al. [17].

Calibration and validation statistics of the models are shown in Tables 5 and 6. Some authors [49] reported calibrations for C16:0 and C18:0 with R^2 of 0.66 and 0.71, respectively, in pork using the SNVD and second-order derivatives, while Sierra, et al. [53] and Cecchinato, et al. [54] found that for beef C16:0 and C18:0 FA had an R^2 of 0.8 and 0.7, respectively. These values were similar to those using lamb and beef mixed together [14] and for rabbit meat [55]. An R^2 equal to or greater than 0.9. has been reported for freeze-dried beef [17] and a high R^2 for predicting C16:0 and C18:0 has also been found for broiler breast using SNVD and first-order derivatives [48]. Therefore, most of the authors reported similar or slightly better results than our results when using beef, pork and lamb with similar mathematical pre-treatments such as SNVD.

Table 5. Calibration and validation statistics for the phospholipids fraction of fatty acid composition (mg FA/100 g of meat).

Fatty Acid	n	SEC	R^2_c	SEP	R^2_p	RPD	Consistency
12:0	206	0.04	0.48	0.15	0.07	1.06	26.67
14:0	215	0.46	0.77	0.73	0.52	1.44	63.01
16:0	210	7.52	0.67	10.45	0.48	1.43	71.96
16:0ald	210	5.31	0.24	6.70	0.13	1.08	79.25
16:1	205	1.35	0.44	1.85	0.32	1.24	72.97
18:0	211	5.72	0.50	6.92	0.36	1.27	82.66
18:0ald	210	3.10	0.40	5.17	0.10	1.06	59.96
t918:1	200	1.56	0.19	2.89	0.03	1.01	53.98
9c18:1	205	14.91	0.50	18.81	0.37	1.29	79.27
11c18:1	212	2.66	0.20	3.51	0.09	1.05	75.78
18:2n-6	200	18.81	0.42	33.94	0.04	0.87	55.42
20:1	215	0.15	0.49	0.20	0.27	1.16	75.00
18:3n-3	207	2.26	0.69	3.11	0.53	1.46	72.67
9c11tCLA	200	0.25	0.36	0.43	0.06	1.04	58.14
20:3n-6	201	1.56	0.26	2.08	0.11	1.07	75.00
20:4n-6	201	5.88	0.55	7.59	0.29	1.28	77.47
20:5n-3	202	1.34	0.56	1.74	0.41	1.30	77.01
22:4n-6	207	1.10	0.59	1.58	0.29	1.23	69.62
22:5n-3	200	1.73	0.58	2.47	0.41	1.33	70.04
22:6n-3	198	0.28	0.12	0.38	0.05	1.07	73.68
Total FA	199	39.14	0.67	63.88	0.44	1.40	61.27
SFA	202	14.80	0.65	20.68	0.57	1.50	71.57
MUFA	210	17.46	0.53	18.85	0.50	1.50	92.63
PUFA	198	19.77	0.65	31.84	0.14	1.24	62.09

n, number of samples used in validation; SEC, standard error of calibration; R^2c, coefficient of determination of calibration; SEP, standard error of validation; R^2p, coefficient of determination of validation; RPD = SD/SEP; Consistency (%) = SEC*100/SEP.

Table 6. NIRS calibration and validation statistics for the total fatty acid composition (mg FA/100 g of meat).

Fatty Acid	n	SEC	R^2_c	SEP	R^2_p	RPD	Consistency
12:0	222	1.02	0.70	0.97	0.72	1.9	105.15
14:0	219	26.05	0.77	27.97	0.74	2.0	93.14
16:0	222	233.56	0.71	234.99	0.72	1.9	99.39
16:0ald	222	5.98	0.13	6.66	0.16	1.0	89.79
16:1	218	32.17	0.76	30.31	0.78	2.1	106.14
18:0	222	158.17	0.61	129.39	0.70	1.8	122.24
18:0ald	222	3.79	0.53	4.85	0.21	1.1	78.14
t918:1	219	41.01	0.47	50.16	0.47	1.4	81.76
9c18:1	217	244.72	0.80	274.92	0.77	2.1	88.59
11c18:1	218	12.14	0.77	12.69	0.74	2.0	95.67
18:2n-6	222	41.63	0.56	41.32	0.43	1.3	100.75
20:1	220	1.40	0.80	1.10	0.71	1.8	127.27
18:3n-3	222	7.73	0.60	7.07	0.65	1.7	109.34
9c11t CLA	222	4.00	0.61	3.25	0.62	1.6	123.08
20:3n-6	217	1.45	0.64	2.09	0.39	1.3	69.38
20:4n-6	222	4.89	0.74	8.60	0.26	1.2	56.86
20:5n-3	210	1.71	0.39	1.90	0.38	1.3	90.00
22:4n-6	217	1.00	0.78	1.74	0.39	1.3	57.47
22:5n-3	211	2.47	0.37	2.79	0.36	1.2	88.53
22:6n-3	222	0.61	0.01	0.60	0.01	1.0	101.67
Total FA	222	908.22	0.74	730.79	0.79	2.2	124.28
SFA	222	412.56	0.73	355.68	0.77	2.1	115.99
MUFA	222	393.64	0.77	340.36	0.77	2.1	115.65
PUFA	222	53.35	0.54	53.81	0.45	1.3	99.15

n, number of samples used in validation; SEC, standard error of calibration; R^2c, coefficient of determination of calibration; SEP, standard error of validation; R^2p, coefficient of determination of validation; RPD = SD/SEP; Consistency (%) = SEC*100/SEP.

Conjugated linoleic acid (CLA) describes a group of 18-carbon fatty acids with two conjugated double bonds. These isomers, of which c9,t11 and t10,c12 are the predominant members in beef, are beneficial for human health [56]. Some authors reported models that are for the entire CLA group, others report models for individual components, in this case, c9t11 and other isomers are used because major CLA isomers can coelute during GC analysis [57]. Hence, the comparison of results is not easy. The model we used (MSC, 1st derivative and $R^2_p = 0.62$) is consistent with those described in the literature because most used SNVD together 1st or 2nd derivative [14,17,53]. Prieto, et al. [51] used finely ground beef to predict several groups of CLA isomers with R^2_c ranging from 0.77 to 0.84, confirming that the finer the grinding, the higher the accuracy [47]. Finally, other authors did not report CLA [24,50,58].

Most FA with more than 19 carbons were not well predicted in our study, which has also been found by other authors [14,17,48,53,55]. Therefore, these are often not reported and published data tend to focus on the main groups of FA [24,50,52,58]. Most authors reported worse statistics for PUFA than for SFA and MUFA [14,48,53,55,59]. This could be explained because long-chain PUFA are mainly located in the membrane phospholipids which are quite constant because they are controlled by a complex enzymatic system, providing low variability among animals and have relatively low concentrations [58].

The models developed for the phospholipid fraction of the FA had RPD lower than 2 being useful just for screening purposes. However, some models for the estimation of FA were adequate for analytical purposes such as 14:0, 16:1, 11c18:1, total FA, SFA and MUFA. The other models remained below RPD = 2. The plots of those models are shown in Figure 3. The main weakness of NIRS to predict the FA composition is the inconsistency. While gas chromatography can identify all the FA that are important for meat science, NIRS does not. The reasons for the low-quality calibrations of some FA of meat include low concentration and variability, presence of water, and comparison of intact vs ground meat. The poor performance of NIRS in prediction equations for FA is due to this low variability and because some FA absorbs at the same wavelengths [14,47]. In our data set the range of variability may result in a complex relationship between the spectra and the response variables that are not predicted under a PLS model [59]. Using the NIRS technique to predict fatty acids is hampered by the absorption of light by the C–H bonds in certain wavelengths.

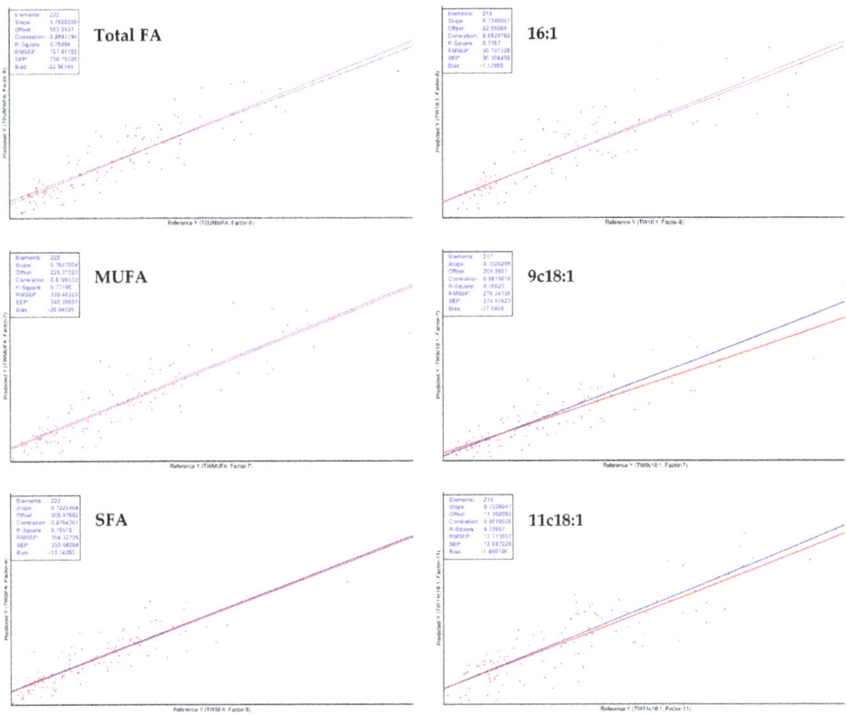

Figure 3. Scatter plots of models of 16:1, 11c18:1, total FA, SFA and MUFA of the total fraction.

Therefore, a C–H bond together with a cis bond modifies the absorption at the same wavelengths as a double cis bond [60]. This means that some individual fatty acids are not determined accurately, which could be related to similarities in the NIR absorption spectra among FA [48,53]. Other authors that have studied the PLS method, used in our study, fail when the relationship between spectra and the analyte of interest is non-linear [59,61]. Spectra collected in the reflectance mode are influenced not only by the main components of meat (water, fat, protein, etc.) but also the particle size, which is affected by the sample homogenization method and has to be accounted for using the right mathematical preprocessing [62]. Our results, using spectra from milled freeze-dried meat, suggest that this type of sample requires little or no preprocessing.

4. Conclusions

This study indicates that NIRS is a feasible and useful tool for screening purposes and has the potential to predict most of the FA of beef. The use of freeze-dried samples, thus reducing the water absorption bands and increasing the concentration of analytes, improved the accuracy of calibrations. Minimal mathematical pre-treatments were required to obtain good results. Using 15 breeds ensured that there was a large variation in the samples, which enabled us to develop good models. However, these improvements were not enough to achieve good calibrations for the phospholipid fraction, mainly due to the low concentrations of the FA in this fraction.

Author Contributions: Conceptualization, G.R. and S.F.; methodology, G.R., S.F., J.-F.H., B.P., S.D., J.L.O., M.C. (Mette Christensen), P.E., I.R., M.C. (Michela Contò), P.A., C.S. and J.L.W.; formal analysis, G.R. and S.F.; investigation, G.R., S.F., J.-F.H., B.P., S.D., J.L.O., M.C. (Mette Christensen), P.E., I.R., M.C. (Michela Contò), P.A., C.S. and J.L.W.; resources, S.F., J.-F.H., S.D., M.C. (Mette Christensen), P.E., I.R., P.A., C.S. and J.L.W.; data curation, G.R. and S.F.; writing—original draft preparation, G.R. and S.F.; writing—review and editing, G:R., S.F., J.-F.H., B.P., S.D., J.L.O., M.C. (Mette Christensen), P.E., I.R., M.C. (Michela Contò), P.A., C.S. and J.L.W.; project administration, S.F., J.-F.H., S.D., M.C. (Mette Christensen), P.E., I.R.,, P.A., C.S. and J.L.W..; funding acquisition, S.F., J.-F.H., S.D., M.C. (Mette Christensen), P.E., I.R., P.A., C.S. and J.L.W. All authors have read and agreed to the published version of the manuscript.

Funding: This research was funded by EU Commission Sixth Framework Programme Project "GemQual" (QLK5-CT-2000-0147) and the Government of Aragón (Grant Research Group Funds, Group A14_20R).

Institutional Review Board Statement: Not applicable.

Informed Consent Statement: Not applicable.

Data Availability Statement: Data presented in this study are available on request due from the corresponding author. The data are not publicly available due to privacy.

Acknowledgments: The authors acknowledge the staff of the different experimental stations and abattoirs that were involved in the rearing and slaughtering of the animals.

Conflicts of Interest: The authors declare no conflict of interest.

References

1. Denoyelle, C.; Cartier, P. Measurement of ground beef composition by near infrared spectrophotometry. *Viandes Prod. Carnes* **1996**, *17*, 191–196.
2. Prieto, N.; Andres, S.; Giraldez, F.J.; Mantecon, A.R.; Lavin, P. Potential use of near infrared reflectance spectroscopy (NIRS) for the estimation of chemical composition of oxen meat samples. *Meat Sci.* **2006**, *74*, 487–496. [CrossRef]
3. Olivan, M.; de la Roza, B.; Mocha, M.; Martinez, M.J. Prediction of physico-chemical and texture characteristics of beef by near infrared transmittance spectroscopy. In Proceedings of the 10th International Conference on Near Infrared Spectroscopy, Kyongju, Korea, 10–15 June 2002; pp. 197–202.
4. Ripoll, G.; Lobon, S.; Joy, M. Use of visible and near infrared reflectance spectra to predict lipid peroxidation of light lamb meat and discriminate dam's feeding systems. *Meat Sci.* **2018**, *143*, 24–29. [CrossRef]
5. Ripoll, G.; Albertí, P.; Panea, B.; Olleta, J.L.; Sañudo, C. Near-infrared reflectance spectroscopy for predicting chemical, instrumental and sensory quality of beef. *Meat Sci.* **2008**, *80*, 697–702. [CrossRef]
6. Aït-Kaddour, A.; Andueza, D.; Dubost, A.; Roger, J.-M.; Hocquette, J.-F.; Listrat, A. Visible and Near-Infrared Multispectral Features in Conjunction with Artificial Neural Network and Partial Least Squares for Predicting Biochemical and Micro-Structural Features of Beef Muscles. *Foods* **2020**, *9*, 1254. [CrossRef] [PubMed]
7. Williams, C.M. Dietary fatty acids and human health. *Ann. Zootech.* **2000**, *49*, 165–180. [CrossRef]
8. Galli, C.; Calder, P.C. Effects of fat and fatty acid intake on inflammatory and immune responses: A critical review. *Ann. Nutr. Metab.* **2009**, *55*, 123–139. [CrossRef] [PubMed]
9. Enser, M.; Richardson, R.; Wood, J.; Gill, B.; Sheard, P. Feeding linseed to increase the n-3 PUFA of pork: Fatty acid composition of muscle, adipose tissue, liver and sausages. *Meat Sci.* **2000**, *55*, 201–212. [CrossRef]
10. De Souza, R.J.; Mente, A.; Maroleanu, A.; Cozma, A.I.; Ha, V.; Kishibe, T.; Uleryk, E.; Budylowski, P.; Schünemann, H.; Beyene, J. Intake of saturated and trans unsaturated fatty acids and risk of all cause mortality, cardiovascular disease, and type 2 diabetes: Systematic review and meta-analysis of observational studies. *Bmj* **2015**, *351*, h3978. [CrossRef]
11. Elmore, J.S.; Cooper, S.L.; Enser, M.; Mottram, D.S.; Sinclair, L.A.; Wilkinson, R.G.; Wood, J.D. Dietary manipulation of fatty acid composition in lamb meat and its effect on the volatile aroma compounds of grilled lamb. *Meat Sci.* **2005**, *69*, 233–242. [CrossRef]
12. Mottram, D.S. Flavour formation in meat and meat products: A review. *Food Chem.* **1998**, *62*, 415–424. [CrossRef]
13. Warren, H.E.; Scollan, N.D.; Nute, G.R.; Hughes, S.I.; Wood, J.D.; Richardson, R.I. Effects of breed and a concentrate or grass silage diet on beef quality in cattle of 3 ages. II: Meat stability and flavour. *Meat Sci.* **2008**, *78*, 270–278. [CrossRef] [PubMed]
14. Mourot, B.P.; Gruffat, D.; Durand, D.; Chesneau, G.; Prache, S.; Mairesse, G.; Andueza, D. New approach to improve the calibration of main fatty acids by near-infrared reflectance spectroscopy in ruminant meat. *Anim. Prod. Sci.* **2014**, *54*, 1848. [CrossRef]
15. Prieto, N.; Dugan, M.E.; Lopez-Campos, O.; Aalhus, J.L.; Uttaro, B. At line prediction of PUFA and biohydrogenation intermediates in perirenal and subcutaneous fat from cattle fed sunflower or flaxseed by near infrared spectroscopy. *Meat Sci.* **2013**, *94*, 27–33. [CrossRef] [PubMed]

16. Pullanagari, R.R.; Yule, I.J.; Agnew, M. On-line prediction of lamb fatty acid composition by visible near infrared spectroscopy. *Meat Sci.* **2015**, *100*, 156–163. [CrossRef] [PubMed]
17. Andueza, D.; Listrat, A.; Durand, D.; Normand, J.; Mourot, B.P.; Gruffat, D. Prediction of beef meat fatty acid composition by visible-near-infrared spectroscopy was improved by preliminary freeze-drying. *Meat Sci.* **2019**, *158*, 107910. [CrossRef] [PubMed]
18. Blanco, M.; Peguero, A. Determination of organic additives in mortars by near-IR spectroscopy. A novel approach to designing a sample set with high-variability components. *Anal. Bioanal. Chem.* **2007**, *387*, 1525–1532. [CrossRef] [PubMed]
19. Núñez-Sánchez, N.; Serradilla, J.; Ares, J.; Garrido-Varo, A. Effect of moisture uptake on the repeatability of near infrared spectroscopy analyses of ewe milk using the dry extract system for infrared (DESIR) method. *J. Near Infrared Spectrosc.* **2008**, *16*, 381. [CrossRef]
20. Maeda, H.; Ozaki, Y.; Tanaka, M.; Hayashi, N.; Kojima, T. Near Infrared Spectroscopy and Chemometrics Studies of Temperature-Dependent Spectral Variations of Water: Relationship between Spectral Changes and Hydrogen Bonds. *J. Near Infrared Spectrosc.* **1995**, *3*, 191–201. [CrossRef]
21. Coppa, M.; Ferlay, A.; Leroux, C.; Jestin, M.; Chilliard, Y.; Martin, B.; Andueza, D. Prediction of milk fatty acid composition by near infrared reflectance spectroscopy. *Int. Dairy J.* **2010**, *20*, 182–189. [CrossRef]
22. Meurens, M.; Eydne, O.; Vanbelle, M. *Fine Analysis of Liquids by NIR Reflectance Spectroscopy of Dry Extract on Solid Support (DESIR)*; Hollow, J., Kaffka, K.J., Gonczy, J.L., Eds.; Akademiai Kiado: Budapest, Hungary, 1987; pp. 297–302.
23. Ripoll, G.; Casasús, I.; Joy, M.; Molino, F.; Blanco, M. Fat color and reflectance spectra to evaluate the β-carotene, lutein and α-tocopherol in the plasma of bovines finished on meadows or on a dry total mixed ration. *Anim. Feed Sci. Technol.* **2015**, *207*, 20–30. [CrossRef]
24. Giaretta, E.; Mordenti, A.; Palmonari, A.; Brogna, N.; Canestrari, G.; Belloni, P.; Cavallini, D.; Mammi, L.; Cabbri, R.; Formigoni, A. NIRs calibration models for chemical composition and fatty acid families of raw and freeze-dried beef: A comparison. *J. Food Compos. Anal.* **2019**, *83*, 103257. [CrossRef]
25. Beak, S.H.; Lee, Y.; Lee, E.B.; Kim, K.H.; Kim, J.G.; Bok, J.D.; Kang, S.K. Study on the fatty acid profile of phospholipid and neutral lipid in Hanwoo beef and their relationship to genetic variation. *J. Anim. Sci. Technol.* **2019**, *61*, 69–76. [CrossRef] [PubMed]
26. Bourre, J. Where to find omega-3 fatty acids and how feeding animals with diet enriched in omega-3 fatty acids to increase nutritional value of derived products for human: What is actually useful. *J. Nutr. Health Aging* **2005**, *9*, 232–242. [PubMed]
27. Scollan, N.D.; Dannenberger, D.; Nuernberg, K.; Richardson, I.; MacKintosh, S.; Hocquette, J.F.; Moloney, A.P. Enhancing the nutritional and health value of beef lipids and their relationship with meat quality. *Meat Sci.* **2014**, *97*, 384–394. [CrossRef]
28. Blesso, C.N. Egg phospholipids and cardiovascular health. *Nutrients* **2015**, *7*, 2731–2747. [CrossRef] [PubMed]
29. Gandemer, G. Lipides du muscle et qualité de la viande. Phospholipides et flaveur. *OCL. OléagineuxCorps GrasLipides* **1997**, *4*, 19–25.
30. European Union. Directive 2010/63/EU of the European Parliament and of the Council of 22 September 2010 on the protection of animals used for scientific purposes. *Off. J. Eur. Union* **2010**, L276/233-L276/279. Available online: https://eur-lex.europa.eu/LexUriServ/LexUriServ.do?uri=OJ:L:2010:276:0033:0079:en:PDF (accessed on 19 June 2021).
31. Albertí, P.; Panea, B.; Sañudo, C.; Olleta, J.L.; Ripoll, G.; Ertbjerg, P.; Christensen, M.; Gigli, S.; Failla, S.; Concetti, S.; et al. Live weight, body size and carcass characteristics of young bulls of fifteen European breeds. *Livest. Sci.* **2008**, *114*, 19–30. [CrossRef]
32. Folch, J.; Lees, M.; Stanley, G. A simple method for the isolation and purification of lipids from animal tissues. *J. Biol. Chem.* **1957**, *226*, 497–509. [CrossRef]
33. Scollan, N.D.; Choi, N.-J.; Kurt, E.; Fisher, A.V.; Enser, M.; Wood, J.D. Manipulating the fatty acid composition of muscle and adipose tissue in beef cattle. *Br. J. Nutr.* **2001**, *85*, 115–124. [CrossRef]
34. Geesink, G.H.; Schreutelkamp, F.H.; Frankhuizen, R.; Vedder, H.W.; Faber, N.M.; Kranen, R.W.; Gerritzen, M.A. Prediction of pork quality attributes from near infrared reflectance spectra. *Meat Sci.* **2003**, *65*, 661–668. [CrossRef]
35. Barlocco, N.; Vadell, A.; Ballesteros, F.; Galietta, G.; Cozzolino, D. Predicting intramuscular fat, moisture and Warner-Bratzler shear force in pork muscle using near infrared reflectance spectroscopy. *Anim. Sci.* **2006**, *82*, 111–116. [CrossRef]
36. Williams, P.; Norris, K. *Near-Infrared Technology in the Agricultural and Food Industries*; American Association of Cereal Chemists, Inc.: St. Paul, Minnesota, 1987; p. 330.
37. MacKintosh, S.B.; Richardson, I.; Kim, E.J.; Dannenberger, D.; Coulmier, D.; Scollan, N.D. Addition of an extract of lucerne (Medicago sativa L.) to cattle diets–Effects on fatty acid profile, meat quality and eating quality of the M. longissimus muscle. *Meat Sci.* **2017**, *130*, 69–80. [CrossRef]
38. Wood, J.D.; Enser, M.; Fisher, A.V.; Nute, G.R.; Sheard, P.R.; Richardson, R.I.; Hughes, S.I.; Whittington, F.M. Fat deposition, fatty acid composition and meat quality: A review. *Meat Sci.* **2008**, *78*, 343–358. [CrossRef] [PubMed]
39. Sevane, N.; Nute, G.; Sañudo, C.; Cortes, O.; Cañon, J.; Williams, J.L.; Dunner, S. Muscle lipid composition in bulls from 15 european breeds. *Livest. Sci.* **2014**, *160*, 1–11. [CrossRef]
40. Sevane, N.; Levéziel, H.; Nute, G.R.; Sañudo, C.; Valentini, A.; Williams, J.; Dunner, S. Phenotypic and genotypic background underlying variations in fatty acid composition and sensory parameters in European bovine breeds. *J. Anim. Sci. Biotechnol.* **2014**, *5*, 20. [CrossRef]
41. Khan, M.; Khan, M.I.; Sahar, A.; Jamil, A. Predicting authenticity and physicochemical characteristics of meat through FT-IR spectroscopy coupled with multivariate analysis. *Pak. J. Agric. Sci.* **2020**, *57*, 1141–1147.

42. Warren, H.E.; Scollan, N.D.; Enser, M.; Hughes, S.I.; Richardson, R.I.; Wood, J.D. Effects of breed and a concentrate or grass silage diet on beef quality in cattle of 3 ages. I: Animal performance, carcass quality and muscle fatty acid composition. *Meat Sci.* **2008**, *78*, 256–269. [CrossRef] [PubMed]
43. Lucarini, M.; Durazzo, A.; Sanchez Del Pulgar, J.; Gabrielli, P.; Lombardi-Boccia, G. Determination of fatty acid content in meat and meat products: The FTIR-ATR approach. *Food Chem* **2018**, *267*, 223–230. [CrossRef] [PubMed]
44. Siemens, B.J.; Daun, J.K. Determination of the fatty acid composition of canola, flax, and solin by near-infrared spectroscopy. *J. Am. Oil Chem. Soc.* **2005**, *82*, 153–157. [CrossRef]
45. Tøgersen, G.; Arnesen, J.; Nilsen, B.; Hildrum, K. On-line prediction of chemical composition of semi-frozen ground beef by non-invasive NIR spectroscopy. *Meat Sci.* **2003**, *63*, 515–523. [CrossRef]
46. Cozzolino, D.; Murray, I. Effect of sample presentation and animal muscle species on the analysis of meat by near infrared reflectance spectroscopy. *J. Near Infrared Spectrosc.* **2002**, *10*, 37–44. [CrossRef]
47. Tao, F.; Ngadi, M. Recent advances in rapid and nondestructive determination of fat content and fatty acids composition of muscle foods. *Crit. Rev. Food Sci. Nutr.* **2018**, *58*, 1565–1593. [CrossRef] [PubMed]
48. Zhou, L.J.; Wu, H.; Li, J.T.; Wang, Z.Y.; Zhang, L.Y. Determination of fatty acids in broiler breast meat by near-infrared reflectance spectroscopy. *Meat Sci.* **2012**, *90*, 658–664. [CrossRef]
49. Ortiz, A.; Parrini, S.; Tejerina, D.; Pinto de Araújo, J.P.; Čandek-Potokar, M.; Crovetti, A.; Garcia-Casco, J.M.; González, J.; Hernández-García, F.I.; Karolyi, D.; et al. Potential Use of Near-Infrared Spectroscopy to Predict Fatty Acid Profile of Meat from Different European Autochthonous Pig Breeds. *Appl. Sci.* **2020**, *10*, 5801. [CrossRef]
50. Pieszczek, L.; Czarnik-Matusewicz, H.; Daszykowski, M. Identification of ground meat species using near-infrared spectroscopy and class modeling techniques—Aspects of optimization and validation using a one-class classification model. *Meat Sci.* **2018**, *139*, 15–24. [CrossRef] [PubMed]
51. Prieto, N.; Lopez-Campos, O.; Aalhus, J.L.; Dugan, M.E.; Juarez, M.; Uttaro, B. Use of near infrared spectroscopy for estimating meat chemical composition, quality traits and fatty acid content from cattle fed sunflower or flaxseed. *Meat Sci.* **2014**, *98*, 279–288. [CrossRef]
52. Davies, A.M.C.; Grant, A. Review: Near-Infrared analysis of food. *Int. J. Food Sci. Technol.* **1987**, *22*, 191–207. [CrossRef]
53. Sierra, V.; Aldai, N.; Castro, P.; Osoro, K.; Coto-Montes, A.; Olivan, M. Prediction of the fatty acid composition of beef by near infrared transmittance spectroscopy. *Meat Sci.* **2008**, *78*, 248–255. [CrossRef] [PubMed]
54. Cecchinato, A.; De Marchi, M.; Penasa, M.; Casellas, J.; Schiavon, S.; Bittante, G. Genetic analysis of beef fatty acid composition predicted by near-infrared spectroscopy. *J. Anim. Sci.* **2012**, *90*, 429–438. [CrossRef] [PubMed]
55. Zomeño, C.; Juste, V.; Hernandez, P. Application of NIRS for predicting fatty acids in intramuscular fat of rabbit. *Meat Sci.* **2012**, *91*, 155–159. [CrossRef] [PubMed]
56. Hargrave-Barnes, K.M.; Azain, M.J.; Miner, J.L. Conjugated linoleic acid-induced fat loss dependence on Delta6-desaturase or cyclooxygenase. *Obesity* **2008**, *16*, 2245–2252. [CrossRef]
57. Cruz-Hernandez, C.; Deng, Z.; Zhou, J.; Hill, A.R.; Yurawecz, M.P.; Delmonte, P.; Mossoba, M.M.; Dugan, M.E.; Kramer, J.K. Methods for analysis of conjugated linoleic acids and trans-18: 1 isomers in dairy fats by using a combination of gas chromatography, silver-ion thin-layer chromatography/gas chromatography, and silver-ion liquid chromatography. *J. AOAC Int.* **2004**, *87*, 545–562. [CrossRef]
58. Scollan, N.; Hocquette, J.F.; Nuernberg, K.; Dannenberger, D.; Richardson, I.; Moloney, A. Innovations in beef production systems that enhance the nutritional and health value of beef lipids and their relationship with meat quality. *Meat Sci.* **2006**, *74*, 17–33. [CrossRef] [PubMed]
59. Barragan-Hernandez, W.; Mahecha-Ledesma, L.; Burgos-Paz, W.; Olivera-Angel, M.; Angulo-Arizala, J. Using near-infrared spectroscopy to determine intramuscular fat and fatty acids of beef applying different prediction approaches. *J. Anim Sci.* **2020**, *98*. [CrossRef]
60. Sato, T.; Kawano, S.; Iwamoto, M. Near infrared spectral patterns of fatty acid analysis from fats and oils. *J. Am. Oil Chem. Soc.* **1991**, *68*, 827–833. [CrossRef]
61. Balabin, R.M.; Lomakina, E.I. Support vector machine regression (SVR/LS-SVM)–an alternative to neural networks (ANN) for analytical chemistry? Comparison of nonlinear methods on near infrared (NIR) spectroscopy data. *Analyst* **2011**, *136*, 1703–1712. [CrossRef]
62. Geladi, P.; MacDougall, D.; Martens, H. Linearization and scatter-correction for near-infrared reflectance spectra of meat. *Appl. Spectrosc* **1985**, *39*, 491–500. [CrossRef]

sensors

MDPI

Review

Potential of Near-Infrared Spectroscopy for the Determination of Olive Oil Quality

Juan Francisco García Martín [1,2]

[1] Departamento de Ingeniería Química, Facultad de Química, Universidad de Sevilla, 41012 Seville, Spain; jfgarmar@us.es

[2] University Institute of Research on Olive Groves and Olive Oils, GEOLIT Science and Technology Park, University of Jaén, 23620 Mengíbar, Spain

Abstract: The analysis of the physico-chemical parameters of quality of olive oil is still carried out in laboratories using chemicals and generating waste, which is relatively costly and time-consuming. Among the various alternatives for the online or on-site measurement of these parameters, the available literature highlights the use of near-infrared spectroscopy (NIRS). This article intends to comprehensively review the state-of-the-art research and the actual potential of NIRS for the analysis of olive oil. A description of the features of the infrared spectrum of olive oil and a quick explanation of the fundamentals of NIRS and chemometrics are also included. From the results available in the literature, it can be concluded that the four most usual physico-chemical parameters that define the quality of olive oils, namely free acidity, peroxide value, K232, and K270, can be measured by NIRS with high precision. In addition, NIRS is suitable for the nutritional labeling of olive oil because of its great performance in predicting the contents in total fat, total saturated fatty acids, monounsaturated fatty acids, and polyunsaturated fatty acids in olive oils. Other parameters of interest have the potential to be analyzed by NIRS, but the improvement of the mathematical models for their determination is required, since the errors of prediction reported so far are a bit high for practical application.

Keywords: chemometrics; olive oil; near-infrared spectroscopy; quality parameters

Citation: García Martín, J.F. Potential of Near-Infrared Spectroscopy for the Determination of Olive Oil Quality. *Sensors* **2022**, *22*, 2831. https://doi.org/10.3390/s22082831

Academic Editors: Mercedes Del Río Celestino and Rafael Font Villa

Received: 18 February 2022
Accepted: 4 April 2022
Published: 7 April 2022

1. Introduction

The International Olive Council defines olive oil as the oil obtained solely from fruits of the olive tree (*Olea europaea* L.), with the exclusion of oils obtained by solvents or by re-esterification procedures and any mixture with oils of another nature. As stated by this international intergovernmental organisation, the olive oils with the highest quality (so-called virgin oil oils) are those obtained 'solely by mechanical or other physical means under conditions, particularly thermal conditions, that do not lead to alterations in the oil, and which have not undergone any treatment other than washing, decantation, centrifugation and filtration' [1]. Virgin olive oils are classified, in turn, into extra virgin olive oils (EVOO), virgin olive oils (VOO), ordinary virgin olive oil, and lampante virgin olive oil [1,2], where EVOO is the olive oil of the highest quality. While the first three virgin olive oils are fit for consumption, lampante virgin olive oil must undergo processing prior to consumption.

Olive oil is regarded as one of the healthiest food oils due to its high content in triglycerides with unsaturated acids, mainly oleic acid, and its phenolic composition. The former is related to a decrease in LDL-cholesterol fraction, while the latter is responsible for the antioxidant properties and the bitter taste of olive oil. Triglycerides account for almost all the saponifiable fraction of the olive oil (roughly 98 wt.%). On the contrary, the phenolic compounds belong to the unsaponifiable fraction, which represents about 2 wt.% of total olive oil. The most representative phenolic compounds in olive oils are oleuropein and hydroxytyrosol. In addition to phenolic compounds, the unsaponifiable fraction is composed of a wide variety of compounds, such as triterpenic alcohols, α-tocopherol (vitamin E), γ-tocopherol, β-carotene (precursor substance of vitamin A and responsible for the yellow–orange colour of olive oil), phytosterols, sterols, hydrocarbons, chlorophylls (responsible for the green colour of olive oil), and volatile compounds responsible for the aroma of olive oil.

Due to the current popularity of the Mediterranean diet and its use in a wide range of different recipes, the nutritional value of olive oil is internationally recognized today. EVOO is mainly used as a salad dressing and food to be eaten cold due to its flavour and taste. The rest of the edible olive oils are used mostly for cooking.

The most common physico-chemical parameters that define the quality of olive oils are the free acidity (FA), the peroxide value (PV), and the absorbency in ultraviolet (at 232 and 270 nm). These three physico-chemical parameters, along with the organoleptic characteristics (odour and taste, defects, fruity attributes, and colour), are used by producers for the determination of the quality of virgin olive oils. Notwithstanding, the International Olive Council establishes additional quality criteria for the designation of olive oils (both edible and non-edible), namely moisture and volatile matter (wt.%), insoluble impurities in light petroleum (wt.%), flash point (°C), trace metal content (mg/kg of iron and copper), fatty acid ethyl esters content (mg/kg), and biophenols content (mg/kg) [1]. Generally, olive oil producers do not regard them as quality parameters, but as composition parameters. Additional physico-chemical parameters such as oxidative stability (h), chlorophyll and carotenoid pigment profiles, and the bitterness index are often included [2]. Regarding organoleptic characteristics, the absence (EVOO) or weak presence (VOO) of sensory defects and the presence of three positive attributes, namely fruitiness, bitterness, and pungency, must be evaluated by skilled tasters.

The determination of the aforementioned physico-chemical parameters is currently carried out in a laboratory using chemicals and generating waste, which is relatively costly and time-consuming. In addition, the online determination of the quality parameters of olive oil during the olive oil extraction process in olive mills is not possible using conventional methods. Among the various alternative, non-destructive methods for these analyses, the use of near-infrared (NIR) spectroscopy stands out. Its aim is to correlate the signal of the olive oils in the NIR spectrum with the quality parameters through the use of chemometrics. This article intends to show the state-of-the-art research and the actual potential of near-infrared spectroscopy (NIRS) for the analysis of olive oil, not only its main four quality parameters, but also other parameters of interest for the olive oil industry. For a better understanding, the following three sections include, in the following order, the fundamentals of NIRS, a description of the main features of the NIR spectrum of olive oil, and a brief explanation of what chemometrics is and how it is applied to NIRS, while the last three sections illustrate the results obtained by various authors on the quality parameters, other compounds of interest and sensory attributes, respectively.

2. Near-Infrared Spectroscopy (NIRS)

NIR spectroscopy is a vibrational spectroscopy, like Raman spectroscopy. A molecule absorbs NIR radiation (from 800 to 2500 nm) if the energy of the radiation corresponds to the energy difference between two vibrational levels and, in addition, a change occurs in the dipole moment of the molecule [3]. This is similar to what happens in the mid-infrared region. However, the bands of fundamental vibrations ($\Delta n = \pm 1$, where n is the vibrational quantum number) do not appear in the NIR spectrum, while absorptions due to the overtones and combination bands are observed. The overtone bands are due to $\Delta n > \pm 1$. Depending on the type of bonds, only the first ($\Delta n = \pm 2$) and second ($\Delta n = \pm 3$) overtones are likely to be observed. Combination bands occur only in polyatomic molecules and are due to simultaneous changes in the energy of two or more modes of vibration [3,4]. Therefore, the near-infrared spectrum is the result of the change in the molecular dipole moment during vibration. For example, the stretches of C=O in the CO_2 molecule and of O–H in the water molecule, which are polar functional groups, have great absorption in the

NIR spectrum [5]. Since the NIR spectrum of an analysed sample is the result of the combinations and overtones of the functional groups of its chemical constituents, the absorption peaks and bands found in the NIR spectrum are generally broad and weak. This makes NIRS more suitable for quantitative analysis than for compound identification (although NIRS can provide some information on functional groups). Hence, NIRS is regarded as a powerful analytical technique for the non-destructive, low-cost, rapid determination of compounds and parameters in food. Since NIR spectroscopy neither requires reagents nor generates waste, other advantages are providing a safe working environment and a huge potential for online measurement.

An NIR spectrometer is composed of a radiation source (the most common is a tungsten–filament–fire halogen lamp with quartz window), a wavelength selector (generally a dispersive equipment), a sample holder, and a detector (generally built with semiconductors such as InGaAs and PbS). There are many sample holders depending on how the NIR spectrum is acquired [4]: transmittance, reflectance, and transflectance (Figure 1).

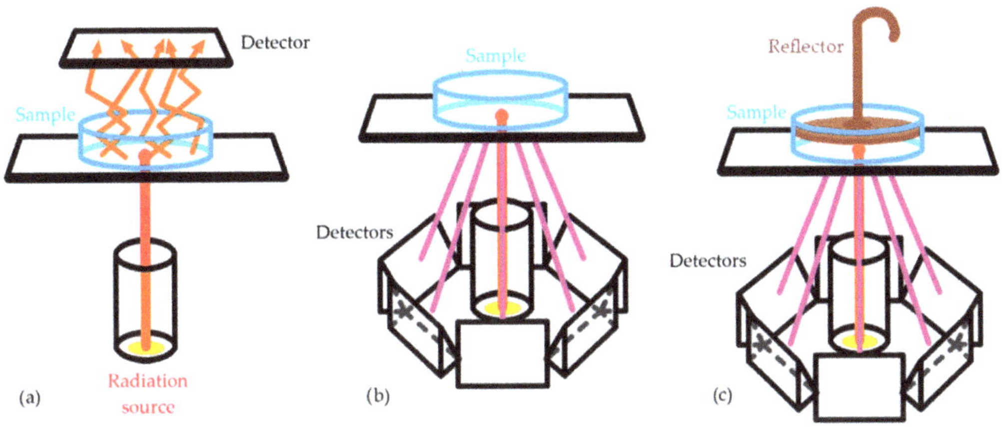

Figure 1. Main configurations to acquire NIR spectra: (**a**) transmittance; (**b**) reflectance; (**c**) transflectance [4].

The use of cuvettes of different path lengths for transmittance and probes for transflectance is best for homogeneous liquids, while reflectance is generally used for solid, heterogeneous samples.

FTIR (Fourier-transform infrared) spectroscopy is an analytical technique generally used to identify functional groups in organic and inorganic compounds by obtaining their infrared spectra in the range of 2500–25,000 nm [6]. FTIR spectrometers acquire infrared spectra from solid, liquid, or gaseous samples in absorption, total, attenuated, and diffuse reflectance, and photoacoustic modes [6]. The raw signal is first Fourier-transformed by the equipment. FTIR spectrometers generate a unique type of signal called an interferogram that has all of the infrared wavelengths encoded into it [7]. Some authors regard the Fourier transform as a type of wavelength selector [4]. Although it is not the most common, FTIR spectroscopy can be applied to the NIR region, i.e., FTNIR spectroscopy, resulting in a faster NIR spectra acquisition with a higher signal-to-noise ratio than conventional NIRS [8]. Some works can be found in the literature on the use of FTNIR with olive oils for quantitative or discriminating purposes [9–14].

Since the 1980s, several works have addressed the determination of the main properties of olives of interest to the farmer. An industrial development of NIRS equipment to measure the internal properties of intact olives occurred about 15 years ago, so nowadays it is easy to find commercial equipment to non-destructively measure the moisture and fat content of olives, such as the OliveScan™2 and Olivia™ equipment (FOSS, Hilleroed,

Denmark), the Luminar 5030 olive and olive paste analyser (Soluciones Integrales de Laboratorio, S.L., El Casar de Talamanca, Spain), and the NIT-38 olive analyser (NIR Technology Systems, Sidney, Australia). In addition, portable NIR spectrophotometers can be purchased for roughly EUR 6000 (e.g., Flame-NIR spectrometers, Ocean Optics, Inc., Orlando, FL, USA) and could be used at any stage of the olive oil production process. Although they have not been recognized as official methods by the International Olive Council, the determinations of fat content and moisture in olives by NIRS have been accredited as official methods by the pertinent authority of diverse countries. Thus, the accreditations 684/LE937 and 1335/LE2481 issued by the ENAC (Spanish Accreditation Bureau, Madrid, Spain) to various laboratories according to the criteria included in the UNE-EN ISO/IEC 17025:2017 standard [15], officially allow the determination of total fat and moisture in intact olives by NIRS following an internal method based on the manufacturer's method FOSS for the Olivia™ equipment (FOSS, Hilleroed, Denmark). Notwithstanding, and despite the large available literature, such industrial development does not exist for the measurement of the quality parameters of olive oil.

At the beginning of the twentieth century, several works have addressed the use of NIRS to determine the olive oil quality's parameters at different points in the oil extraction process in olive mills [16–18]. Thus, NIRS equipment was installed on an olive oil production line, in order to take samples at the exit of the clarifying centrifuge and from the hopper where the oil is continuously weighed, as shown in Figure 2 [18]. Furthermore, NIRS has been applied to design a quality control system for the identification of adulterated olive oils with other oils such as sunflower oil, corn oil, and raw olive waste [19], and to the differentiation of olive oils that belong to different denominations of origin [20]. Comprehensive reviews on these latter topics can be found elsewhere [21].

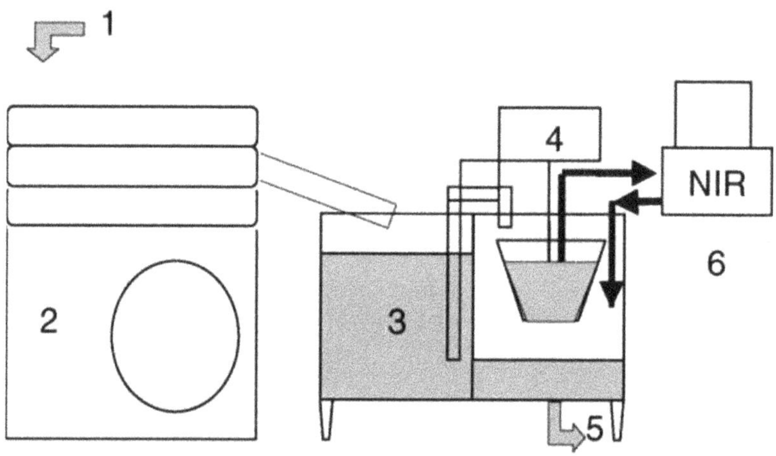

Figure 2. Schematic diagram of the proposed NIRS sensor in the last pass in the olive oil extraction process: (1) oil from horizontal centrifuge decanter; (2) vertical centrifuge for oil clarification; (3) tank for oil sedimentation; (4) continuous oil weigher; (5) to oil storage container; (6) NIRS equipment [18].

Once experts in the olive oil production process have been consulted, three points within the process, which can be found in Figure 2, seem to be the most suitable for the sampling of olive oils and the on-site determination of their quality parameters by NIRS. The first is at the exit of the horizontal decanter (2- or 3-outlet decanter). However, the samples taken at this point would be more intended for experimental purposes and the enhancement of the process (assessment of temperature of the olive oil in the decanter, etc.), and the values of the quality parameters could not match those of the oil once bottled. The second and third would be at the exit of the vertical centrifuge for oil clarification (or the sedimentation tank if available in the olive mill) and at the olive oil storage containers, respectively. Nevertheless, considering that NIRS is a rapid, non-destructive, technique that requires minimal sample preparation (no reagent) and relatively small amounts of a. sample (a few mL of olive oil), experts consider that the most suitable location for the NIRS equipment would be at the bottling plant. In this way, after filling a bottle with olive oil, a small sample would be immediately taken and its NIR spectrum acquired, which would provide the actual values of the quality parameters of the olive oil contained in the bottle. This would also speed up and make the olive oil labelling process more precise, as long as the International Olive Council (or the national quality bureaus) accepts NIRS as an official method for the determination of the olive oil's quality parameters.

However, such types of online proposals have not been, or have been installed only to a limited extent, in olive mills. This is because the development of robust mathematical models is the key to the industrial application of NIRS for online monitoring. These models, previously obtained by using chemometrics, could be the starting point for designing and installing an online tool for the determination of the quality parameters of olive oil on process lines at olive mills, but a full-scale application requires a huge number of samples, not only from the different varieties of olives that the olive mill works with but also over several harvestings in order to develop mathematical models that can be used in a production context. In addition, the chemometric tools of NIRS equipment should provide self-learning model calibration systems. That is to say, the just-acquired spectra directly from the oils in the process line should be automatically included in the calibration data set to strengthen the models by expanding the data sets over time [16]. In this sense, FOSS annually updates the calibration models of their NIRS equipment to measure properties in intact olives and olive pastes, and customers have to pay to update their equipment if they apply for it.

3. Near-Infrared Spectrum of Olive Oils

The sample temperature has a great influence on the NIR radiation that it reflects and absorbs, which makes temperature a parameter of paramount importance in NIRS. For olive oils (and other oils), a sample temperature of approximately 32 °C is usually chosen [22–28]. At this temperature, olive oil is a homogeneous liquid, with a non-important loss of volatile compounds occurring. Therefore, the only sample preparation required in NIRS is heating and maintaining olive oil at that temperature during spectrum acquisition. On the other hand, near-infrared radiation penetrates deeper into organic samples than other electromagnetic radiations, such as ultraviolet (UV), visible, far-infrared, and mid-infrared radiations [29]. Therefore, the optical path length chosen when acquiring NIR spectra has a significant influence on the radiation absorption intensity at different wavelengths. Figures 3–6 show the visible-NIR spectra of 127 olive oils from the variety 'Picual' acquired using a Labspec Pro 350-2500P visible/NIR spectrophotometer (Analytical Spectral Devices Inc., Boulder, CO, USA) equipped with three detectors and an operating in transmittance mode. This equipment was used for the spectral acquisition of olive oils in the range 350–2500 nm using quartz cuvettes with different path lengths (from 0.5 to 10 mm) as sample holders. The reflectance was transformed into absorbance. As shown in these figures, the use of cuvettes with higher path lengths for spectral acquisition leads to higher absorbance in the NIR region, thus providing smoother NIR spectra that are more suitable for further building mathematical models for the determination of quality parameters [22]. In this sense, small differences in path length (0.2 and 0.5 mm) when acquiring the spectra of South African EVOO in the transflectance mode with quartz cuvettes as sample holders have been reported not to have a significant effect on regression model performance [9].

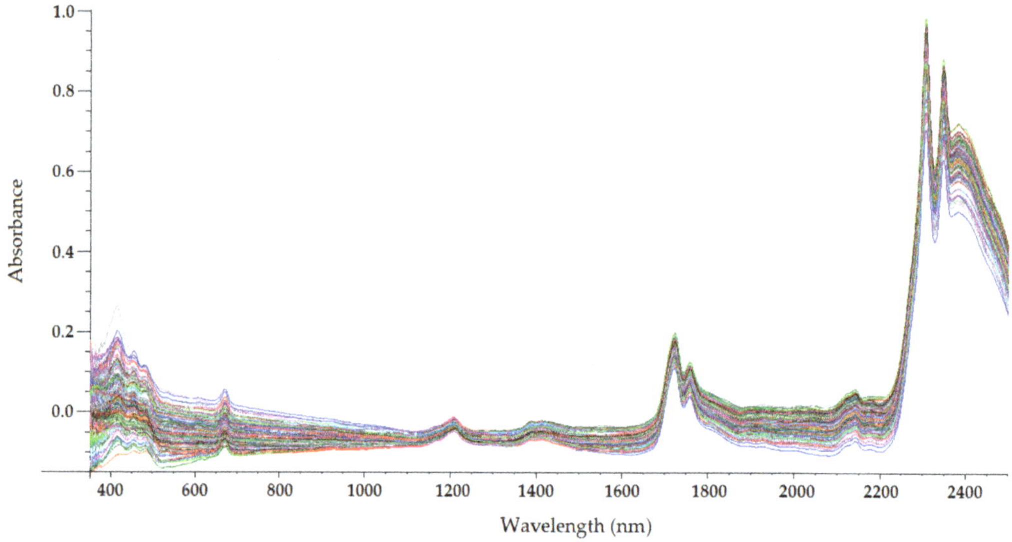

Figure 3. Mean-normalized visible/NIR spectra of 127 olive oils obtained with 0.5-mm path-length quartz cuvette [22].

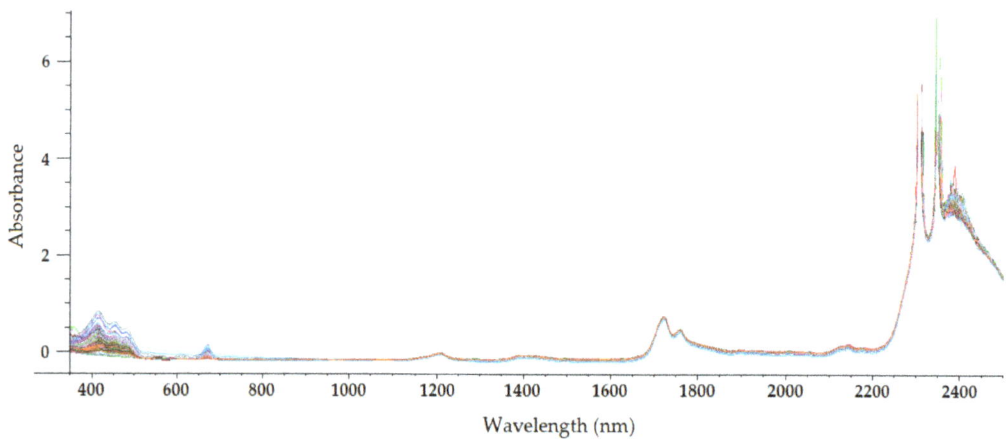

Figure 4. Mean-normalized visible/NIR spectra of 127 olive oils obtained with 2-mm path-length quartz cuvette [22].

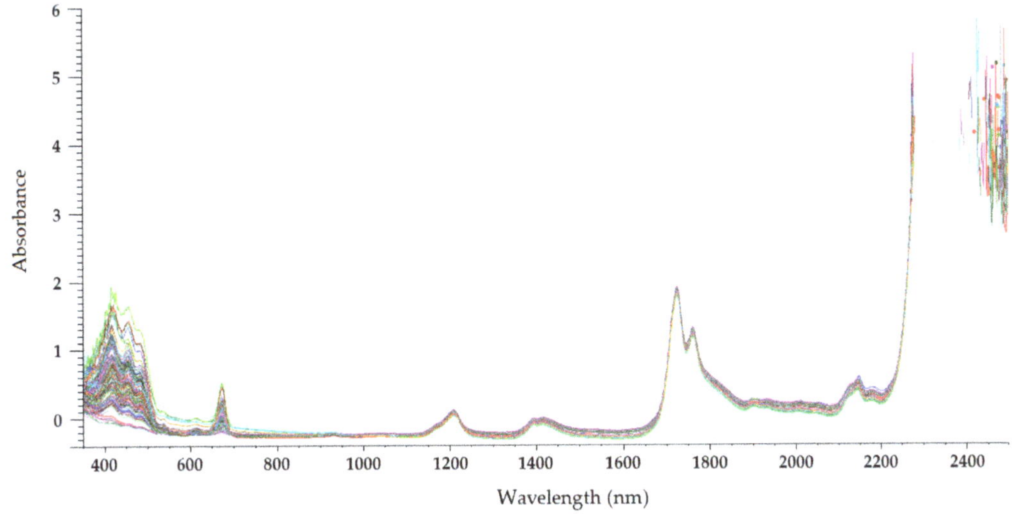

Figure 5. Mean-normalized visible/NIR spectra of 127 olive oils obtained with 5-mm path-length quartz cuvettes [22].

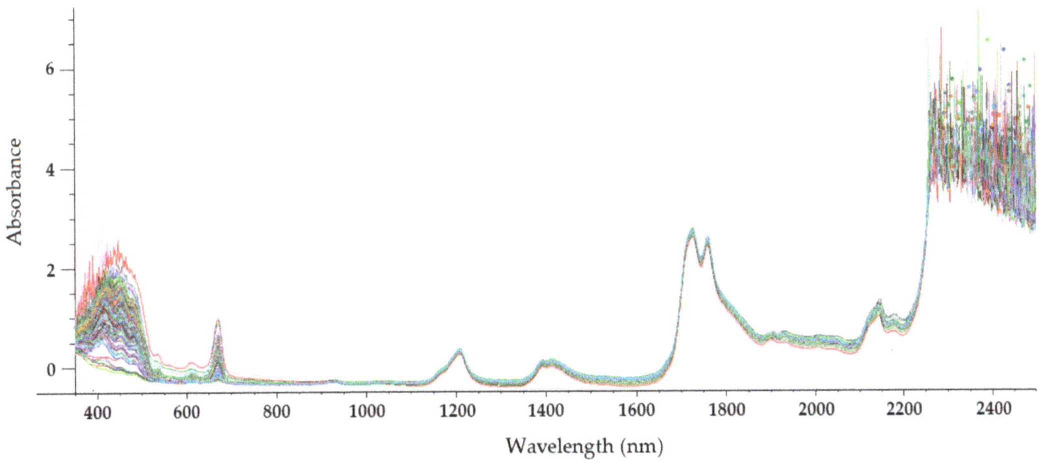

Figure 6. Mean-normalized visible/NIR spectra of 127 olive oils obtained with 10-mm path-length quartz cuvette [22].

The NIR spectrum of olive oil has previously been described in the literature by various authors [21,22,27,29,30]. In fact, the NIR spectrum of olive oil is quite similar to that of triglycerides, as olive oil is mainly composed of triglycerides [31]. What is more, as triolein is the main triglyceride and therefore the major component of olive oil, the highest absorption band in the NIR spectrum of olive oil is the same as that of the triolein spectrum, which is observed at 1725 nm [29]. The two regions of the NIR spectrum that are of great importance [31] can be clearly observed in Figures 3–6. One is the absorption band near 1720 nm, which is related to the first overtone of the C-H vibration of several chemical groups such as $-CH_3$, $-CH_2$ and $=CH_2$, and the other is the absorption peaks at 1660 and 2145 nm, which are related to the C-H vibration of *cis*-unsaturation. When the degree of *cis*-unsaturation increases, the absorption peak at 1725 nm (*cis*-C18:1) shifts to lower wavelengths, i.e., to 1717 nm and 1712 nm for *cis*-C18:2 and *cis*-C18:3, respectively [31]. The high adsorption peak at 2145 nm makes the main peaks related to saturated and *trans* fatty acids, usually observed at 2128 and 2131 nm, respectively, hardly noticeable in the olive oil spectrum [29]. Wavelengths close to 1800 nm have also been related to the saturated fatty acids [29]. Finally, a broad absorbance band can be observed at 1210 nm as a result of second overtones of C–H and CH=CH– stretching vibrations [27].

Besides the bands and peaks corresponding to molecules that compose the fatty acids of the olive oil, a broad band at 1400 nm and a broader one at around 1950 nm are also observed in these figures. These bands have been related to the presence of water, to be specific to its first overtone, and to a combination band, respectively [25].

As observed in Figures 3–6, the absorbance in the 2300–2500 nm region is out of the range of the detector used (a holographic fast scanner InGaAs detector, cooled at 25 °C, and coupled with a high-order blocking filter) when increasing the path length of the cuvette (lower radiation transmission and therefore higher absorbance by the olive oil). This problem has also been reported when disposable borosilicate vials were used for the spectral acquisition of olive oil between 400 and 2500 nm in the transmittance mode [32]. This problem was attributed to the high absorbance of this material. For this reason, quartz cuvettes are the most used and appropriate sample holders for NIRS, since quartz does not absorb radiation in the NIR region [21]. However, since neither of the two regions of major importance in the NIR spectrum of olive oil falls in this region, the absorbance at wavelengths between 2200 and 2500 nm can be discarded when working with olive oil NIR spectra without losing important information on the sample.

Regarding the visible spectrum, it is sometimes used together with the NIR spectrum for the determination of olive oil's quality parameters. There are three main absorption peaks of olive oil in the visible spectrum. The first is found at 415 nm (dark blue coloured light) and is related to carotenoids, pheophytin *a*, pheophorbide *a*, and pyropheophytin *a* [33]. The second absorption peak can be observed at 450 nm (blue light), which is characteristic of carotenoids [33]. The third absorption peak is found at 670 nm, and is related to chlorophylls [27]. It is worth noting that the two former peaks (between 350 and 500 nm) were easier to differentiate with the 0.5-mm and 2-mm quartz cuvettes (Figures 3 and 4, respectively) than with the 5-mm path-length cuvette (Figure 5). They could not be clearly distinguished using the 10-mm path-length cuvette (Figure 6), which could indicate that increasing the path length when working in the transmittance mode reduces the quality of the visible spectrum of olive oil. This is contrary to what was found in the NIR spectrum.

The features of the visible and NIR spectra of olive oil have been exploited in different ways. For example, the absorbances in the 470–690 nm, 1145–1265 nm, and 1355–1500 nm visible/NIR ranges have been related to olive pomace oil, so these spectral ranges have been used to determine the amount of olive pomace adulterating EVOO with a low standard error of prediction (SEP = 3.27 wt.%) [5]. Besides, two minor carbonyl absorptions at 1894 and 1930 nm have been used to assess the authenticity of EVOO based on the ratio of absorption intensity at these wavelengths, which are related to the loss of volatiles from EVOO, and therefore to the loss of quality of olive oils [11]. On the other hand, the use of wavelengths in which the absorption of NIR radiation is related to the structure of fatty acids (aliphatic chains), and therefore responsible for the free acidity of olive oil, resulted in more reliable mathematical models for the determination of free acidity in edible olive oils [22].

4. Chemometrics Coupled with NIRS

Chemometrics is defined as 'the science of relating measurements made on a chemical system or process to the state of the system via application of mathematical or statistical methods', according to the International Chemometrics Society [34]. It started to be applied to spectroscopic data about five decades ago. Chemometrics coupled with NIRS can be defined as the application of statistics and mathematical models to extract the desired information from the NIR spectra. The NIR spectra of olive oils are difficult to interpret since they are the result of overlapped overtones and combination bands, which can contain different baselines or noise. The combination of NIRS and chemometrics provides calibration models for olive oil spectra analysis and both classification and discrimination tools. Chemometrics coupled with NIRS are also suitable to handle the dimensional overload, collinearity, spectral interferences, and spectral noise on olive oil NIR spectra. To do this, several specific software has been developed, such as The Unscrambler (CAMO Software AS, Oslo, Norway) or the Chemometrics Toolbox (Eigenvector Research, Inc., Manson, WA, USA) for MatLab (The MathWorks, Inc., Natick, MA, USA), which allow obtaining results with great precision, speed, and comfort.

To speed up data evaluation and to increase the precision of the mathematical models, pre-treatments are generally applied to raw spectra, consisting of classical methods for spectral normalization, smoothing, and differentiation [35,36]. Spectra pre-treatments include data spectra derivatization, normalization, baseline correction, standard normal variate, mean centring, Savitzky and Golay smoothing, first and second derivatives and multiplicative scatter corrections [8,36–38]. The use of spectra pre-treatments, which at first is an advantage for the use of NIRS for the determination of quality parameters of olive oil, can result in a huge hindrance to the implementation of NIRS for online monitoring or on an industrial scale. For example, when applying a normalization (generally maximum normalization or mean normalization) to olive oil spectra, all available spectra are selected for that normalization, and the normalized spectra are subsequently used to build a calibration method for the determination of one or more olive oil properties via chemometrics. As mentioned above, the chemometric tools coupled with NIRS should provide self-learning calibration models. That is to say, spectra acquired later (e.g., olive oils from next harvestings) must be included in the calibration data set to expand data sets and strengthen models over time [16]. The problem is that the current set of spectra has already been normalized. The new added spectra cannot be normalized in the same way. At most, all the spectra (old and newly acquired) could be normalized together, but this normalization would be different from the normalization done with the old spectra, thus affecting the later selection of outliers, the developed calibration model, etc. As a result, this kind of pre-treatments would be difficult to implement for an online measurement of olive oil's quality parameters during olive oil extraction at the olive mills.

For olive oils, chemometrics coupled with NIRS are generally used for oil classification (including adulterations) or property quantification. To do this, there are mathematical algorithms that explore the correlation structure within a single data block. For olive oil classification, unsupervised pattern recognition such as principal component analysis (PCA) and supervised pattern recognition such as partial least squares (PLS) combined with discriminant analysis (DA) is the most used chemometric technique [36,39]. Many works can be found in the literature for the detection of adulteration in olive oils using NIRS. Thus, PCA has been applied to detect corn, sunflower, or raw olive residue oils in the range 0–30 wt.% in VOO and EVOO [19], to detect between 5 and 50 wt.% sunflower, soybean, and sesame oils in VOO [40], and to detect corn, sunflower, soybean, and canola oils in EVOO, with lower limits of adulteration detection of approximately 20, 20, 15, and 10 wt.%, respectively [41], all of them in the laboratory. The good results obtained in the determination of adulteration in EVOO using PCA and NIRS have led to testing the use of portable NIR spectrometers, which could provide in situ information on adulteration. In this sense, it was proven that the use of PCA and a portable spectrometer, which collected spectra in the range 908−1676 nm, resulted in a reliable tool to identify, classify, and quantify the content of different vegetable oils (canola, corn, soybean, and sunflower oil) in EVOO at a confidence level of 95% [42]. On the other hand, PLS-DA has been applied, for example, to detect corn, hazelnut, soya, and sunflower oils in olive oils [43]. Furthermore, PCA and PLS-DA of olive oil NIR spectra have also been applied to predict the geographical origins of olive oils. For example, 57 EVOO were successfully classified according to their geographical origin (Chianti Classico or Maremma) using different pre-treatments and chemometric methods; among them, PCA stood out [44]. Both PCA and PLS-DA were used to discriminate between 135 VOO (10 commercial VOO and 125 VOO from 5 French Protected Designation of Origin) based on their NIR spectra features [45].

In order to correlate the NIR or visible/NIR spectra of olive oils with the quantifiable parameters of interest, multivariate calibration methods are applied, namely multiple linear regression (MLR), principal component regression (PCR), and partial least squares (PLS) regression. Regarding the determination of olive oil quality parameters by NIRS, few papers can be found in the literature that apply MLR or PCR [46]. In contrast, in almost all the published articles available in the literature dealing with NIRS and the determination of the quality parameters of olive oil, the building of predictive models is based on PLS regression [9,18,22–25,28,30,46,47].

The parameter of interest (acidity, peroxide value, etc.) must be previously analysed by the traditional, official method (i.e., the reference method according to the International Olive Council standard), to use the obtained values for building the mathematical model with which this parameter will be measured in the future by NIRS. That is to say, the spectra of the olive oils will be correlated with the values of the parameter of interest measured with the reference method.

For a quick explanation of these three regression methods, R will be defined as the matrix $i \times j$ of the absorbances of the i samples at the j wavelengths of the NIR spectrum and C as the matrix $i \times 1$ of the different values of the olive oil's parameter to be analysed by the NIRS for each sample.

Multiple linear regression (MLR) is a method that directly establishes a linear combination of the variables of R (absorbances at different wavelengths) that reproduces the values of C (values of the olive oil's parameter measured by the reference method) minimising the error (Equation (1)).

$$C = (R \times S) + E \tag{1}$$

where S stands for the matrix of coefficients that, multiplied by the values of R, provides the matrix of values of the analysed parameter (C), and E is the residual error matrix [4,38,48]. This method is the least used and is applied when the number of samples is greater than the number of variables [37].

Principal component regression (PCR) is a method in which the matrix V of the principal components (PC) of R is first determined. The first principal component (PC1) is the vector in the column space of R that describes the maximum amount of variation within the spectra of the olive oils. The second principal component (PC2) describes the maximum residual variation not described by PC1, and so on. The minimum number of PC that minimises the information not explained is selected. Then, the projection of R in V is performed, thus obtaining the matrix of scores U (Equation (2)). Finally, a linear combination of U provides the values of C that minimise the error (Equation (3)).

$$U = (R \times V) \tag{2}$$

$$C = (U \times S) + E \tag{3}$$

Thus, to determine by NIRS the value of the parameter C of an olive oil sample, different from those used for PCR, the scores matrix U_{unk} is obtained from the absorbance matrix R_{unk} by multiplying it by the matrix of principal components V. Then, the value of the parameter C of that sample is obtained by introducing U_{unk} in Equation (3) [4,38,48].

Finally, partial least squares (PLS) regression is the most used method, and the most suitable when the number of samples is smaller than the number of variables [37,49]. Furthermore, PLS regression provides a better approach to quantitative modelling than MLR, because the correlations among the noise in R are more realistic [49].

In this method, the projection of both R and C is performed in the space V defined by the PC, i.e., the projection of R in V provides a matrix of scores U, and the projection of C in V leads to the score matrix T (Equation (4)).

$$T = C \times V \quad (4)$$

From these score matrices, the following equations are obtained:

$$R = (U \times P) + E \quad (5)$$

$$C = (T \times Q) + F \quad (6)$$

$$T = (b \times U) + G \quad (7)$$

where P stands for the loadings matrix of R, Q is the loadings matrix of C, b is a constant and E, F and G are the residual matrices (error matrices). The ideal situation to relate R to C is when U and T are very similar. That is, b should be close to 1.

Therefore, for an olive sample not used in the PLS regression of which the value of the parameter C is unknown, the scores matrix U_{unk} is calculated from the values of its NIR spectrum matrix R_{unk} using Equation (2), which in turn will allow one to obtain the scores matrix T_{unk} using Equation (7). Once T_{unk} has been calculated, the matrix C, that is, the parameter of olive oil to be calculated by NIRS, is obtained [4,38,48].

Once the calibration model is built by MLR, PCR, or PLS, it is necessary to assess its predictive capacity when applied to samples not used in the calibration process. In other words, validation is necessary to determine the extent to which the results obtained can be extrapolated from samples different from those used to build the calibration method, so the model can be used to determine the parameter desired by NIRS in olive oils from, for example, future harvestings [4,38]. Therefore, in the research papers available in the literature for the determination of the olive oil's quality parameters by NIRS, the samples are usually divided into calibration and validation sets, so that some of the well-characterized samples are reserved to validate the accuracy of the model. In most cases, the calibration set is made up of two thirds of the samples and the validation set of the remaining third, the selection from the samples of each set being random [9,23–25,47]. Other authors have selected one out of four olive oils for the validation set, the remaining olive oils forming the calibration set [30]. However, this does not guarantee a good spread of spectral variability within both sets, so samples for the calibration set should not be selected primarily as a function of their number, but rather for their variability [34]. This means that increasing the number of samples for the calibration set does not always result in a more accurate and robust model.

The simplest solution is to distribute samples uniformly within both calibration and validation sets, taking into account the highest and the lowest values of the parameter of interest of olive oil (measured in the samples with the reference method) to be analysed by NIRS. However, with this solution, only the variability in the analysed parameter is distributed, while the distribution of the variability in the spectral information remains uncertain. The most used method in NIRS that takes into account the variability among spectra is the Kennard–Stone method [50]. The Kennard–Stone algorithm is applied to the spectra (not to the values of the parameter of interest). To select the samples for the calibration set, the algorithm starts by searching for the two samples with the largest Euclidean distance. The following samples for this set will be those that maximise the Euclidean distance from previously selected samples, and so on. This will guarantee that all the variation within the spectral information is contained in the calibration sample set. The Kennard–Stone method has been applied, for example, in the determination of the acidity of olive oil by NIRS [22]. When there is not a validation set of samples, an internal validation method is used, which uses the same samples of the calibration set to validate the mathematical model. The most commonly used internal validation method is full cross-validation (CV). It consists of creating models using all samples except one and validating the model with the excluded sample (leave-one-out method). Therefore, *n* calibration models are built from *n* samples. The standard error of cross validation is obtained from the arithmetic mean of the error values obtained in the *n* models [51].

The robustness of the PLS calibration models is usually evaluated by the multiple correlation coefficient of calibration (r^2_c), while their ability to predict the parameter of interest is assessed by the standard error of prediction (SEP) or the root mean square error of prediction (RMSEP). Both SEP and RMSEP describe the error between the results from the reference method and the results from the NIRS equipment for a set of unknown samples not used for the building of the PLS calibration model. SEP is related to the precision of the model, while RMSEP is related to its accuracy. If the samples were not divided into calibration and validation sets, then the standard error of cross-validation (SECV) or the root mean square error of cross-validation (RMSECV) is used instead.

Generally, an ideal PLS model should have a very high r^2_c and a value of SEP close to the standard error of laboratory (SEL) of the reference method. The closer SEP is to SEL, the greater the precision of the PLS model and the probability of this to provide roughly the same values of the parameter of interest as the reference method. Table 1 summarises the criteria proposed by Shenk and Westerhaus to assess the statistical results of the PLS calibration models and their validations.

Table 1. Criteria for the assessment of PLS models in NIRS [52].

	Calibration (r^2_c)		Prediction (SEP)
$r^2_c \geq 0.90$	Excellent precision	SEP = 1–1.5 SEL	Excellent precision
r^2_c = 0.70–0.89	Good precision	SEP = 2–3 SEL	Good precision
r^2_c = 0.50–0.69	Good separation between low, medium, and high values	SEP = 4 SEL	Medium precision
r^2_c = 0.30–0.49	Correct separation between low and high values	SEP = 5 SEL	Low precision
r^2_c = 0.05–0.29	It is better than no analysing		

r^2_c: correlation coefficient of calibration; SEP: standard error of prediction; SEL: standard error of laboratory.

Usually, the larger the number of samples used for building the calibration model, the better the predictive capacity of the model, and the smaller the error of prediction. There is not a rule about how many samples should contain the calibration set, but it is informally accepted that at least 100 samples should be used for building the calibration models. However, this number of samples is not mandatory and robust calibration models can be built with fewer samples.

The number of principal components used in the PLS model is also related to the performance of the model. The lower the PC number, the better. Normally, the minimum number of PC that maximises the explained information of the PLS model is chosen.

The performance of the PLS models is also assessed by the ratio of performance to deviation (RPD), also called the residual predictive deviation. This parameter is defined as the ratio of the standard deviation (σ) of the reference data from the validation set to the SEP. It is assumed that PLS models with RPD values higher than 3 can be suitable for routine analysis. This parameter is very popular in the literature but, in the opinion of the author, is a tricky parameter. For example, consider that a parameter of food must have a value less than 1 unit to be accepted for human consumption. Imagine that the samples to validate a PLS model have values in this parameter from 0 to 10 units used, the average value of the samples is 5 units, the standard deviation is 2.5 units, and the achieved SEP is 0.5 units. As RPD is defined as σ/SEP, then RPD = 2.5/0.5 = 5, the method will thus be regarded as a method of great precision. In the opinion of the author, a new PLS model for determining a parameter in food (based on the data obtained from a reference method) which must be less than 1 unit and of which the SEP is 0.5 units (i.e., the average difference between the values provided by the new method and the reference method is 0.5 units), is not a very good one, regardless of its acceptable RPD value. The next section will provide some examples on this matter.

Unfortunately, although PLS regression is a powerful tool for building calibration models from NIR full-spectrum, even noise, background and uninformative wavelengths have the possibility of being included in the models [22,37,53]. In the literature, several mathematical methods can be found to remove these wavelengths and only let those wavelengths that actually contribute to the PLS model remain, such as Monte Carlo uninformative variable elimination (MCUVE) [22,53–55], moving window variable importance in projection [56,57], the successive projections algorithm (SPA) [22,55,58], etc. Other authors perform the selection of the spectral variables involved in the models by consecutive cycles, removing those which contribution to the model (regression coefficient) close to zero in each cycle [24,28].

Another interesting option to improve the performance of PLS models is to remove outliers. If the prediction sample is inconsistent with the calibration data, it is regarded as a prediction outlier [59]. They can be removed manually or by applying multivariate outlier detection methods. However, wavelength selection and outlier removal must be carefully performed or avoided at early stages due to the risk of eliminating important spectral information related to the quality parameter of interest. As mentioned above, NIRS equipment should provide self-learning model calibration systems, i.e., spectra from new samples (new harvestings, different geographical origins, etc.) should be automatically included in the calibration data set to strengthen the PLS models by expanding the data sets over time [8]. Only once a robust PLS model is created for determining a quality parameter from hundreds (or thousands) of olive oils of different varieties, harvestings, geographical origins, etc., should the selection of variables and removal of outliers be performed, and the resulting PLS validated with new samples from next harvestings, etc.

5. Determination of Olive Oil's Quality Parameters by NIRS

5.1. Free Acidity (FA)

The acidity value or free acidity of an oil is a measurement of its free fatty acids content, which is released from the hydrolysis of oil triglycerides by lipolytic enzymes. These enzymes are normally present in the seed and pulp cells of olives. When the integrity of the fruit is damaged, the enzymes react with the oil contained in vacuoles. Unhealthy, damaged, or bruised olives, along with unsuitable storage conditions, are responsible for olive oils with high acidity values [2].

FA is expressed as a percentage of grams of oleic acid per 100 g of oil. The conventional determination of FA is carried out in the laboratory using chemicals according to the Official Methods of Analysis of the European Commission [60], being relatively costly and time-consuming. Briefly, the method consists of placing a few grams of olive oil into wide-mouth Erlenmeyer flasks, along with an ethyl alcohol:ethyl ether solution (1:1 *v*/*v*) and a few drops of phenolphthalein, to neutralize the free fatty acids with NaOH until pink in colour [22].

Olive oils with FA greater than 2% are not regarded as fit for consumption and must be refined prior to consumption [1]. With regard to edible olive oils, according to the European Regulation, the maximum levels of free acidity for EVOO and VOO are 0.8% and 2%, respectively, while the FA threshold for olive oils (blends of refined olive oil and VOO fit for consumption) and olive pomace oils (obtained by treating olive pomace with solvents) is 1%.

The estimation of FA by NIRS has been previously assayed by several authors (Table 2), achieving significantly good results in general. Thus, the average FA values for 'Arbequina' and 'Picual' olive oils were 0.49 ± 0.01 and 0.33 ± 0.00, respectively, by means of the reference method [60], while the average values were 0.54 ± 0.15 and 0.37 ± 0.16, respectively, using the 1100–2500 nm NIR spectrum [18]. For the calibration set, these authors used olive oils with acidity between 0.12 and 15.1%, while for the validation set, the olive oils had FAs ranging between 0.16 and 12.2%. Using 15 PC, these authors achieved a R^2_{cal} of 0.998 and a SEP of 0.16% (Table 2). This error was very close to the SEL estimated by the authors (0.1%), which accounts for the robustness of the PLS model. As illustrated in Table 2, by reducing the free acidity range of olive oils for creating the PLS models, lower SEP was achieved. It is worth noting that an SEP of 0.35% [27] and an RMSEP of 0.34% [61] led to an RPD greater than 3. As indicated in the previous section, these values of RPD could make one think that these PLS models have good precision. However, these predictive errors seem to be slightly too high to be suitable for measuring FA or discriminating between edible olive oils, of which the maximum allowed FA is 2% (0.8% for EVOO). Besides, these errors are much higher than SEL for the reference method reported by several authors: 0.1% [18], 0.048% [22], and 0.032% [9]. In Table 2, it can be observed that PLS models with low RPD such as [47] showed low prediction errors, because the FA range chosen to build the calibration model was more appropriate.

Table 2. PLS statistics obtained for free acidity (FA) of olive oils using different FA ranges, visible/NIR spectral intervals and optical path lengths.

FA Range (%)	Spectral Acquisition	Spectrum (nm)	n_{cal}	n_{val}	Path Length (mm)	PC	r^2_{cal}	SEP	RPD	Reference
0.25–4.5	Transmittance	1100–2500	72	35	0.2	7	0.69	0.12	1.8	[9]
0.25–4.5	Transflectance	978–2500	72	35	0.6	8	0.58	0.15	1.5	[9]
0.15–1.3	Reflectance	1961–2212	62	17	8.0	12	0.99	0.060	-	[12]
0.12–15.1	Transmittance	1100–2500	131	45	1.0	15	0.99	0.16	-	[18]
0.10–1.3	Transmittance	800–2200	87	40	10.0	15	0.94	0.075	2.6	[22]
0.16–0.5	Reflectance	1100–2300	38	19	-	-	0.89	0.023	-	[24]
0.10−8.7	Transmittance	350–2500	222	47	5.0	-	0.86	0.35	3.1	[27]
0.10–5.7	Transflectance	400–2500	359	100	0.1	-	0.99	0.060 [1]	7.7	[30]
0.06–8.0	Transmittance	400–2250	208	-	-	-	0.97	0070 [2]	8.4	[32]
0.10–1.1	Transmittance	800–2500	60	37	8.0	-	0.76	0.080 [3]	1.5	[47]
0.36–3.3	Transmittance	400–1100	34	14	3.0	2	0.88	0.34 [3]	3.1	[61]
0.11–1.7	Transmittance	800–2500	14	10	6.5	13	0.99	0.048 [3]	-	[62]
0.15–2.2	Transmittance	1000–2222	49	11	-	8	0.98	0.088 [4]	4.9	[63]

n_{cal} = number of samples in the calibration set; n_{val} = number of samples in the validation set; PC = number of principal components; r^2_c = multiple correlation coefficient of calibration; SEP = standard error of validation; RPD = ratio of performance to deviation. [1] bias-corrected standard error of prediction; [2] standard error of cross validation; [3] root mean square error of prediction; [4] root mean square error of cross validation.

All in all, it can be concluded that the free acidity of olive oils can be measured by NIRS with great precision. This precision can be graphically observed when plotting the predicted values against the FA measured by the reference method (Figure 7).

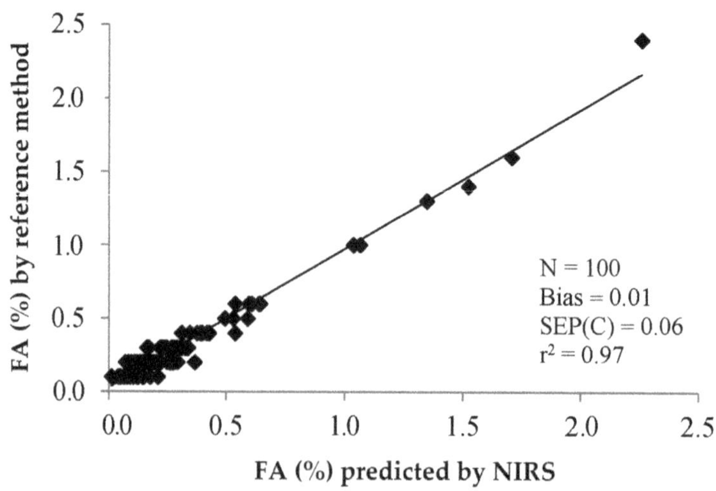

Figure 7. Validation exercise for the determination of the free acidity of olive oil by PLS-NIRS using all the wavelengths between 400 and 2500 nm [30].

In a previous work, the elimination of noise and uninformative spectral variables affecting a PLS model for the determination of FA of olive oils by NIRS was assayed by the Monte Carlo uninformative variable elimination (MCUVE) method and the successive projections algorithm (SPA) [22]. When using the 1401 wavelengths from 800 to 2200 nm, the achieved SEP was 0.75%. The PLS model built with the 314 wavelengths selected by MCUVE led to SEP = 0.064%, while the MLS model built with the 85 wavelengths selected by SPA was 0.051%, quite close to the SEL (0.048%) reported by the author. This improvement in the goodness in the prediction can be visually observed when plotting the FA values predicted by the PLS calibration model built with the full NIR spectrum (Figure 8) and by the MLS calibration model built with the 85 selected wavelengths by SPA (Figure 9) against the FA values obtained using the reference method.

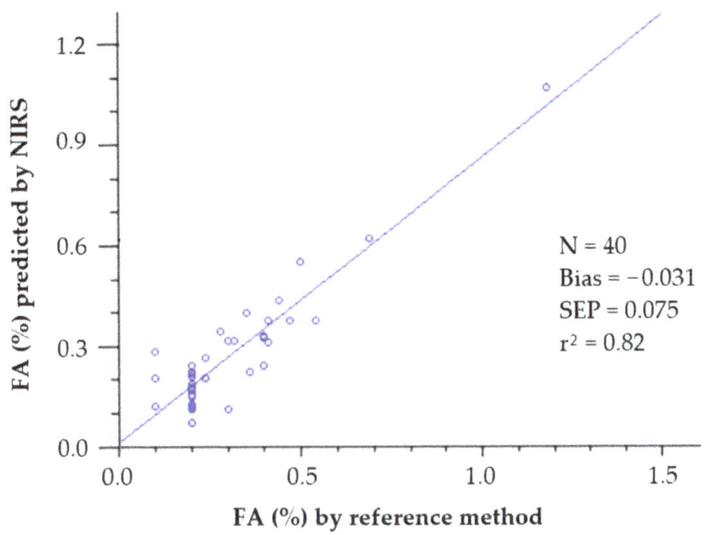

Figure 8. Validation exercise for the determination of the free acidity of olive oil by PLS-NIRS using all the wavelengths between 800 and 2200 (FA values retrieved from [22]).

Interestingly, only 12 of the 80 wavelengths selected by SPA were among the 314 wavelengths selected by MCUVE. This accounted for the difficulty of interpreting the NIR spectra and PLS models obtained from them. It was found that most of the selected wavelengths by MCUVE were related to the main NIR absorption bands of free fatty acids. On the contrary, most of the wavelengths selected by SPA were correlated with triacylglycerols [22]. A tentative assignment of the wavelength ranges selected by each method has been carried out by the author (data not previously published) and illustrated in Table 3. Several years later, the author tried to predict the FA of waste cooking oils with the PLS models obtained for olive oils, with the wavelengths selected by MCUVE and SPA. The statistics were quite poor (hence the prediction error was very high), which could be due to impurities in the waste cooking oils used or the premature removal of wavelengths when creating the PLS models. In the end, the author had to build a specific PLS model to determine the free acidity of waste cooking oils by NIRS [64].

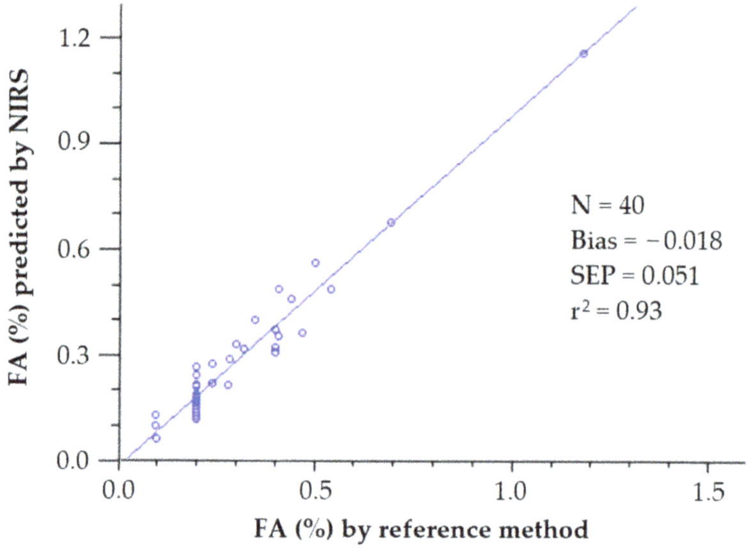

Figure 9. Validation exercise for the determination of the free acidity of olive oil by PLS-NIRS using 85 wavelengths from the NIR spectrum (FA values retrieved from [22]).

Table 3. Spectral bands' tentative assignments correlated to FA of olive oil obtained by the Monte Carlo uninformative variable elimination (MCUVE) method and the successive projections algorithm (SPA).

MCUVE (nm)	SPA (nm)	Bond	Vibration	Molecule/Compound	Reference
1202–1221	-	C–H C–H	First overtone Second overtone	–CH=CH– $-CH_2$	[29] [65]
1484–1506	-	O–H stretch	First overtone	Cellulose	[65]
1531–1569	-	O–H stretch	First overtone	Starch	[65]
1582–1603	-	-	-	-	-
1613–1644	-	C–H stretch	First overtone	$=CH_2$	[65]
-	1717–1729	C–H stretch C–H stretch	First overtone First overtone	$-CH_3$	[29,65]
-	1751–1763	C–H stretch	First overtone	$-CH_2$	[29,65]
1915–1934	-	C=O stretch	Second overtone	CONH	[65]
1957–1973	-	O–H stretch	O–H bend combination	Starch and cellulose	[65]
2154–2192	-	C–H Asym C–H stretch C–H stretch	Combination C–H deformation combination C=O stretch combination Combination bands	$-CH_2$ -HC=CH- Protein -COOH	[29] [65] [65] [66]

Finally, some works can be found in the literature on the determination of FA in other IR spectral regions different from the NIR range. For example, the use of FTIR spectroscopy in the infrared spectral region from 5800 to 6075 nm and the wavelength 3308 nm resulted in an R^2_{cal} of 0.99 and a root mean square error of cross-validation of 0.0107% [13]. However, the calibration model was built with solely a set of 15 samples with FA between 0 and 1%, which were prepared by the gravimetric addition of oleic acid to deodorised olive oil.

5.2. Peroxide Value (PV)

Peroxides are the primary products of the oxidation of olive oil. The peroxide value is a measure of the total peroxides in olive oil expressed as mEq O_2/kg oil, and therefore a major quality guide. EVOO and VOO cannot exceed the maximum value of 20 mEq O_2/kg, the limit fixed by the International Olive Council [1].

The reference method consists of dissolving the oil sample in acetic acid and chloroform, adding potassium iodide and subsequent titration with sodium thiosulphate of the liberated iodine [60]. The precision of the reference method was determined from the results of collaborative tests by the International Olive Council [28], the reproducibility and repeatability coefficients of variation being 7.1% and 1.9%, respectively, for EVOO, and 13.8% and 3.4% for ordinary olive oils. The standard error of laboratory was reported to be 1.41 meq O_2/kg [9].

Table 4 illustrates the errors of prediction of PV by NIRS achieved by several authors. As can be seen, most of the SEP values are close to the reported SEL, even though the RPD values are not too high. Therefore, it can be concluded that PV is another olive oil's quality parameter that can be predicted by NIRS.

Table 4. PLS statistics obtained for peroxide value (PV) of olive oils using different PV ranges, visible/NIR spectral intervals, and optical path lengths.

PV Range (meq O_2/kg)	Spectral Acquisition	Spectrum (nm)	n_{cal}	n_{val}	Path Length (MM)	PC	R^2_{cal}	SEP	RPD	Reference
2.2–74.0	Transmittance	1100–2500	90	44	0.2	6	0.92	4.15	3.5	[9]
2.2–74.0	Transflectance	1100–2500	90	44	0.6	8	0.87	5.28	2.8	[9]
3.0–32.0	Reflectance	1333–1587	65	14	8.0	12	0.98	1.0	-	[12]
5.6–43.9	Transmittance	350–2500	199	46	5.0	-	0.87	3.82	2.8	[27]
1.6–44.5	Transflectance	400–2500	359	100	0.1	-	0.83	1.31 [1]	2.0	[30]
2.6–18.0	Transmittance	400–2250	125	-	-	-	0.92	1.34 [2]	2.7	[32]
7.1–75.4	Transmittance	800–2500	60	37	8.0	-	0.92	2.65 [3]	1.6	[47]
3.6–8.0	Transmittance	400–1100	34	14	3.0	2	0.83	2.25 [3]	3.1	[61]
2.5–17.2	Transmittance	800–2500	14	10	6.5	10	0.94	1.87 [3]	-	[62]
0.0–26.7	Transmittance	1000–2222	49	11	-	8	0.84	3.0 [4]	1.8	[63]

n_{cal} = number of samples in the calibration set; n_{val} = number of samples in the validation set; PC = number of principal components; r^2_c, = multiple correlation coefficient of calibration; SEP = standard error of validation; RPD = ratio of performance to deviation. [1] bias-corrected standard error of prediction; [2] standard error of cross validation; [3] root mean square error of prediction; [4] root mean square error of cross validation.

However, when comparing Figure 7 with Figure 10, it can be observed that the precision of the determination of the peroxide value by NIRS seems to be lower than that of the free acidity.

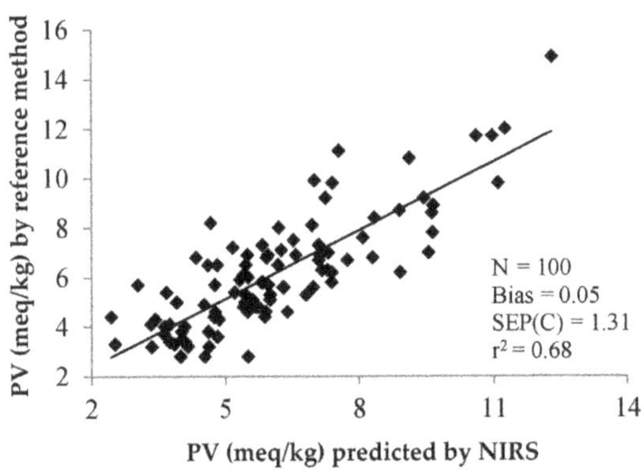

Figure 10. Validation exercise for the determination of the peroxide value of olive oil by PLS-NIRS using all the wavelengths between 400 and 2500 nm [30].

5.3. K270 and K232

The determination of UV-specific extinctions permits an approximation of the oxidation process in unsaturated oils. At 232 nm, primary oxidation products show an absorption (conjugated dienes) that increases due to the defective storage of olive fruits or faulty oil extraction. Secondary oxidation products, such as carbolynic compounds (aldehydes and ketones), are detected at 270 nm, indicating an advanced oxidation process. The maximum permitted values are 2.5 for K232 and 0.20 for K270 [1]. The extinction coefficients K232 and K270 are measured by UV spectrophotometric analysis at the specific wavelengths of 232 and 270 nm and are expressed in absorbance units (AU). Notwithstanding, their determination has been assayed by NIR and visible/NIR spectroscopy (Tables 5 and 6).

Table 5. PLS statistics obtained for K232 of olive oils using different K232 ranges, visible/NIR spectral intervals, and optical path lengths.

K232 (AU)	Spectral Acquisition	Spectrum (nm)	n_{cal}	n_{val}	Path Length (mm)	PC	r^2_{cal}	SEP	RPD	Reference
1.7–20.4	Transmittance	1100–2500	70	34	0.2	6	0.94	0.94	3.6	[9]
1.7–20.4	Transflectance	978–2500	70	34	0.6	4	0.87	1.3	2.6	[9]
0.9−5.0	Transmittance	350–2500	223	55	5.0	-	0.82	0.32	2.6	[27]
1.4–5.4	Transflectance	400–2500	359	100	0.1	-	0.75	0.10 [1]	1.5	[30]
1.2–2.0	Transmittance	800–2500	60	37	8.0	-	0.40	0.090 [2]	1.2	[47]
1.5–3.5	Transmittance	1000–2222	49	11	-	8	0.84	0.27 [2,3]	1.6	[63]

n_{cal} = number of samples in the calibration set; n_{val} = number of samples in the validation set; PC = number of principal components; r^2_c = multiple correlation coefficient of calibration; SEP = standard error of validation; RPD = ratio of performance to deviation. [1] bias-corrected standard error of prediction; [2] root mean square error of prediction; [3] root mean square error of cross validation.

Table 6. PLS statistics obtained for K270 of olive oils using different K270 ranges, visible/NIR spectral intervals and optical path lengths.

K270 (AU)	Spectral Acquisition	Spectrum (nm)	n_{cal}	n_{val}	Path Length (mm)	PC	r^2_{cal}	SEP	RPD	Reference
0.10–2.0	Transmittance	1100–2500	70	34	0.2	6	0.87	0.094	2.5	[9]
0.10–2.0	Transflectance	978–2500	70	34	0.6	3	0.71	0.13	1.8	[9]
0.07–0.41	Transflectance	400–2500	359	100	0.1	-	0.67	0.012 [1]	2.2	[30]
0.06–0.17	Transmittance	800–2500	60	37	8.0	-	0.54	0.020 [2]	1.2	[47]
0.08–0.21	Transmittance	1000–2222	49	11	-	10	0.74	0.019 [3]	1.6	[63]

n_{cal} = number of samples in the calibration set; n_{val} = number of samples in the validation set; PC = number of principal components. [1] bias-corrected standard error of prediction; [2] root mean square error of prediction; [3] root mean square error of cross-validation.

The standard errors of laboratory for these methods have been reported to be 0.42 and 0.048 for K232 and k270, respectively [9]. These SEL have been calculated for ranges of values exceeding, by far, the limits established by the International Olive Council. Thus, SEL for K232 (0.42) was provided for samples in the range 1.7–20.4, while the maximum permitted value is 2.5. Similarly, SEL for K270 (0.048) was calculated for samples in the range 0.10–2.0, while the maximum permitted value is 0.2. This could make the comparison between the errors of prediction and SEL difficult. In any case, some of the statistics illustrated in Tables 5 and 6 show the feasibility of using NIRS to determine K232 and K270. Reference [30] and, to a lesser extent, reference [47] show bias-corrected SEP and RMSEP, respectively, suitable for predicting the extinction coefficients by visible/NIR or NIR spectroscopy, respectively. Figures 11 and 12 show the relation between K232 and 270 predicted by NIRS and K232 and 270 analysed by the reference method, as reported in Reference [30]. Similarly to what was observed for PV, the determination of the specific extinction coefficients by NIRS has lower precision than the determination of FA. It is noteworthy that all but one of the RPD values reported in Tables 5 and 6 are less than 3, which accounts for the poor practical application of this parameter.

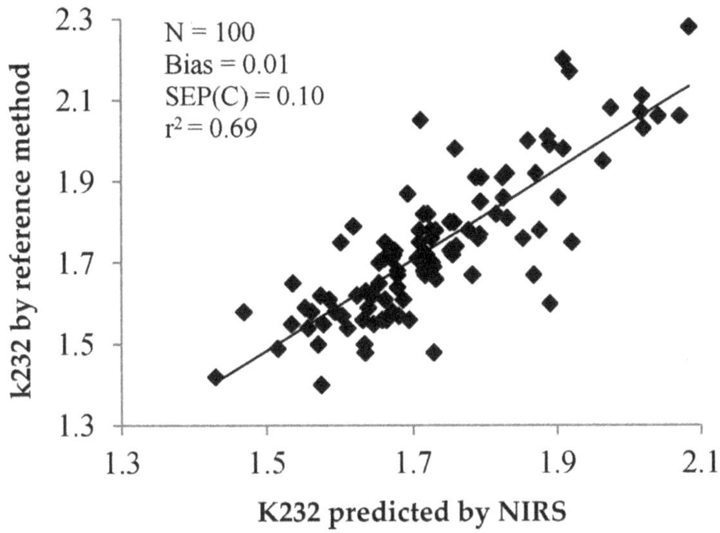

Figure 11. Validation exercise for the determination of K232 of olive oil by PLS-NIRS using all the wavelengths between 400 and 2500 nm [30].

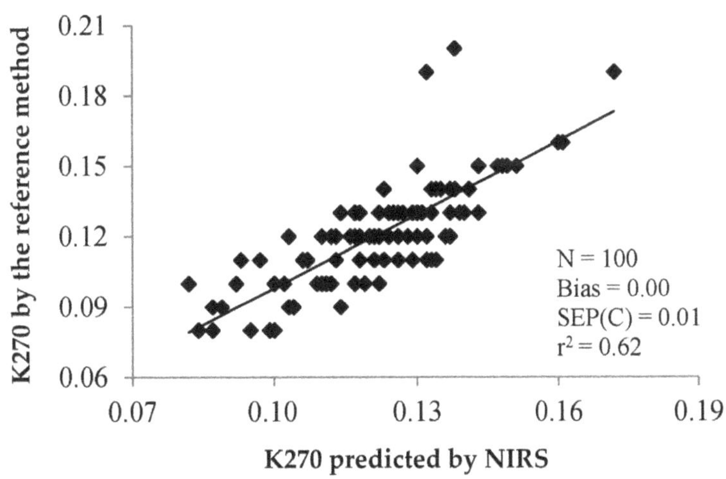

Figure 12. Validation exercise for the determination of K270 of olive oil by PLS-NIRS using all the wavelengths between 400 and 2500 nm [30].

6. Other Compounds

In addition to the four basic quality physico-chemical parameters of olive oil, several other compounds and parameters of olive oil have been assayed to be quantified by NIRS (Table 7). The RPD values reported by the different authors on the various parameters are generally low, but RPD is not considered in the discussion of the statistics collected in Table 7 to the problems that its interpretation presents, as has been pointed out in the previous sections. For most of these parameters, the authors did not provide SEL, so the most suitable approach to assess the feasibility of NIRS to predict these parameters is then to compare the error of prediction with the unit range of these parameters for olive oils.

Table 7. PLS statistics obtained for different compounds and parameters of olive oils using different unit ranges, visible/NIR spectral intervals, and optical path lengths.

Parameter	Units	Range	Spectral Acquisition	Spectrum (nm)	n_{cal}	n_{val}	Path Length (mm)	PC	r^2_{cal}	SEP	RPD	Reference
K225	AU	0.06–0.66	Transmittance	1100–2500	149	30	1.0	13	0.87	0.058	-	[18]
Carotenoids	mg/kg	1.6–18.1	Transmittance	450–2500	151	32	1.0	4	0.985	0.66	-	[17]
Carotenoids	mg/kg	0.12–13.1	Transmittance	1100–2500	64	32	0.2	5	0.66	1.1	1.7	[9]
Carotenoids	mg/kg	0.12–13.1	Transflectance	978–2500	64	32	0.6	3	0.52	1.4	1.4	[9]
Carotenoids	mg/kg	2.1–38.5	Transmittance	1100–2500	205	50	5.0	-	0.62	-	-	[46]
Carotenoids	mg/kg	2.1–38.5	Transmittance	350–2500	205	50	5.0	-	0.95	1.8	3.9	[46]
Chlorophylls	mg/kg	0.7–27.5	Transmittance	450–2500	151	32	1.0	4	0.993	0.96	-	[17]
Chlorophylls	mg/kg	0.082–25.2	Transmittance	1100–2500	65	32	0.2	8	0.56	3.6	1.5	[9]
Chlorophylls	mg/kg	0.082–25.2	Transflectance	978–2500	65	32	0.6	3	0.31	4.4	1.2	[9]
Chlorophylls	g/kg	0–14.5	Transmittance	400–2250	168	-	-	-	0.98	0.51 [1]	5.6	[32]
Chlorophylls	mg/kg	1.4–88.1	Transmittance	1100–2500	205	53	5.0	-	0.56	-	-	[46]
Chlorophylls	mg/kg	1.4–88.1	Transmittance	350–2500	205	53	5.0	-	0.96	3.5	4.1	[46]
Alkyl esters	mg/kg	3–610	Transflectance	400–2500	359	100	0.1	-	0.79	19.5 [2]	1.9	[30]
Ethyl esters	mg/kg	1–461	Transflectance	400–2500	359	100	0.1	-	0.80	14.2 [2]	1.9	[30]
Moisture	wt.%	0.01−0.63	Transflectance	400–2500	283	66	0.1	-	0.71	0.04 [2]	1.5	[30]
Total polyphenols	mg/kg	44.5–738.8	Transmittance	1100–2500	67	31	0.2	2	0.21	89.7	1.1	[9]
Total polyphenols	mg/kg	44.5–738.8	Transflectance	978–2500	67	31	0.6	2	0.34	82.1	1.2	[9]
Total polyphenols	mg/kg	110.7–594.0	Transmittance	800–2500	60	37	0.8	-	0.85	44.5 [3]	1.7	[47]
Squalene	g/kg	1.0−10.1	Transflectance	1100–2300	118	59	-	-	0.86	1.2	2.3	[23]
Squalene	g/kg	1.0−10.1	Transmittance	350–2500	118	59	10.0	-	0.76	1.0	1.9	[23]
α-tocopherol	mg/kg	54.5–755.9	Transflectance	1100–2300	218	109	10.0		0.95	47.2	2.4	[25]
α-tocopherol	mg/kg	54.5–755.9	Transmittance	350–2500	218	109	10.0		0.94	58.3	1.9	[25]
α-tocopherol	mg/kg	91.0–249.3	Transmittance	800–2500	60	37	0.8	-	0.71	15.2 [3]	1.3	[47]
β-tocopherol	mg/kg	0.5–14.1	Transflectance	1100–2300	218	109	10.0		0.64	1.4	1.0	[25]
β-tocopherol	mg/kg	0.5–14.1	Transmittance	350–2500	218	109	10.0		0.66	1.3	1.1	[25]
β-tocopherol	mg/kg	9.11–17.2	Transmittance	800–2500	60	37	0.8	-	0.42	1.5 [3]	1.0	[47]
γ-tocopherol	mg/kg	1.8–103.8	Transflectance	1100–2300	218	109	10.0		0.92	6.3	1.9	[25]
γ-tocopherol	mg/kg	1.8–103.8	Transmittance	350–2500	218	109	10.0		0.87	8.1	1.5	[25]
γ-tocopherol	mg/kg	10.7–36.6	Transmittance	800–2500	60	37	0.8	-	0.63	2.2 [3]	1.2	[47]
Total tocopherol	mg/kg	63.1–1078.0	Transflectance	1100–2300	218	109	10.0		0.92	61.8	2.0	[25]
Total tocopherol	mg/kg	63.1–1078.0	Transmittance	350–2500	218	109	10.0		0.91	76.2	1.6	[25]
Total tocopherol	mg/kg	110.8–278.8	Transmittance	800–2500	60	37	0.8	-	0.61	19.3 [3]	1.2	[47]
Oxidative stability	h	15.2−90.6	Transmittance	350–2500	133	43	5.0	-	0.94	7.4	3.0	[27]

n_{cal} = number of samples in the calibration set; n_{val} = number of samples in the validation set; PC = number of principal components; r^2_c = multiple correlation coefficient of calibration; SEP = standard error of validation; RPD = ratio of performance to deviation. [1] standard error of cross validation; [2] bias-corrected standard error of prediction; [3] root mean square error of prediction.

The UV absorbance at K225 is an index of oil bitterness. High bitterness in olive oils is not well accepted by consumers [18]. Using 13 PC, an R^2_{cal} of 0.870 and an SEP of 0.058 were obtained for this parameter (Table 7), with SEL = 0.026 [18].

For carotenoid and chlorophyll pigments in VOO, Jiménez Marquez [17] concluded that his results showed similarities between visible-near infrared transmittance spectroscopy and reference laboratory methods. The SEL for chlorophylls was 0.25 mg/kg, while SEL for carotenes was 0.35 mg/kg for the ranges indicated in Table 7, with SEP being slightly superior to SEL [17]. The standard error of the laboratory depends on many factors. The range of concentrations used can be highlighted. In this sense, other authors have found that SEL was 0.23 mg/kg for carotenoids in the range 0.12–13.13 mg/kg, and 0.47 mg/kg for chlorophylls in the range 0.082–25.23 mg/kg [9]. Of note is that β-carotene is the precursor substance of vitamin A and is responsible for the yellow–orange colour of olive oil, while chlorophylls are responsible for the green colour of olive oil. Therefore, both absorb radiation mainly in the visible spectrum.

One could ask why one would determine these compounds by NIRS, since they absorb mainly in the visible spectrum and, as for the K225, K232 and K270 parameters, ultraviolet radiation. As stated in Section 3, the peaks observed at 420 and 460 nm in the olive oil spectrum correspond mainly to carotenoids, while the peak at approximately 670 nm corresponds to chlorophyll absorption [33]. This was the reason why the PLS models built solely with the NIR spectrum (Table 7) achieved maximum R^2_{cal} of 0.66 and 0.56 for carotenoids and chlorophylls, respectively, which are too low for practical use [9,46]. For this reason, NIRS (or visible/NIR spectroscopy) should be implemented as a multiparametric tool, i.e., not only to determine a property of olive oil, but as many parameters as possible from its NIR (or visible/NIR) spectral information. The idea is to find out the composition and quality parameters of olive oil by simply acquiring its NIR or visible/NIR spectrum in a few seconds. This is the main advantage of NIRS when compared to the laborious, time-consuming reference methods that have to be individually carried out in the laboratory for each quality parameter of olive oil.

Alkyl esters in olive oils are derived from the non-desired fermentation of the fruit, normally when overripe or incorrectly stored, thus suffering damage in the cell structure prior to entering the olive oil processing. The most important quality of olive oil is the number of ethyl esters, which is regarded as a quality criterion by the International Olive Council. The content of fatty acid ethyl esters must be $\leq$35 mg/kg for an oil to be classified as EVOO [1]. The SEP illustrated in Table 7 for ethyl esters (14.2 mg/kg) [30] seems to be a bit excessive to meet the requirements of the International Olive Council.

Moisture, which can promote the rancidification of olive oil, leading to an unpleasant taste and an unpleasant odour, has been determined by PLS-NIRS to achieve an r^2_{cal} of 0.71 and a bias-corrected SEP of 0.04 wt.% [30]. Taking into account that the moisture and volatile matter content is another quality criterion of the International Olive Council, and it must be $\leq$0.2 wt.% for edible olive oils [1], this prediction error should be lowered a bit. The relation between analysed and predicted values obtained by these authors is illustrated in Figure 13. It is worth noting that the determination of water content in olive oils by NIRS has not been assayed to date by using only the wavelengths where the broad absorption bands of water are found (at 1400 and 1950 nm).

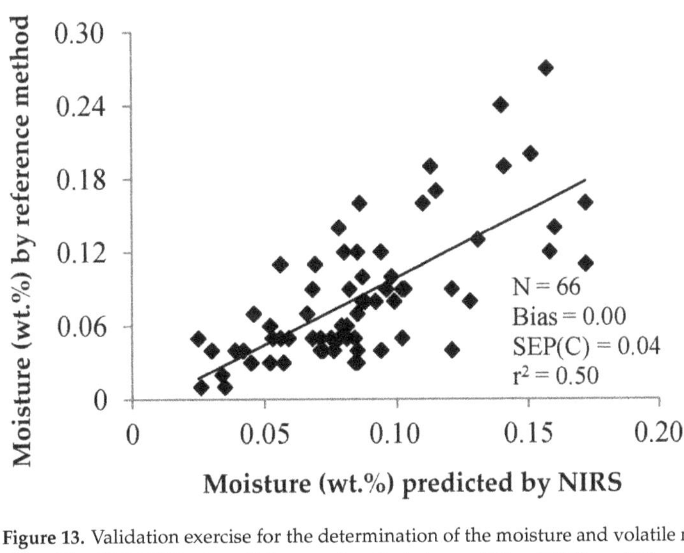

Figure 13. Validation exercise for the determination of the moisture and volatile matter content of olive oil by PLS-NIRS using all the wavelengths between 400 and 2500 nm [30].

Parameters such as total polyphenols have not been successfully predicted by NIRS. In this sense, SEPs of 82.10 and 89.66 mg/kg were obtained when analysing total polyphenols in EVOO using two spectrometers, SEL being 9.24 mg/kg for samples in the range 44.49–738.76 mg/kg [9]. Other authors achieved a good correlation coefficient of calibration

(r^2_{cal} = 0.85) and a lower error of prediction (RMSEP = 44.5 mg/kg) [47], but these were still a bit high for practical use.

Squalene is a hydrocarbon that can be found in relatively high quantities (between 60 and 75 wt.%) within the unsaponifiable fraction of olive oil, accounting for between 0.2 and 7.5 g/kg of olive oil [67]. In spite of the multiple pieces of scientific evidence of the beneficial effects of squalene on human health, its determination is generally not performed in the olive oil industry, as squalene is neither considered a quality nor a purity parameter in olive oil regulation [1]. The only attempt found in the literature to determine squalene in olive oil by NIRS or visible/NIR spectroscopy used EVOO, VOO, ordinary oil oils, pomace oils, and lampante oils of different varieties for the calibration and validation exercises [23]. The best results were obtained with the NIR spectra (Table 7). However, the SEP achieved (1 g/kg) is too high for its use in the olive oil industry since, as aforementioned, the concentration of squalene in olive oils ranges between 0.2 and 7.5 g/kg olive oil.

Olive oil is a notorious source of vitamin E (α-tocopherol). EVOO and VOO contain about 207.3 mg α-tocopherol per kg of olives. Pomace olive oils contain higher amounts of vitamin E, up to 981.6 mg/kg [68]. The determination of α-tocopherol, β-tocopherol, γ-tocopherol and total tocopherols of olive oils has been assayed using their NIR and visible/NIR spectra [25]. In that work, lampante and pomace olive oils were used in the calibration PLS models along with EVOO and VOO to increase the diversity of tocopherols, so that the range of concentrations of α-tocopherol (Table 7) was much higher than the content of α-tocopherol reported for EVOO and VOO [68]. Models using only NIR wavelengths predicted the content in α-, γ- and total tocopherols better than those using all wavelengths from the visible/VIR spectrum [25]. The PLS-NIR model for α-tocopherol achieved a good correlation coefficient of calibration (0.95), but SEP (47.2 mg/kg) seems to be quite high for practical application, taking into account that the average content of vitamin E in olive oils is 207.3 mg/kg [68]. The statistics found by other authors did not improve the ability of NIRS to determine α-tocopherol in olive oils [47].

Finally, other parameters of interest for the quality of olive oil, such as the oxidative stability, for which the units are time-based, have been predicted by visible/NIR spectroscopy with relatively good precision [27], as illustrated in Table 7.

On the other hand, olive oil is practically composed of fat (the saponifiable fraction accounts for roughly 98 wt.% olive oil). The fatty acid profile of olive oils is one of the most suitable and with the highest precision analysis that NIRS can perform [10,18,31,32,69]. The current European regulation settles the obligation of food manufacturers to include nutritional information on their product labels [26]. Mandatory information on food labels includes energy value, total fat content, total saturated fatty acids (TSFA), and other compounds that olive oil does not contain, such as carbohydrates, sugars, proteins, and salt. As voluntary nutritional information, the European label can contain other nutritional information, such as monounsaturated fatty acids (MUFA) and polyunsaturated fatty acid (PUFA) content. Furthermore, food labelling regulations in the USA and Canada also require a declaration of TSFA content on product labels [10]. Regarding olive oil, the most frequently included information on its nutritional label is total fat, saturated fat, monounsaturated fat, and polyunsaturated fat [26]. It has been reported that the first overtone of MUFA can be observed at 1724 and 1766 nm, with the combination bands at 2358 nm [29]. As for PUFA, 1660, 1698, and 1730 nm wavelengths have correlated with the first overtone, 1162 and 1212 nm with the second overtone, and 2136, 2176, 2224, 2310, 2348, and 2434 nm with combination bands [29]. Some works available in the literature have shown the feasibility of NIRS for determining TSFA, MUFA, and PUFA in the American, Canadian, Spanish, and Portuguese EVOO, VOO, and ordinary olive oils [10,14,26,47]. Hence, NIRS is suitable for the nutritional labelling of olive oil.

7. Sensory Attributes

The sensory parameters of olive oil are of equal importance as the physico-chemical quality parameters described in Section 5. Notwithstanding, scarce information can be found in the literature on the use of NIRS for the determination of the sensory parameters of olive oil. The prediction of the minor composition of VOO, in particular its phenolic and volatile compounds, as well as its organoleptic attributes, has been assayed in the 800–2500 nm NIR spectrum. Acceptable multivariate algorithms based on the multiple coefficient of determination were obtained for some minor components, such as hydroxytyrosol derivatives (r^2 = 0.86–0.88) and C6 alcohols (r^2 = 0.69–0.80), and for positive sensory attributes such as 'fruity' (r^2 = 0.87) and 'bitter' (r^2 = 0.85) [47]. More research is needed to correlate the NIR spectra of olive oil with its sensory parameters before regarding NIRS as a potential tool for the determination of these parameters.

8. Conclusions

The information available in the literature illustrates that the application of NIRS to olive oil could undergo an industrial development similar to that of olives and olive pastes, which have commercial, available NIRS equipment for assessing some of its main parameters of interest. A sampling system of olive oils and NIRS equipment for the acquisition of their NIR spectra could be implemented in the olive oil mill or in the bottling plant, thus allowing the on-site determination of their main quality parameters.

The four primary olive oils' quality parameters (FA, PV, K232 and K270) can be accurately determined by NIRS spectroscopy, based on promising results reported by different authors. In addition, NIRS is suitable for the nutritional labelling of olive oil, since its feasibility for determining TSFA, MUFA, and PUFA has been demonstrated. Therefore, all these parameters in an olive oil could be measured by NIRS, as a multiparametric analytical technique, simply by acquiring the NIR spectrum of the oil and using the PLS model developed for each parameter.

Other parameters such as α-tocopherol (vitamin E), fatty acid ethyl esters, squalene and K225 show potential to be determined by NIRS, but the prediction errors reported by the various authors are still a bit high for practical application. Furthermore, by expanding the wavelength range to which spectra are acquired to the visible region, other compounds such as carotenoids and chlorophylls, which absorb radiation mainly in the visible spectrum, could be quantified by visible/NIR spectroscopy. However, visible/NIR spectrophotometers are more expensive than NIR spectrophotometers, so the implementation of one or another will depend on the robustness of the PLS models for each parameter and, hence, their practical application.

NIRS equipment at olive oil mills or bottling plants should provide self-learning model calibration systems, so that samples from new harvestings, different designations, geographical origin, and varieties, etc. are automatically added to the calibration set to strengthen the PLS models over time. Validation exercises with samples not used to build the PLS models are mandatory to assess their performance.

Spectra pre-treatments (derivatisation, normalisation, baseline correction, standard normal variate, mean centring, Savitzky and Golay smoothing, first and second derivatives, multiplicative scatter corrections) enhance the handling of the spectra and the building of the PLS calibration models. Similarly, the selection of actual contributing spectral variables and the removal of outliers can improve the performance of the PLS models. Notwithstanding, these latter two procedures must be carefully performed or avoided at early stages of the model building (when there is not a large calibration sample set) due to the risk of removing important spectral information related to the quality parameter of interest.

The ultimate goal is to achieve acceptance of NIRS as an official method for the determination of the quality parameters and the nutritional parameters for the labelling of olive oil by the relevant national authorities and, as a priority, the International Olive Council, which will greatly contribute to the industrial development of NIRS equipment for the olive oil industry.

Funding: This research received no external funding.

Institutional Review Board Statement: Not applicable.

Acknowledgments: The author would like to earnestly thank José M. García-Martos and Fernando Martínez-Román, both from the Instituto de la Grasa of the Spanish National Research Council, for their insightful suggestions and reassurance.

Conflicts of Interest: The author declares no conflict of interest.

References

1. *Trade Standard Applying to Olive Oils and Olive Pomace Oils. Rev. 13*; International Olive Council: Madrid, Spain, 2019.
2. Plasquy, E.; García Martos, J.M.; Florido, M.C.; Sola-Guirado, R.R.; García Martín, J.F. Cold storage and temperature management of olive fruit: The impact on fruit physiology and olive oil quality—A review. *Processes* **2021**, *9*, 1543. [CrossRef]
3. Siesler, H.W. Basic Principles of Near-Infrared Spectroscopy. In *Handbook of Near-Infrared Analysis*; Burns, D.A., Ciurczak, E.W., Eds.; CRC Press: Boca Raton, FL, USA, 2007.
4. Peguero Gutiérrez, A. La Espectroscopía NIR en la Determinación de Propiedades Físicas y Composición Química de Intermedios de Producción y Productos Acabados. Ph.D. Thesis, Universitat Autònoma de Barcelona, Barcelona, Spain, 2010.
5. Yang, H.; Irudayaraj, J. Comparison of near-infrared, Fourier transform-infrared, and Fourier transform-Raman methods for determining olive pomace oil adulteration in extra virgin olive oil. *J. Am. Oil Chem. Soc.* **2001**, *78*, 889–895. [CrossRef]
6. Berna, F. Fourier transform infrared spectroscopy (FTIR). In *Encyclopedia of Earth Sciences Series*; Gilbert, A.S., Ed.; Springer: Dordrecht, The Netherlands, 2017; pp. 285–286.
7. Thermo Fisher Scientific. *Advantages of Fourier-Transform Near-Infrared Spectroscopy*; Thermo Fisher Scientific Inc.: Madison, WI, USA, 2006.
8. Rohman, A. Infrared spectroscopy for quantitative analysis and oil parameters of olive oil and virgin coconut oil: A review. *Int. J. Food Prop.* **2017**, *20*, 1447–1456. [CrossRef]
9. Manley, M.; Eberle, K. Comparison of Fourier transform near infrared spectroscopy partial least square regression models for South African extra virgin olive oil using spectra collected on two spectrophotometers at different resolutions and path lengths. *J. Near Infrared Spectrosc.* **2006**, *14*, 111–126. [CrossRef]
10. Mossoba, M.M.; Azizian, H.; Tyburczy, C.; Kramer, J.K.G.; Delmonte, P.; Kia, A.R.F.; Rader, J.I. Rapid FT-NIR analysis of edible oils for total SFA, MUFA, PUFA, and trans FA with comparison to GC. *J. Am. Oil Chem. Soc.* **2013**, *90*, 757–770. [CrossRef]
11. Azizian, H.; Dugan, M.E.R.; Kramer, J.K.G. Clarifications of the carbonyl and water absorptions in Fourier transform near infrared spectra from extra virgin olive oil. *Eur. J. Lipid Sci. Technol.* **2021**, *123*, 2000288. [CrossRef]
12. Lagardere, L.; Lechat, H.; Lacoste, F. Détermination de l'acidité et de l'indice de peroxyde dans les huiles d'olive vierges et dans les huiles raffinées par spectrométrie proche infrarouge à transformée de Fourier. *Oléagineux Corp. Gras Lipides* **2004**, *11*, 70–75. [CrossRef]
13. Tarhan, İ.; Ismail, A.A.; Kara, H. Quantitative determination of free fatty acids in extra virgin olive oils by multivariate methods and Fourier transform infrared spectroscopy considering different absorption modes. *Int. J. Food Prop.* **2017**, *20*, S790–S797. [CrossRef]
14. Milinovic, J.; Garcia, R.; Rato, A.E.; Cabrita, M.J. Rapid assessment of monovarietal Portuguese extra virgin olive oil's (EVOO's) fatty acids by Fourier-transform near-infrared spectroscopy (FT-NIRS). *Eur. J. Lipid Sci. Technol.* **2019**, *121*, 1800392. [CrossRef]
15. *ISO/IEC 17025*; Requisitos Generales Para la Competencia de los Laboratorios de Ensayo y Calibración. Asociación Española de Normalización: Madrid, Spain, 2017.
16. Giovenzana, V.; Beghi, R.; Romaniello, R.; Tamborrino, A.; Guidetti, R.; Leone, A. Use of visible and near infrared spectroscopy with a view to on-line evaluation of oil content during olive processing. *Biosyst. Eng.* **2018**, *172*, 102–109. [CrossRef]
17. Márquez, A.J. Monitoring Carotenoid and Chlorophyll Pigments in Virgin Olive Oil by Visible-Near Infrared Transmittance Spectroscopy. On-Line Application. *J. Near Infrared Spectrosc.* **2003**, *11*, 219–226. [CrossRef]
18. Marquez, A.J.; Díaz, A.M.; Reguera, M.I.P.P. Using optical NIR sensor for on-line virgin olive oils characterization. *Sens. Actuators B Chem.* **2005**, *107*, 64–68. [CrossRef]
19. Wesley, I.J.; Barnes, R.J.; McGill, A.E.J. Measurement of adulteration of olive oils by near infra-red spectroscopy. *J. Am. Oil Chem. Soc.* **1995**, *72*, 289–292. [CrossRef]
20. Bertran, E.; Blanco, M.; Coello, J.; Iturriaga, H.; Maspoch, S.; Monteliu, I. Near infrared spectrometry and pattern recognition as screening methods for the authentication of virgin olive oils of very close geographical origins. *J. Near Infrared Spectrosc.* **2000**, *8*, 45–52. [CrossRef]
21. Casale, M.; Simonetti, R. Review: Near infrared spectroscopy for analysing olive oils. *J. Near Infrared Spectrosc.* **2014**, *22*, 59–80. [CrossRef]
22. García Martín, J.F. Optical path length and wavelength selection using Vis/NIR spectroscopy for olive oil's free acidity determination. *Int. J. Food Sci. Technol.* **2015**, *50*, 1461–1467. [CrossRef]
23. Cayuela, J.A.; García, J.F. Nondestructive measurement of squalene in olive oil by near infrared spectroscopy. *LWT Food Sci. Technol.* **2018**, *88*, 103–108. [CrossRef]

24. Cayuela, J.A.; García, J.M.; Caliani, N. NIR prediction of fruit moisture, free acidity and oil content in intact olives. *Grasas Aceites* **2009**, *60*, 194–202. [CrossRef]
25. Cayuela, J.A.; García, J.F. Sorting olive oil based on alpha-tocopherol and total tocopherol content using near-infra-red spectroscopy (NIRS) analysis. *J. Food Eng.* **2017**, *202*, 79–88. [CrossRef]
26. Cayuela-Sánchez, J.A.; Palarea-Albaladejo, J.; García-Martín, J.F.; Del Carmen Pérez-Camino, M. Olive oil nutritional labeling by using Vis/NIR spectroscopy and compositional statistical methods. *Innov. Food Sci. Emerg. Technol.* **2019**, *51*, 139–147. [CrossRef]
27. Cayuela Sánchez, J.A.; Moreda, W.; García, J.M. Rapid determination of olive oil oxidative stability and its major quality parameters using vis/NIR transmittance spectroscopy. *J. Agric. Food Chem.* **2013**, *61*, 8056–8062. [CrossRef] [PubMed]
28. Cayuela, J.A. Assessing olive oil peroxide value by NIRS, and on reference methods. *NIR News* **2017**, *28*, 12–16. [CrossRef]
29. García-González, D.L.; Baeten, V.; Fernández Pierna, J.A.; Tena, N. Infrared, raman, and fluorescence spectroscopies: Methodologies and applications. In *Handbook of Olive Oil: Analysis and Properties*; Springer: Boston, MA, USA, 2013; pp. 335–393; ISBN 9781461477778.
30. Garrido-Varo, A.; Sánchez, M.T.; De la Haba, M.J.; Torres, I.; Pérez-Marín, D. Fast, low-cost and non-destructive physico-chemical analysis of virgin olive oils using near-infrared reflectance spectroscopy. *Sensors* **2017**, *17*, 2642. [CrossRef] [PubMed]
31. Sato, T.; Kawano, S.; Iwamoto, M. Near infrared spectral patterns of fatty acid analysis from fats and oils. *J. Am. Oil Chem. Soc.* **1991**, *68*, 827–833. [CrossRef]
32. Mailer, R.J. Rapid evaluation of olive oil quality by NIR reflectance spectroscopy. *J. Am. Oil Chem. Soc.* **2004**, *81*, 823–827. [CrossRef]
33. Moyano, M.J.; Meléndez-Martínez, A.J.; Alba, J.; Heredia, F.J. A comprehensive study on the colour of virgin olive oils and its relationship with their chlorophylls and carotenoids indexes (I): CIEXYZ non-uniform colour space. *Food Res. Int.* **2008**, *41*, 505–512. [CrossRef]
34. Moros, J.; Garrigues, S.; De la Guardia, M. Vibrational spectroscopy provides a green tool for multi-component analysis. *Trends Anal. Chem.* **2010**, *29*, 578–591. [CrossRef]
35. Gabrielsson, J.; Trygg, J. Recent developments in multivariate calibration. *Crit. Rev. Anal. Chem.* **2006**, *36*, 243–255. [CrossRef]
36. Miller, J.N.; Miller, J.C. *Statistics and Chemometrics for Analytical Chemistry*, 6th ed.; Pearson Education Limited: Harlow, UK, 2010; ISBN 978-0-273-73042-2.
37. Beebe, K.R.; Kowalski, B.R. An introduction to multivariate calibration and analysis. *Anal. Chem.* **1987**, *59*, 1007A–1017A. [CrossRef]
38. Guzmán Jiménez, E. Desarrollo de Técnicas Ópticas de Análisis Rápido Para el Control de Calidad en el Proceso de Elaboración de Aceite de Oliva Virgen. Ph.D. Thesis, Universidad de Córdoba, Córdoba, Spain, 2013.
39. Kaufmann, K.C.; Sampaio, K.A.; García-Martín, J.F.; Barbin, D.F. Identification of coriander oil adulteration using a portable NIR spectrometer. *Food Control* **2022**, *132*, 108536. [CrossRef]
40. Weng, X.-X.; Lu, F.; Wang, C.-X.; Qi, Y.-P. Discriminating and quantifying potential adulteration in virgin olive oil by near infrared spectroscopy with BP-ANN and PLS. *Spectrosc. Spectr. Anal.* **2009**, *29*, 3283–3287. [CrossRef]
41. Vanstone, N.; Moore, A.; Martos, P.; Neethirajan, S. Detection of the adulteration of extra virgin olive oil by near-infrared spectroscopy and chemometric techniques. *Food Qual. Saf.* **2018**, *2*, 189–198. [CrossRef]
42. Borghi, F.T.; Santos, P.C.; Santos, F.D.; Nascimento, M.H.C.; Corrêa, T.; Cesconetto, M.; Pires, A.A.; Ribeiro, A.V.F.N.; Lacerda, V.; Romão, W.; et al. Quantification and classification of vegetable oils in extra virgin olive oil samples using a portable near-infrared spectrometer associated with chemometrics. *Microchem. J.* **2020**, *159*, 105544. [CrossRef]
43. Kasemsumran, S.; Kang, N.; Christy, A.; Ozaki, Y. Partial least squares processing of near-infrared pectra for discrimination and quantification of adulterated olive oils. *Spectrosc. Lett.* **2005**, *38*, 839–851. [CrossRef]
44. Casale, M.; Oliveri, P.; Casolino, C.; Sinelli, N.; Zunin, P.; Armanino, C.; Forina, M.; Lanteri, S. Characterisation of PDO olive oil Chianti Classico by non-selective (UV–visible, NIR and MIR spectroscopy) and selective (fatty acid composition) analytical techniques. *Anal. Chim. Acta* **2012**, *712*, 56–63. [CrossRef]
45. Galtier, O.; Dupuy, N.; Le Dréau, Y.; Ollivier, D.; Pinatel, C.; Kister, J.; Artaud, J. Geographic origins and compositions of virgin olive oils determinated by chemometric analysis of NIR spectra. *Anal. Chim. Acta* **2007**, *595*, 136–144. [CrossRef]
46. Cayuela, J.A.; Yousfi, K.; Carmen Martínez, M.; García, J.M. Rapid determination of olive oil chlorophylls and carotenoids by using visible spectroscopy. *J. Am. Oil Chem. Soc.* **2014**, *91*, 1677–1684. [CrossRef]
47. Inarejos-García, A.M.; Gómez-Alonso, S.; Fregapane, G.; Salvador, M.D. Evaluation of minor components, sensory characteristics and quality of virgin olive oil by near infrared (NIR) spectroscopy. *Food Res. Int.* **2013**, *50*, 250–258. [CrossRef]
48. López Barrera, M.D.C. Caracterización de Aceites Usados de HORECA Mediante Métodos Tradicionales y Espectroscopía de Infrarrojo Cercano. Bachelor's Thesis, Universidad de Sevilla, Sevilla, Spain, 2017.
49. Wold, S.; Sjöström, M.; Eriksson, L. PLS-regression: A basic tool of chemometrics. In *Chemometrics and Intelligent Laboratory Systems*; Elsevier: Amsterdam, The Netherlands, 2001; Volume 58, pp. 109–130.
50. Kennard, R.W.; Stone, L.A. Computer Aided Design of Experiments. *Technometrics* **1969**, *11*, 137–148. [CrossRef]
51. Picard, R.R.; Cook, R.D. Cross-validation of regression models. *J. Am. Stat. Assoc.* **1984**, *79*, 575–583. [CrossRef]
52. Shenk, J.S.; Westerhaus, M. Calibration the ISI way. In *Near Infrared Spectroscopy: The Future Waves*; Davies, A.M.C., Williams, P.C., Eds.; NIR Publications: Chichester, UK, 1996; pp. 198–202. ISBN 978-0952866602.

53. Han, Q.J.; Wu, H.L.; Cai, C.B.; Xu, L.; Yu, R.Q. An ensemble of Monte Carlo uninformative variable elimination for wavelength selection. *Anal. Chim. Acta* **2008**, *612*, 121–125. [CrossRef] [PubMed]
54. Cai, W.; Li, Y.; Shao, X. A variable selection method based on uninformative variable elimination for multivariate calibration of near-infrared spectra. *Chemom. Intell. Lab. Syst.* **2008**, *90*, 188–194. [CrossRef]
55. Hao, Y.; Sun, X.; Zhang, H.; Liu, Y. Application of effective wavelength selection methods to determine total acidity of navel orange. *Sens. Lett.* **2011**, *9*, 1229–1234. [CrossRef]
56. Lu, B.; Castillo, I.; Chiang, L.; Edgar, T.F. Industrial PLS model variable selection using moving window variable importance in projection. *Chemom. Intell. Lab. Syst.* **2014**, *135*, 90–109. [CrossRef]
57. Jiang, J.H.; James, R.; Siesler, B.H.W.; Ozaki, Y. Wavelength interval selection in multicomponent spectral analysis by moving window partial least-squares regression with applications to mid-infrared and near-infrared spectroscopic data. *Anal. Chem.* **2002**, *74*, 3555–3565. [CrossRef]
58. Araújo, M.C.U.; Saldanha, T.C.B.; Galvão, R.K.H.; Yoneyama, T.; Chame, H.C.; Visani, V. The successive projections algorithm for variable selection in spectroscopic multicomponent analysis. *Chemom. Intell. Lab. Syst.* **2001**, *57*, 65–73. [CrossRef]
59. Fernández Pierna, J.A.; Wahl, F.; De Noord, O.E.; Massart, D.L. Methods for outlier detection in prediction. *Chemom. Intell. Lab. Syst.* **2002**, *63*, 27–39. [CrossRef]
60. European Union Commission (EEC). Regulation EEC 2568/91 on the characteristics of olive oil and olive-residue oil and on the relevant methods of analysis. *Off. J. Eur. Communities* **1991**, *248*, 1–15.
61. Abu-Khalaf, N.; Hmidat, M. Visible/Near Infrared (VIS/NIR) spectroscopy as an optical sensor for evaluating olive oil quality. *Comput. Electron. Agric.* **2020**, *173*, 105445. [CrossRef]
62. Armenta, S.; Garrigues, S.; De la Guardia, M. Determination of edible oil parameters by near infrared spectrometry. *Anal. Chim. Acta* **2007**, *596*, 330–337. [CrossRef]
63. Peršurić, Ž.; Saftić, L.; Mašek, T.; Kraljević Pavelić, S. Comparison of triacylglycerol analysis by MALDI-TOF/MS, fatty acid analysis by GC-MS and non-selective analysis by NIRS in combination with chemometrics for determination of extra virgin olive oil geographical origin. A case study. *LWT Food Sci. Technol.* **2018**, *95*, 326–332. [CrossRef]
64. García Martín, J.F.; López Barrera, M.D.C.; Torres García, M.; Zhang, Q.A.; Álvarez Mateos, P. Determination of the acidity of waste cooking oils by near infrared spectroscopy. *Processes* **2019**, *7*, 304. [CrossRef]
65. Shenk, J.S.; Workman, J.J.J.; Westerhaus, M.O. Application of NIR spectroscopy to agricultural products. In *Handbook of Near-Infrared Analysis*; Burns, D.A., Ciurczak, E.W., Eds.; Dekker Inc.: New York, NY, USA, 2001; pp. 383–431.
66. Kandala, C.V.; Sundaram, J.; Puppala, N. Analysis of moisture content, total oil and fatty acid composition by NIR reflectance spectroscopy: A review. In *Lecture Notes in Electrical Engineering*; Springer: Berlin/Heidelberg, Germany, 2012; Volume 146, pp. 59–80. [CrossRef]
67. Tiscornia, E.; Forina, M.; Evangelisti, F. Composizione chimica dell'olio di oliva e sue variazioni indotte dal processo di rettificazione. *Riv. Ital. Sostanze Grasse* **1980**, *59*, 519–556.
68. Perona, J.S.; Arcemis, C.; Ruiz-Gutierrez, V.; Catalá, A. Effect of dietary high-oleic-acid oils that are rich in antioxidants on microsomal lipid peroxidation in rats. *J. Agric. Food Chem.* **2005**, *53*, 730–735. [CrossRef]
69. García-Martín, J.F.; Alés-Álvarez, F.J.; Del Carmen López-Barrera, M.; Martín-Domínguez, I.; Álvarez-Mateos, P. Cetane number prediction of waste cooking oil-derived biodiesel prior to transesterification reaction using near infrared spectroscopy. *Fuel* **2019**, *240*, 10–15. [CrossRef]

sensors

MDPI

Article

Detecting Bacterial Biofilms Using Fluorescence Hyperspectral Imaging and Various Discriminant Analyses

Ahyeong Lee, Saetbyeol Park, Jinyoung Yoo, Jungsook Kang, Jongguk Lim, Youngwook Seo, Balgeum Kim and Giyoung Kim *

Rural Development Administration, 310 Nongsaengmyeong-ro, Deokjin-gu, Jeonju 54875, Korea; lay117@korea.kr (A.L.); psb0911@korea.kr (S.P.); young3233@korea.kr (J.Y.); js72kang@korea.kr (J.K.); limjg@korea.kr (J.L.); yws25@korea.kr (Y.S.); bgkim8732@korea.kr (B.K.)

* Correspondence: giyoung@korea.kr; Tel.: +82-63-238-4103; Fax: +82-63-238-4105

Abstract: Biofilms formed on the surface of agro-food processing facilities can cause food poisoning by providing an environment in which bacteria can be cultured. Therefore, hygiene management through initial detection is important. This study aimed to assess the feasibility of detecting *Escherichia coli* (*E. coli*) and *Salmonella typhimurium* (*S. typhimurium*) on the surface of food processing facilities by using fluorescence hyperspectral imaging. *E. coli* and *S. typhimurium* were cultured on high-density polyethylene and stainless steel coupons, which are the main materials used in food processing facilities. We obtained fluorescence hyperspectral images for the range of 420–730 nm by emitting UV light from a 365 nm UV light source. The images were used to perform discriminant analyses (linear discriminant analysis, *k*-nearest neighbor analysis, and partial-least squares discriminant analysis) to identify and classify coupons on which bacteria could be cultured. The discriminant performances of specificity and sensitivity for *E. coli* (1–4 log CFU·cm^{-2}) and *S. typhimurium* (1–6 log CFU·cm^{-2}) were over 90% for most machine learning models used, and the highest performances were generally obtained from the *k*-nearest neighbor (*k*-NN) model. The application of the learning model to the hyperspectral image confirmed that the biofilm detection was well performed. This result indicates the possibility of rapidly inspecting biofilms using fluorescence hyperspectral images.

Keywords: *E. coli*; *S. typhimurium*; biofilm; hyperspectral imaging; discriminant analysis

Citation: Lee, A.; Park, S.; Yoo, J.; Kang, J.; Lim, J.; Seo, Y.; Kim, B.; Kim, G. Detecting Bacterial Biofilms Using Fluorescence Hyperspectral Imaging and Various Discriminant Analyses. *Sensors* **2021**, *21*, 2213. https://doi.org/10.3390/s21062213

Academic Editor: Mercedes Del Río Celestino

Received: 18 January 2021
Accepted: 18 March 2021
Published: 22 March 2021

1. Introduction

Despite the increasing social interest in safe agro-food, food poisoning occurs frequently. Food poisoning due to foodborne illnesses is one of the major public health problems worldwide [1]. The Centers for Disease Control and Prevention (CDC) estimates that food poisoning causes 48 million people to get sick, 128,000 to be hospitalized, and 3000 to die each year [2]. Although the cause of food poisoning varies, most food poisoning incidents are caused by ingestion of food contaminated with germs or viruses [3,4]. In the case of food contamination occurring in the process of food processing and distribution, cross-contamination is one of the main routes of contamination caused by the surface of germ-contaminated facilities, machinery, and containers and contaminated washing water [5,6].

Food safety accidents caused by cross-contamination in food processing facilities [7–11] or home kitchens [12–14] are related to biofilms found in a wide range of environments. Biofilms are composed of an extracellular polymeric substance, which comprises mostly polysaccharides secreted by microorganisms. Biofilms are firmly attached to the surface of living organisms (vegetables, meat, etc.) or non-living objects (stainless steel, plastics, etc.) [15]. Sessile bacteria attached to biofilms are known to withstand stress better than planktonic bacteria in stressful environments such as in the presence of antibiotics, disinfectants, high temperatures, light, and dryness [16,17]. Cross-contamination occurs when food passes through a surface contaminated with biofilms or when cells are separated from the biofilm structure through an aerosol derived from a contaminated food processing facility [1,11]. According to the US National Institutes of Health, about 80% of human infections are related to biofilms [18]. These biofilms cause serious hygiene problems and economic losses due to food spoilage, equipment damage, and food poisoning (microbiological infection). Therefore, this is an important issue for the food industry, including the agricultural and livestock production and processing industries, to tackle [19].

Soon [4] conducted a factor analysis through cases of food safety incidents and recalls of food safety accidents that occurred around the world during 2008–2018. He revealed that 40.11% of food safety hazards were caused by microbiological hazards, and the major hazards were *Listeria monocytogenes* (32.91%), *Salmonella spp.* (29.85%), and *Escherichia coli* (17.86%). Among them, *Escherichia coli* (*E. coli*) infection can occur by eating contaminated agricultural products (sprouts, spinach, lettuce, etc.) or undercooked meat and is highly likely to lead to hospitalization in disease outbreaks [2]. Food poisoning due to *E. coli* infection is associated with serotype O157:H7. In particular, Shiga toxin-producing *Escherichia coli* causes hemolytic uremic syndrome (also known as Hamburger's disease), which leads to acute kidney damage; it is a fatal disease in children and the elderly [20]. *Salmonella typhimurium* (*S. typhimurium*) causes salmonellosis, which leads to acute and chronic enteritis symptoms such as diarrhea and abdominal pain [21]. The main sources of disease are contaminated poultry, meat, and eggs [22]. The CDC estimates that *Salmonella* causes approximately 1.2 million diseases and 450 deaths each year in the United States and reports that it is the most common cause of food poisoning in June, July, and August [2,23]. As noted in many studies, *E. coli* [11,24–26] and *S. typhimurium* [27–32] form a strong biofilm on the surface of non-living objects or living organisms. Cross-contamination in agro-food processing lines caused by biofilms is a serious concern in the food industry.

There is growing demand for technology that can quickly and easily detect biological contaminants to prevent cross-contamination [33,34]. Conventional biofilm testing techniques use swabs to collect samples of the suspected area. After that, the bacteria are cultured to check the presence of contamination using various methods such as contact plate system, microbial diagnostic platform like TEMPO® (bioMerieux, Marcy-l'Etoile, France), and adenosine triphosphate (ATP) determination [35,36]. In this case, testing performance is degraded when we collect samples using swabs, and it takes a long time to cultivate bacteria. As an alternative, hyperspectral imaging (HSI) technology with rapid and non-destructive inspection characteristics is drawing attention [37]. HSI technology can not only identify the physical chemistry characteristics of a substance through spectroscopic analysis but also simultaneously obtain information about the spatial distribution of certain components through image analysis [38]. Since hyperspectral imaging has many independent variables, analysis methods that can reduce the number of independent variables have been used rather than using general multiple regression analysis [39–43]. Among them, partial least-squares regression (PLSR) and principal component analysis (PCA) are mainly used. PLSR analysis is suitable for regression modeling under the condition that the number of samples is less than the number of variables. PCA is suitable for classification modeling that reduces the number of independent variables through dimensional transformation [44]. Recently, there are increasing cases of applying various machine learning techniques [45,46] or artificial neural networks [47–49] to increase the performance of the model.

Many studies have demonstrated that HSI technology is a powerful tool for monitoring food safety incidents in relation to biofilms, which cause cross-contamination in the food industry [50–54]. Zhu [54] determined corn contaminated by aflatoxin mycotoxin over 100 ppb with 95.3% performance using fluorescence hyperspectral images. Jun [52] identified a biofilm over 7 log $CFU{\cdot}cm^{-2}$ formed by *E. coli* and *S. typhimurium* with a performance of 95% using the one-wavelength image and the ratio image of the fluorescence hyperspectral image of two wavelength bands. Lee [15] identified a biofilm over 1 log $CFU{\cdot}cm^{-2}$ formed by *E. coli* on a high-density polyethylene (HDPE) coupon. Lee used the single-wavelength method that Jun [52] used for detecting biofilm and machine learning techniques. Then Lee confirmed that the prediction performance of a biofilm by machine learning techniques is higher than that by the single-wavelength method.

This study was conducted to examine the feasibility of rapidly inspecting biofilms using HSI technology by expanding the targets of detection than the previous research [15] we did. First, we confirmed the fluorescence characteristics of *E. coli* and *S. typhimurium* using a microplate reader. Then, we obtained a biofilm fluorescence image between 420 and 730 nm for a 365 nm UV light source using a hyperspectral imaging device. The biofilm is formed by *E. coli* and *S. typhimurium* on the surface of HDPE and stainless steel (SS), which are the main materials in agro-food processing facilities. Consequently, we developed a biofilm discrimination model for hyperspectral images by applying various machine learning algorithms and compared their performances.

2. Materials and Methods

2.1. *Fluorescence Characteristics of Food Poisoning Bacteria*

To examine the fluorescence characteristics of food poisoning bacteria, strains of non-pathogenic *Escherichia coli* (*E. coli*, KCCM11234) and *Salmonella typhimurium* (*S. typhimurium*, KCCM12041) were obtained from the Korean Culture Center of Microorganisms. All reagents and media were sterilized at 121 °C for 15 min using an autoclave (MLS–2420; SANYO, Tokyo, Japan) before use. Each strain was individually grown in tryptic soy broth (TSB; BD, Franklin Lakes, NJ, USA) at 36 °C for 24 h for activation. The cultures were transferred onto the surface of a tryptic soy agar (TSA; BD, Franklin Lakes, NJ, USA) plate by loop and incubated at 36 °C for 24 h. Single colonies, which were formed after incubation, were collected from the plates and were suspended in 0.1 M phosphate-buffered saline (PBS) solution in a microtube. The suspension was centrifuged at 8000 rpm for 3 min (Eppendorf centrifuge 5804 R; Eppendorf, Hamburg, Germany). Then, the supernatant was removed using a pipette, and the pellets were resuspended in 0.1 M PBS solution. The washing step was performed three times. The optical density (OD) of the cell suspension was measured at 600 nm using a microplate reader (Infinite M1000; Tecan, Männedorf, Switzerland), and the final concentration was adjusted to 10^9 CFU·mL^{-1} (OD 0.1 and 100 μL). Then, each strain was serially diluted tenfold from 10^6 to 10^1 with 0.1 M PBS solution. The 0.1 M PBS solution was used as a control group. Each cell suspension with the adjusted number of *E. coli* and *S. typhimurium* cells was placed in a 96-well plate. Fluorescence emission intensity from 400 to 800 nm bands was acquired for excitation light from 350 to 400 nm bands at 5 nm intervals using a microplate reader.

2.2. *Bacterial Biofilm Formation*

The biofilm was formed by using the non-pathogenic *E. coli* and *S. typhimurium* strains that were previously obtained. All HDPE and SS coupons (20 × 50 × 1 mm^3) for the formation of biofilms were washed with an ultrasonic cleaner (WUC-A, DAIHAN Ultrasonic cleaner; Wonju, Korea), which was sterilized at 121 °C for 15 min in an autoclave (MLS–2420; SANYO, Tokyo, Japan) and then completely dried before use. The OD of the cell suspension was measured at 600 nm using a microplate reader, and 100 μL of the cell suspension was adjusted to an OD value of 0.1. To achieve biofilm formation on the surface of HDPE and SS coupons, the cell suspension was inoculated in a 50 mL conical tube containing 15 mL of TSB (approximately 10^2 CFU·mL^{-1}), and coupons were placed in a conical tube and incubated at 36 °C for seven days, as shown in Figure 1a. Every 24 h interval, the culture medium of the conical tube was removed and 0.1 M PBS solution was added and washed twice with gentle stirring. Then, 15 mL of sterilized TSB medium was added to supply nutrients for biofilm formation. For the control, the process proceeded under the same conditions, but the bacteria were not inoculated into the TSB medium.

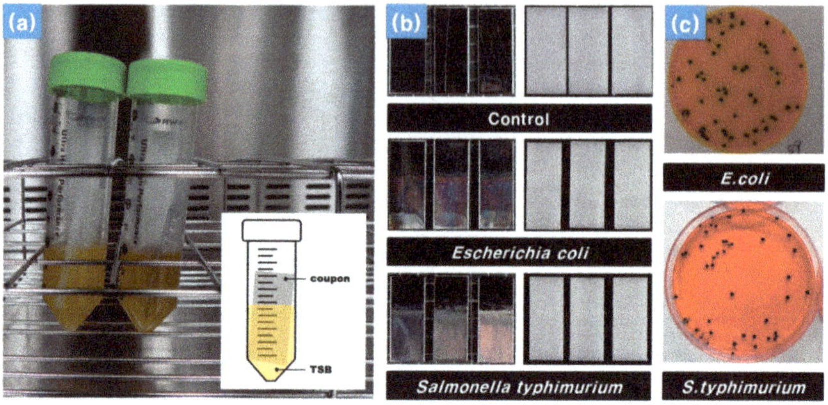

Figure 1. (**a**) Bacterial biofilm formation using a conical tube, (**b**) stainless steel (SS) coupons (left) and high-density polyethylene (HDPE) coupons (right), and (**c**) cultivation of *Escherichia coli* (*E. coli*) and *Salmonella typhimurium* (*S. typhimurium*).

After biofilm incubation, each coupon was rinsed three times with distilled water using a pipette to reduce interference from other substances, such as the medium and loosely attached bacteria cells. Coupon surfaces were dried completely on a sterile workbench, from which hyperspectral images were acquired, as shown in Figure 1b. To enumerate attached cells, the biofilms formed on the surface of the coupon were carefully detached with a cell scraper. Detached biofilms were transferred into 0.1 M PBS solution and diluted sequentially. The diluted culture medium was incubated at 36 °C for 24–48 h. The degree of biofilm formation on the surface of coupons was determined by counting the number of colonies using the standard plate count (SPC) method and the dry-film method (Figure 1c). To determine the number of adherent cells, the SPC method was performed with plate count agar (Plate Count Agar; Becton Dickinson and Company, Franklin Lakes, NJ, USA), and the dry-film method used a rapid-dry film (3M Petri-film *E. coli*/coliform count plates; 3M, St. Paul, MN, USA) for *E. coli*. In the case of *S. typhimurium*, the cell suspension was streaked onto XLT4 selective media (Xylose-Lysine-Tergitol 4 agar; Becton Dickinson and Company, Sparks, MD, USA).

2.3. Hyperspectral Imaging System

Figure 2 shows a fluorescence hyperspectral imaging system using ultraviolet (UV) excitation light. This system was composed of a highly sensitive electron-multiplying charge-coupled device (EMCCD, MegaLuca; Andor Technology Inc., Belfast, Northern Ireland) for obtaining hyperspectral images. The EMCCD camera consisted of 8 μm × 8 μm pixels and received a 14-bit digital image at a rate of 12.5 MHz. The EMCCD camera was thermo-electrically cooled to a temperature of −20 °C using a two-stage Peltier device. The imaging spectrograph (VNIR Hyperspec; HeadwallPhotonics Inc., Fitchburg, MA, USA) and a Schneider–Kreuznach Xenoplan 1.4/23 C-mount lens (f/1.9 35 mm Compact Lens; Schneider Optics, Hauppauge, NY, USA) were positioned in front of the EMCCD. The light source was a 365 nm UV beam (model XX–15A 365 nm; Spectronics Corp., Westbury, NY, USA), and the motorized sample stand was driven by a linear motor (XSlide; Velmax Inc., Bloomfield, NY, USA). The field of view of an image is limited by the size of the slit, which was 25 μm in this study. The fluorescence generated from the sample by the UV light source passed through the lens and slit of the imaging spectrograph. Then, the line scan image acquired through the slit was spectroscopically irradiated onto the EMCCD surface. Consequently, each line scan image was collected with spatial information horizontally and spectral information vertically.

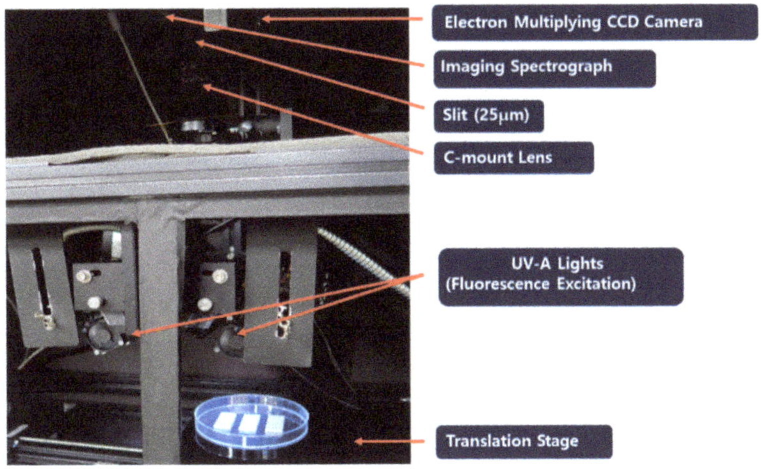

Figure 2. The ultraviolet fluorescence hyperspectral imaging system.

2.4. Acquisition of Hyperspectral Fluorescence Images and Spectra

To investigate the possibility of biofilm detection using hyperspectral images, 7-day-passed coupons after inoculation of *E. coli* and *S. typhimurium* were used in each experiment: 15 HDPE coupons and 15 SS coupons were used as the test group contained in the culture medium inoculated with bacteria, while 9 HDPE coupons and 9 SS coupons were used as the control group treated in the culture medium without bacteria. Hyperspectral images were acquired for both sides (front and back) of the coupons, and 96 hyperspectral images were obtained in the end. In the case of *S. typhimurium*, 96 hyperspectral images were acquired in the same manner from the third to the fifth day after inoculation of *S. typhimurium*. There was a difference in the culture rates of *E. coli* and *S. typhimurium* even after adjusting for the initial number of cultured bacteria.

For each sample, we obtained hyperspectral fluorescence images, dark reference images, and white reference images. The fluorescence hyperspectral images were acquired by the line scanning method, with 340 lines and 1 mm intervals for exposure times of 100 ms using a UV beam. The hyperspectral image included a spatial resolution of 310 × 502 pixels and contained 420–730 nm wavelength images, which were equally divided into 65 bands, with a waveband interval of 4.8 nm. The hyperspectral image was averaged after 4 repeated measurements for the same location to remove the noise.

Dark reference plate images for device noise compensation were measured by blocking the light using the cover of the camera lens. Fluorescence reference plate images were measured using a plate where the fluorescence was uniformly displayed. The white reference plate images were measured by a premium white inkjet paper (Union Camp Co.) exhibiting uniform blue fluorescence [55]. Fluorescence hyperspectral images were transformed for a total of 65 bands using Equation (1):

$$I_{fluorescence}\,(i) = \frac{I_r\,(i) - I_d\,(i)}{I_f\,(i) - I_d\,(i)} \quad (1)$$

where $I_{fluorescence}$ is the corrected relative fluorescence image, I_r is the raw hyperspectral fluorescence image, I_f is the hyperspectral white reference image, and I_d is the hyperspectral dark reference image at the *i*-th wavelength. Before analyzing the hyperspectral fluorescence image of the biofilm, it was preprocessed by a normalization method to remove the effects of non-uniformity of the light source and the electrical noise signal of the hyperspectral imaging equipment. Then, pixel and average fluorescence spectra were extracted from the modified fluorescence hyperspectral image.

2.5. Biofilm Detection Algorithm

Figure 3 shows a flowchart of the biofilm detection algorithm using hyperspectral imaging technology. As a preprocessing step, the original sample fluorescence hyperspectral images were corrected using dark and white reference images. A region of interest (ROI) of the spectra was extracted for the HDPE and SS coupon regions of the test group and the control group from the calibrated hyperspectral images. For the next step, the extracted spectra were discriminated and analyzed through various discriminant analyses. Discriminant analyses included decision trees (DTs), *k*-nearest neighbor (*k*-NN) analysis, linear discriminant analysis (LDA), and partial least-squares discriminant analysis (PLS-DA).

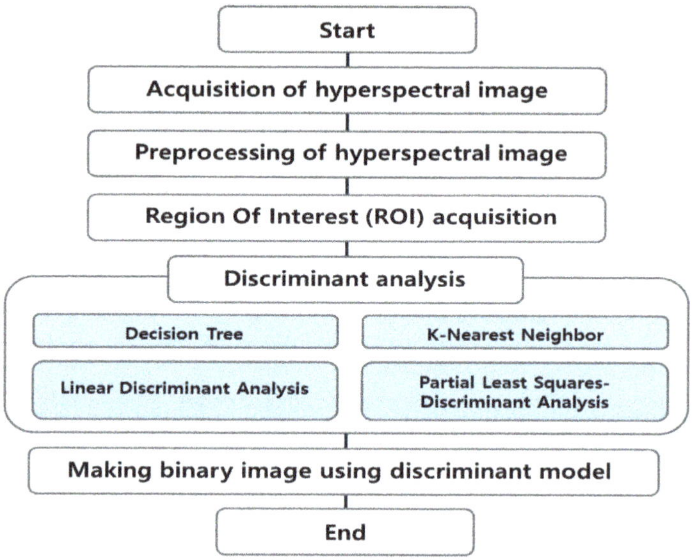

Figure 3. Flowchart of detecting bacterial biofilms using discriminant analyses.

2.5.1. Decision Tree

A DT predicts a class by plotting decision rules in a tree structure and classifying samples into several smaller groups. It is a top-down approach, where classes are divided by the partitioning rule until the stop criterion is met [56]. A DT is a popular supervised learning model for classification and regression because of its easy interpretation. However, it is easy to over-fit the training data, so tuning hyper-parameters (e.g., partitioning rule and stop criteria) is important.

2.5.2. k-Nearest Neighbor

The *k*-NN classifies samples into the closest class based on the distance between samples in the feature space. It is named in that it predicts values from *k* neighbors [57]. The distance between samples is measured through the Euclidean distance. The Euclidean distance between the datum x and y is calculated using Equation (2):

$$d(x,y) = \sqrt{(x_1 - y_1)^2 + (x_2 - y_2)^2 + \cdots + (x_n - y_n)^2} \quad (2)$$

where d represents the distance and n is the number of features. In the *k*-NN for classification, the input data are predicted as the largest number of class of the closest k data. As the value of k increases, the effect of noise can decrease, but the boundary between classes becomes unclear. Therefore, a process of finding proper *k*-value through repeated experiments is required [58].

2.5.3. Linear Discriminant Analysis

LDA reduces the dimension of feature vectors by maximizing the ratio of variance between classes and within classes [59–61]. It means that LDA finds the optimal transformation matrix (w) that maximizes the criterion function W_{lda}, which is the ratio of the within-class scatter (S_w) and between-class scatter (S_B) like Equations (3)–(5):

$$W_{lda} = \frac{|W^T S_B W|}{|W^T S_W W|} \tag{3}$$

$$S_w = \sum_{i=1}^{n} \sum_{x \in n_i} (x - m_i)(x - m_i)^T \tag{4}$$

$$S_B = \sum_{i=1}^{n} k_i (m_i - m)(m_i - m)^T \tag{5}$$

where n is the number of classes, n_i is a set of data belonging to the i-th class, m_i is the mean of the i-th class, and k_i is the size of n_i.

2.5.4. Partial Least Squares Discriminant Analysis

PLS-DA is an analysis technique based on PLSR that classifies predicted regression models using threshold values. Although PLS-DA and PLSR are the same analysis methods, PLSR uses continuous dependent variables (e.g., spectra) to develop and predict regression models, and PLS-DA is a variant version for categorical prediction models [62]. PLSR model is calculated using Equations (6) and (7):

$$X = TP^T + E \tag{6}$$

$$Y = UQ^T + F \tag{7}$$

where n is the number of samples, p is the number of variables, $X = (n \times p)$ matrix, $Y = (n \times 1)$ matrix, T *and* U are $(n \times p)$ score matrices of latent vectors, P and Q are matrices of loading, and E and F are the error terms (residuals). PLS-DA is performed by applying the score obtained through PLSR to discriminant analysis.

2.5.5. Biofilm Detecting Performance

The development of an *E. coli* and *S. typhimurium* biofilm prediction model and detection of biofilm regions through discriminant analyses were performed using the open statistical software R (ver. 2019; R Foundation, Vienna, Austria) and the commercial software MATLAB (ver. 2018; MathWorks Inc., Matick, MA, USA). To develop a biofilm detection model, 80% of the spectra extracted from the hyperspectral images were used to develop a prediction model, and the remaining 20% of the spectra were used to verify the biofilm prediction performance. Fivefold cross-validation was performed to prevent over-fitting, and the prediction performance was calculated using Equations (8) and (9):

$$\text{Control group performance (Specificity)} = \frac{TN}{TN + FP} \tag{8}$$

$$\text{Experimental group performance (Sensitivity)} = \frac{TP}{TP + FN} \tag{9}$$

where TP (true positive) is the frequency of accurately predicting the area where the biofilm was formed, FP (false positive) is the frequency of erroneously predicting the area where the biofilm was not formed, TN (true negative) is the frequency of accurately predicting the region where the biofilm was not formed, and FN (false negative) is the frequency of erroneous prediction of the region where the biofilm was formed.

Additionally, the receiver operating characteristics (ROC) curve was drawn and the area under the curve (AUC) was calculated for choosing the best model. The ROC curve is the plot with the true-positive rate against the false-positive rate, and yjr AUC is the area under the ROC curve. After selecting the model with the highest performance, the biofilm detection result was validated by applying the model to hyperspectral images that were not used for model development.

3. Results and Discussion

3.1. Fluorescence Characteristics of Food Poisoning Bacteria

Figure 4 shows fluorescence emission spectra obtained by the microplate from 400 to 800 nm bands for excitation light from 350 to 400 nm bands. Figure 4a,b is the fluorescence emission spectra of *E. coli* and *S. typhimurium,* respectively. It can be seen that both *E. coli* and *S. typhimurium* cultures exhibited a high fluorescence expression intensity in 400–450 nm wavelength bands.

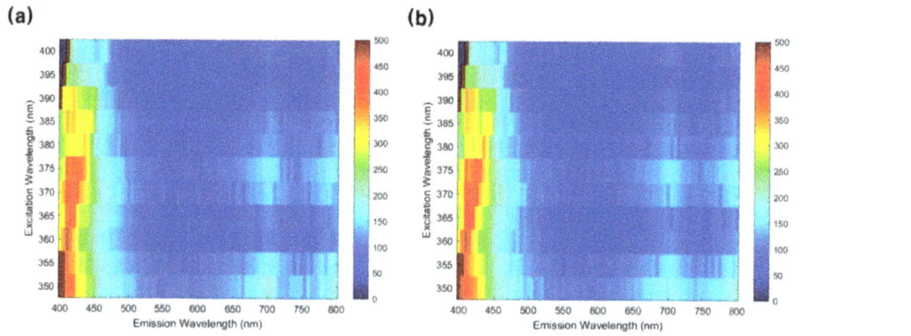

Figure 4. Fluorescence emission results from 400 to 800 nm bands for the excitation light from 350 to 400 nm bands: (**a**) *E. coli* and (**b**) *S. typhimurium*.

Figure 5 shows the fluorescence emission spectra for each concentration of food poisoning bacteria for 365 nm excitation light. Figure 5a,b shows the fluorescence emission spectra of *E. coli* and *S. typhimurium* concentrations controlled at 10^5, 10^4, 10^3, 10^2, 10^1, and 10^0 (PBS) CFU·mL^{-1}, respectively. It can be seen that the fluorescence intensity is high in the 400–450 nm wavelength bands and around 700 nm bands. It was also confirmed that both *E. coli* and *S. typhimurium* showed high fluorescence intensity in the 415 nm wavelength band according to the bacterial concentration. In the case of the band around 700 nm, the bacterial culture medium showed higher fluorescence intensity than the PBS solution, but there was no difference in fluorescence intensity according to the bacterial concentration.

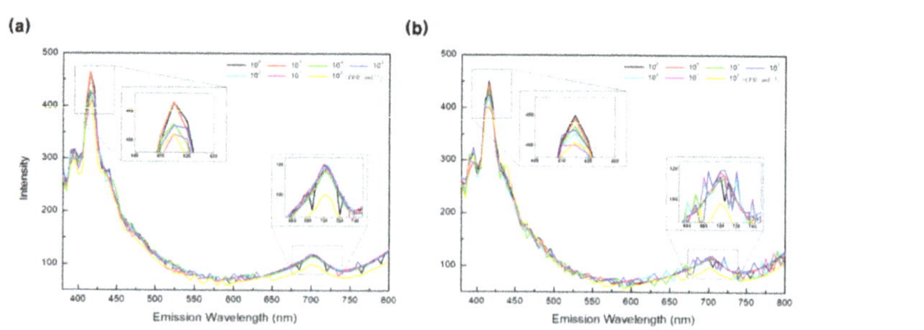

Figure 5. Fluorescence emission spectra for six concentrations (10^5, 10^4, 10^3, 10^2, 10^1, and 10^0 CFU·mL^{-1}) of food poisoning bacteria for 365 nm excitation light: (**a**) *E. coli* and (**b**) *S. typhimurium*.

3.2. Food Poisoning Bacteria Biofilm Formation

As a result of measuring the number of bacteria using the standard plate count method and dry-film method, E. coli was successfully cultured on 16 pieces of HDPE coupons and

26 pieces of SS coupons, while *S. typhimurium* was cultured on 23 pieces of HDPE coupons and 26 pieces of SS coupons.

Figure 6 shows the number of bacteria measured for each culture. The number of *E. coli* per HDPE coupon ranged from 0.78 to 3.94 log CFU·cm^{-2}, whereas the number of *E. coli* per SS coupon ranged from 0.78 to 3.51 log CFU·cm^{-2}. In the case of *S. typhimurium*, 2.4 to 4.99 log CFU·cm^{-2} bacteria were formed on HDPE coupons and 1.93 to 6 log CFU·cm^{-2} bacteria were formed on SS coupons.

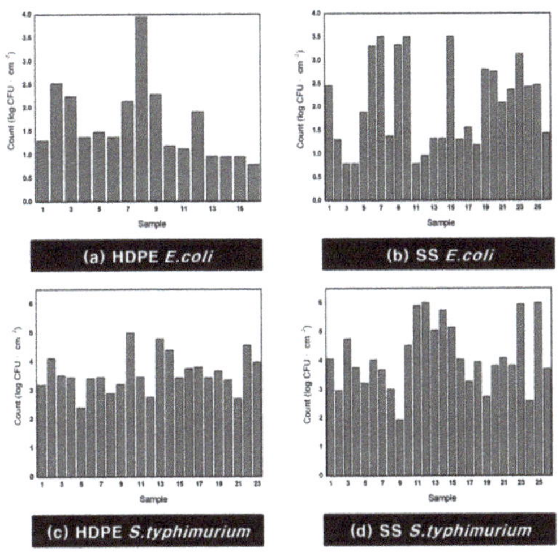

Figure 6. Cultivated *E. coli* and *S. typhimurium* count graph: (**a**) *E. coli* cultivated on high-density polyethylene (HDPE), (**b**) *E. coli* cultivated on stainless steel (SS), (**c**) *S. typhimurium* cultivated on HDPE, and (**d**) *S. typhimurium* cultivated on SS.

Figure 7a,b shows the RGB hyperspectral images of the 551.8 nm band for the test group coupons. In the case of RGB images, areas were not visually separated based on whether the biofilm was formed or not. In the case of hyperspectral images, the intensity difference was not significant, but it is difficult to distinguish a biofilm region using only one wavelength image.

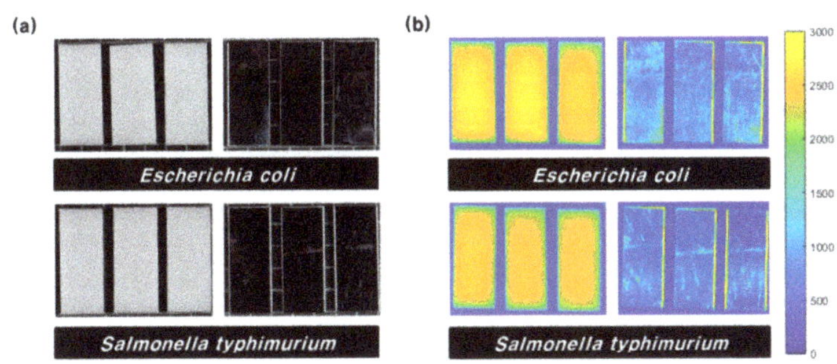

Figure 7. HDPE and SS coupons of bacterial biofilms: (**a**) RGB images and (**b**) hyperspectral images of the 551.8 nm band.

Spectrum extraction from the hyperspectral image was performed using the area contained in the medium. A total of 183,212 spectra were extracted for use in model development. In the case of *E. coli*, 17,185 spectra were extracted from 10 of 16 HDPE coupons and 25,926 spectra were extracted from 15 of 26 SS coupons for model development.

A total of 31,921 spectra were extracted from 18 of 23 HDPE coupons and 34,600 spectra were extracted from 18 of 26 coupons for the *S. typhimurium* model. As a control group, 19,662 and 15,709 spectra were extracted from 12 HDPE and 12 SS coupons in the case of the *E. coli* model, respectively. In the case of the *S. typhimurium* model, 19,053 and 19,156 spectra were extracted from 12 HDPE and 12 SS coupons, respectively. Consequently, 36,847 spectra were used to develop a model to detect *E. coli* on HDPE coupons, and 41,635 spectra were used to detect *E. coli* on SS coupons. In the case of *S. typhimurium*, 50,974 and 53,756 spectra were used for detection on HDPE and SS coupons, respectively.

3.3. Biofilm Detection Model

Figure 8 shows the average value of the extracted pixel spectra to develop a biofilm detection model. Figure 8a,b shows the average spectrum of *E. coli* cultured on HDPE and SS coupons, and Figure 8c,d shows the average spectrum of *S. typhimurium* cultured on HDPE and SS coupons. Both HDPE and SS coupons showed the highest fluorescence signal near 550 nm, and the fluorescence signal of SS coupons was low in all wavelength regions compared to HDPE coupons. It can be seen that the fluorescence signal intensity of the region where the biofilm is formed appears higher than that of the control region for all wavelength ranges. However, the wavelength bands could not indicate a distinct difference between the experimental group and the control group, except for the *E. coli* biofilm formed on SS coupons. Moreover, the spectral deviation was severe for each pixel, so it was be difficult to distinguish the two groups with only part of the wavelengths.

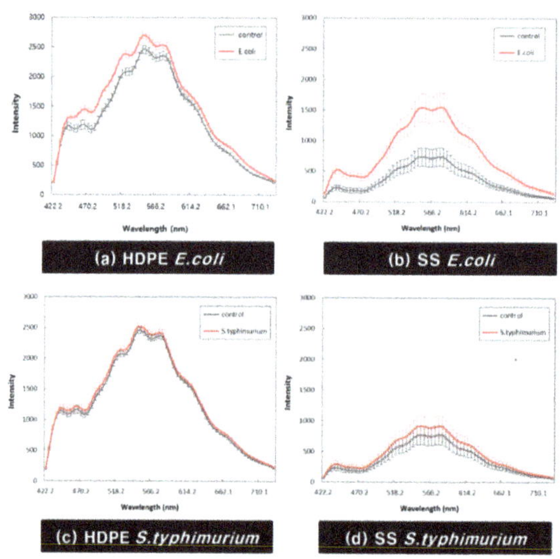

Figure 8. Average spectra of extracted pixel spectra from hyperspectral images: (**a**) *E. coli* cultivated on HDPE, (**b**) *E. coli* cultivated on SS, (**c**) *S. typhimurium* cultivated on HDPE, and (**d**) *S. typhimurium* cultivated on SS.

The biofilm detection model was developed by applying discriminant analyses using multiple wavelength bands. The DT, *k*-NN, LDA, and PLS-DA methods were used, which are widely used in discriminant analysis. After dividing the total amount of data in the 80:20 ratio, the model was trained using 80% of the data and the performance of the model was verified using the remaining 20% of the data. Fivefold cross-validation was performed, and average performances of specificity and sensitivity were calculated. Table 1 shows the results of discriminant analyses after classifying the test group's spectrum by 1 and the control group's spectrum by 0.

Table 1. Discriminant model performance.

				Performance (%)											
				1		2		3		4		5		Average	
				spe	sen	spe	sen	spe	sen	spe	sen	spe	sen	spe	sen
HDPE	*E. coli*	DT	*Train*	97.05	94.09	91.07	95.89	96.64	96.53	94.33	95.04	94.23	95.07	94.66	95.32
			Test	97.06	94.86	91.33	95.79	96.52	96.04	93.53	94.88	94.03	94.61	94.49	95.24
		k-NN	*Train*	99.99	100	100	100	99.99	100	99.99	100	99.99	100	**99.99**	**100**
			Test	99.97	100	100	100	100	100	99.94	100	100	100	**99.98**	**100**
		LDA	*Train*	99.82	100	99.81	100	99.78	99.99	99.79	99.99	99.75	100	99.79	**100**
			Test	99.65	99.97	99.71	100	99.85	99.97	99.8	100	99.91	100	99.78	99.99
		PLS-DA	*Train*	96.76	98	96.67	97.97	96.8	98.02	96.78	97.89	96.63	97.61	96.73	97.90
			Test	96.74	98.19	96.66	97.72	96.67	97.78	96.59	97.91	97.06	97.53	96.74	97.83
HDPE	*S. typhimurium*	DT	*Train*	85.4	76.41	84.46	78.66	87.63	69.57	84.81	76.78	86.69	70.64	85.80	74.41
			Test	84.07	75.89	83.71	76.79	87.48	69.84	85.18	75.89	85.85	70.22	85.26	73.73
		k-NN	*Train*	98.9	97.24	98.86	97.36	98.94	97.21	98.97	97.21	98.85	97.22	**98.90**	**97.25**
			Test	97.85	94.83	97.78	94.7	97.64	95.13	98.19	95.29	97.84	94.85	**97.86**	94.96
		LDA	*Train*	96.47	96.26	96.52	96.46	96.61	96.29	96.5	96.27	96.46	96.42	96.51	96.34
			Test	96.65	96.69	96.47	95.96	96.3	96.31	96.5	96.68	96.53	96.15	96.49	**96.36**
		PLS-DA	*Train*	87.17	44.95	86.92	45.42	87.15	45.11	87.01	44.68	87.23	45.19	87.10	45.07
			Test	87.13	45.25	86.69	46.02	87.1	44.78	88.25	44.76	86.47	44.41	87.13	45.04
SS	*E. coli*	DT	*Train*	95.93	85.35	96.31	84.88	96.24	84.77	95.84	85.33	96.29	84.92	96.12	85.05
			Test	95.55	85.81	96.18	84.26	96.46	84.73	95.9	85.88	96.24	84.13	96.07	84.96
		k-NN	*Train*	98.41	93.43	98.32	93.31	98.38	93.51	98.35	93.37	98.67	94.31	**98.43**	**93.59**
			Test	97.21	92.12	97.51	91.19	97.34	90.84	91.85	97.67	97.35	91.21	**96.25**	92.61
		LDA	*Train*	92.15	92.9	92.09	93.2	92.01	93	92	92.87	92.19	92.93	92.09	92.98
			Test	91.9	93.16	92.46	91.9	92.47	92.62	91.86	93.52	91.94	92.95	92.13	**92.83**
		PLS-DA	*Train*	91.66	89.2	91.58	89.47	91.54	89.41	91.53	89.41	91.43	89.44	91.55	89.39
			Test	90.14	91.39	90.96	89.26	92.05	89.06	91.39	89.48	92.01	88.99	91.31	89.64
SS	*S. typhimurium*	DT	*Train*	99.19	9.17	92.82	23.26	99.18	9.2	99.41	8.43	99.45	8.41	**98.01**	11.69
			Test	99.22	9.17	92.73	21.72	99.28	9.02	99.4	8.44	99.25	8.51	**97.98**	11.37
		k-NN	*Train*	94.37	58.83	94.27	58.26	94.58	58.12	94.58	58.2	94.78	58.41	94.52	**58.36**
			Test	85.65	38.29	85.67	40.77	84.42	40.36	85.26	38.41	85.75	39.5	85.35	**39.47**
		LDA	*Train*	91.05	26.42	90.76	26.84	91.02	26.74	91.15	26.31	91.12	26	91.02	26.46
			Test	91.4	26.12	90.83	26.02	90.86	25.46	90.9	26.86	91.07	27.47	91.01	26.39
		PLS-DA	*Train*	93.89	20.81	93.64	21.27	93.93	20.82	93.93	20.88	94.02	20.55	93.88	20.87
			Test	94.36	20.65	93.69	20.38	93.77	20.87	93.53	20.93	94.01	21.33	93.87	20.83

In general, discriminant model performances of specificity and sensitivity were higher than 90%, except the *S. typhimurium* model for SS coupons. The performances of the detection model for the biofilm formed on HDPE coupons was higher than that of the biofilm formed on SS coupons. It is estimated that the edge part of SS coupons reflected fluorescence, which made model learning difficult. In addition, it was found that the models used on *E. coli* had higher detection performances than those used on *S. typhimurium*. In particular, the *k*-NN model average specificity in the test group was 99.98% and 97.86% and the average sensitivity was 100% and 94.96% for the HDPE coupons of *E. coli* and *S. typhimurium*, respectively. The average specificity of the *k*-NN model in the test group was 96.25% and 85.35% and the average sensitivity was 92.61% and 39.47% for the SS coupons of *E. coli* and *S. typhimurium*, respectively, showing higher classification performances than other discriminant analyses. However, in some cases, the test group showed better performance than the *k*-NN model. In the case of the *S. typhimurium* model for HDPE coupons, the average sensitivity of LDA in the test group was 96.36%, and for the *E. coli* model for SS coupons, the average sensitivity was 92.83%. In the case of the *S. typhimurium* model for SS coupons, the average specificity of the DT analysis in the test group was 97.98%. Nevertheless, the *k*-NN model showed the highest performance under many conditions.

Then, the ROC curve was drawn (Figure 9), and the AUC was calculated for choosing the best model. In the case of the *E. coli* model for HDPE coupons, the AUC of the DT analysis in the test group was 0.968, that of the *k*-NN analysis was 1, that of LDA was 1, and that of PLS-DA was 0.997. In the case of the *S. typhimurium* model for HDPE coupons, the AUC of the DT analysis in the test group was 0.808, that of the *k*-NN analysis was 0.994, that of LDA was 0.994, and that of PLS-DA was 0.769. In the case of the *E. coli* model for SS coupons, the AUC of the DT analysis in the test group was 0.911, that of the *k*-NN analysis was 0.984, that of LDA was 0.975, and that of PLS-DA was 0.967. In the case of the *S. typhimurium* model for SS coupons, the AUC of the DT analysis in the test group was 0.636, that of the *k*-NN analysis was 0.715, that of LDA was 0.671, and that of PLS-DA was 0.667. Therefore, the *k*-NN algorithm is suitable for detecting *E. coli* and *S. typhimurium* biofilms.

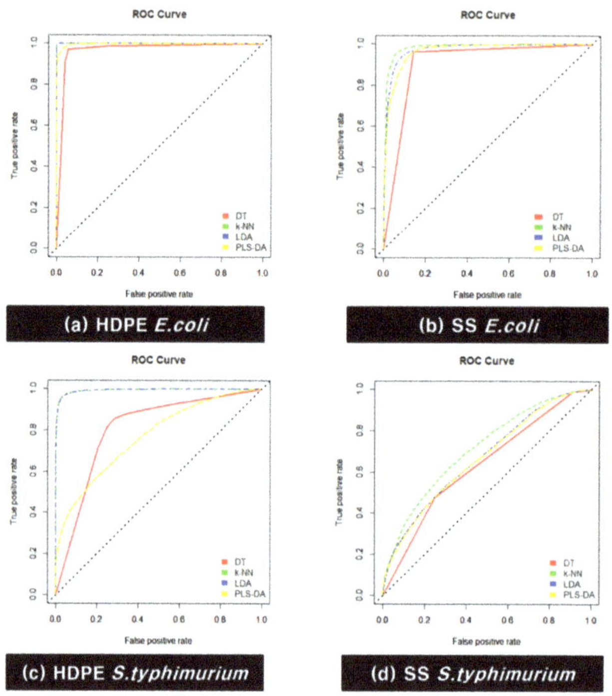

Figure 9. Receiver operating characteristics (ROC) curve of the discriminant model: (**a**) *E. coli* cultivated on HDPE, (**b**) *E. coli* cultivated on SS, (**c**) *S. typhimurium* cultivated on HDPE, and (**d**) *S. typhimurium* cultivated on SS.

3.4. Food Poisoning Bacteria Biofilm Detection Result

Among discriminant analyses, the *k*-NN model, which showed the highest detection performance, was applied to all pixels constituting the hyperspectral image to predict the presence of a biofilm. First, the coupon region area and background area were binarized by setting the ROI in the image. Then, the *k*-NN model was applied to the coupon region area. Figure 10 shows the binarization result of bacterial detection. It was confirmed that the biofilm area, which was difficult to determine with the naked eye, was successfully detected. Jun [52] detected the *E. coli* biofilm on the surface of SS coupons using a short wavelength, and the minimum detection limit of the *E. coli* biofilm was 7 log CFU·cm^{-2}. In addition, the same research group detected the *E. coli* biofilm on the surface of HDPE coupons using a short wavelength and a ratio of two wavelength images in the hyperspectral image. The minimum detection limit of *E. coli* was reported to be 7.56 log CFU·cm^{-2}. Lee [15] detected the *E. coli* biofilm formed on the surface of HDPE coupons using multiple wavelengths and confirmed that the *k*-NN algorithm had the highest detection performance. In this study, biofilms of *E. coli* and *S. typhimurium* were detected using discriminant analyses and the *k*-NN model had the highest detection performance, which agrees with the results reported by Lee [15]. In addition, it was confirmed that the *E. coli* biofilm formed at a concentration of 0.78 log CFU·cm^{-2} and *S. typhimurium* biofilm formed at a concentration of 1.93 log CFU·cm^{-2} could be detected. Using the *k*-NN algorithm, the minimum detection limit was lowered to almost one-tenth of that in previous studies. Recently, as sensor technology has rapidly developed, miniaturization of the hyperspectral imaging system has also progressed. Currently, our research was conducted in a laboratory environment, so there is a limitation that learning models cannot be used directly in the field. However, if a portable hyperspectral system that is capable of implementing our learning model is made, real-time biofilm detection would be possible in the field.

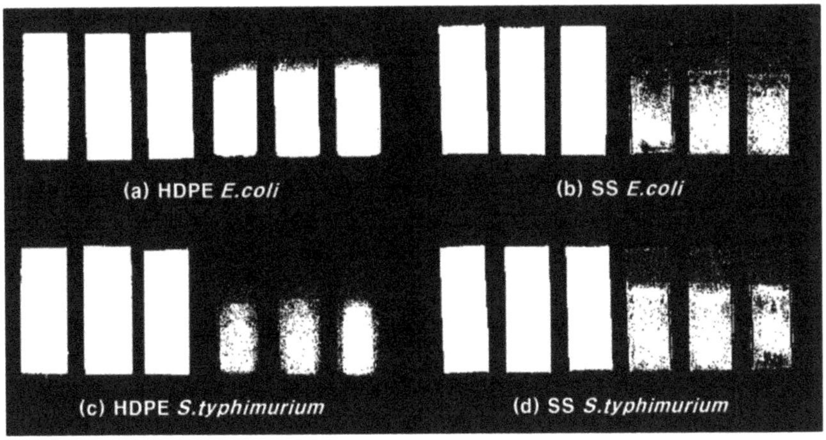

Figure 10. Region of interest of coupons (left) and bacterial biofilm detection result (right): (**a**) *E. coli* cultivated on HDPE, (**b**) *E. coli* cultivated on SS, (**c**) *S. typhimurium* cultivated on HDPE, and (**d**) *S. typhimurium* cultivated on SS.

4. Conclusions

This study was conducted to determine whether biofilms, which are the main cause of cross-contamination of bacteria in agro-food processing facilities, can be quickly inspected using hyperspectral imaging technology and various discriminant analysis techniques. Biofilms were formed by *E. coli* and *S. typhimurium* on the surface of HDPE and SS coupons, which are used as main materials in agro-food processing facilities.

To examine the fluorescence characteristics of food poisoning bacteria, the fluorescence emission intensity of *E. coli* and *S. typhimurium* from 400 to 800 nm bands was acquired for excitation light from 350 to 400 nm bands at 5 nm intervals using a microplate reader. As a result, high fluorescence intensity was confirmed according to the bacterial concentration in the 400–450 nm wavelength bands, and fluorescence expression was also confirmed in the 700 nm band.

Hyperspectral fluorescence images of a 365 nm UV light source in the wavelength range of 420 to 730 nm were acquired using a line scan apparatus. The average spectra of the samples were extracted from the corrected hyperspectral image data. Extracted spectra were analyzed by various discriminant analyses such as DTs, *k*-NN, LDA, and PLS-DA. The *k*-NN algorithm predicted the biofilm region with a high performance of 90% or more. The minimum detection limit of the detected biofilm was 0.78 log CFU$\cdot$cm^{-2} and 1.93 log CFU$\cdot$cm^{-2} for *E. coli* and *S. typhimurium*, respectively. The biofilm detection model using the *k*-NN algorithm was applied to all pixels of the hyperspectral images, and it was found that the biofilm region could be accurately detected. The possibility of real-time biofilm detection in the field using hyperspectral images was confirmed.

In the case of discriminant analysis, there is a disadvantage that the data volume used for model development is large and the model training time is long. These issues can be solved using an optimization technique by selecting wavelength bands essential for model development. In addition, further research is required to secure more samples to reduce the false detection rate.

Author Contributions: Writing—original draft preparation and theoretical analysis, A.L.; conceptualization, S.P. and J.Y.; writing—review and editing, J.K. and J.L.; tests and analysis, Y.S. and B.K.; correction and review of the manuscript, G.K. All authors have read and agreed to the published version of the manuscript.

Funding: This study was carried out with the support of the Research Program for Agricultural Science & Technology Development (Project No. PJ01248701), National Institute of Agricultural Sciences, Rural Development Administration, Republic of Korea.

Institutional Review Board Statement: Not applicable.

Informed Consent Statement: Not applicable.

Data Availability Statement: Data sharing not applicable.

Conflicts of Interest: The authors declare no conflict of interest.

References

1. Srey, S.; Jahid, I.K.; Ha, S.D. Biofilm formation in food industries: A food safety concern. *Food Control* **2013**, *31*, 572–585. [CrossRef]
2. Centers for Disease Control and Prevention. Available online: https://www.cdc.gov/foodsafety/foodborne-germs.html/ (accessed on 14 December 2020).
3. Food and Drug Administration. Available online: https://www.fda.gov/food/outbreaks-foodborne-illness/foodborne-pathogens (accessed on 14 December 2020).
4. Soon, J.M.; Brazier, A.K.; Wallace, C.A. Determining common contributory factors in food safety incidents–A review of global outbreaks and recalls 2008–2018. *Trends Food Sci. Technol.* **2020**, *97*, 76–87. [CrossRef]
5. Jahid, I.K.; Ha, S.-D. A review of microbial biofilms of produce: Future challenge to food safety. *Food Sci. Biotechnol.* **2012**, *21*, 299–316. [CrossRef]
6. Alegbeleye, O.O.; Singleton, I.; San'Ana, A.S. Sources and contamination routes of microbial pathogens to fresh produce during field cultivation: A review. *Food Microbiol.* **2018**, *73*, 177–208. [CrossRef]
7. Guobjoernsdottir, B.; Einarsson, H.; Thorkelsson, G. Microbial adhesion to processing lines for fish fillets and cooked shrimp: Influence of stainless steel surface finish and presence of gram-negative bacteria on the attachment of *Listeria monocytogenes*. *Food Technol. Biotechnol.* **2005**, *43*, 55–61.
8. Teh, K.H.; Flint, S.; Palmer, J.; Andrewes, P.; Bremer, P.; Lindsay, D. Proteolysis produced within biofilms of bacterial isolates from raw milk tankers. *Int. J. Food Microbiol.* **2012**, *157*, 28–34. [CrossRef] [PubMed]
9. Galie, S.; García-Gutiérrez, C.; Miguélez, E.M.; Villar, C.J.; Lombó, F. Biofilms in the food industry: Health aspects and control methods. *Front. Microbiol.* **2018**, *9*, 898. [CrossRef] [PubMed]
10. Abdallah, M.; Benoliel, C.; Drider, D.; Dhulster, P.; Chihib, N.E. Biofilm formation and persistence on abiotic surfaces in the context of food and medical environments. *Arch. Microbiol.* **2014**, *196*, 453–472. [CrossRef]
11. Giaouris, E.; Heir, E.; Hébraud, M.; Chorianopoulos, N.; Langsrud, S.; Møretrø, T.; Habimana, O.; Desvaux, M.; Renier, S.; Nychas, G.-J. Attachment and biofilm formation by foodborne bacteria in meat processing environments: Causes, implications, role of bacterial interactions and control by alternative novel methods. *Meat Sci.* **2014**, *97*, 298–309. [CrossRef]
12. Zore, A.; Bezek, K.; Jevšnik, M.; Abram, A.; Runko, V.; Slišković, I.; Raspor, P.; Kovačević, D.; Bohinc, K. Bacterial adhesion rate on food grade ceramics and Teflon as kitchen worktop surfaces. *Int. J. Food Microbiol.* **2020**, *332*, 108764. [CrossRef]
13. De Jong, A.; Verhoeff-Bakkenes, L.; Nauta, M.; De Jonge, R. Cross-contamination in the kitchen: Effect of hygiene measures. *J. Appl. Microbiol.* **2008**, *105*, 615–624. [CrossRef]
14. El-Liethy, M.A.; Hemdan, B.A.; El-Taweel, G.E. Prevalence of *E. coli*, *Salmonella*, and *Listeria* spp. as potential pathogens: A comparative study for biofilm of sink drain environment. *Orig. Artic.* **2020**, *40*, e12816. [CrossRef]
15. Lee, A.Y.; Seo, Y.W.; Lim, J.G.; Park, S.B.; Yoo, J.Y.; Kim, B.G.; Kim, G.Y. Detection of *E. coli* biofilms with hyperspectral imaging and machine learning techniques. *Korean J. Agric. Sci.* **2020**, *47*, 645–655. [CrossRef]
16. Aryal, M.; Muriana, P.M. Efficacy of Commercial Sanitizers Used in Food Processing Facilities for Inactivation of Listeria monocytogenes, *E. Coli* O157: H7, and *Salmonella* Biofilms. *Foods* **2019**, *8*, 639. [CrossRef]
17. Miao, J.; Liang, Y.; Chen, L.; Wang, W.; Wang, J.; Li, B.; Li, L.; Chen, D.; Xu, Z. Formation and development of Staphylococcus biofilm: With focus on food safety. *J. Food Saf.* **2017**, *37*, e12358. [CrossRef]
18. Davies, D. Understanding biofilm resistance to antibacterial agents. *Nat. Rev. Drug Discov.* **2003**, *2*, 114–122. [CrossRef]
19. Bai, A.J.; Rai, V.R. Bacterial quorum sensing and food industry. *Compr. Rev. Food Sci. Food Saf.* **2011**, *10*, 183–193. [CrossRef]
20. World Health Organization. Available online: https://www.who.int/news-room/fact-sheets/detail/e-coli (accessed on 14 December 2020).
21. Gast, R.K.; Porter, R.E.J. *Salmonella* Infections. In *Diseases of Poultry*, 14th ed.; John Wiley & Sons, Inc.: Hoboken, NJ, USA, 2020; pp. 717–753. [CrossRef]
22. Solomon, E.B.; Niemira, B.A.; Sapers, G.M.; Annous, B.A. Biofilm Formation, Cellulose Production, and Curli Biosynthesis by *Salmonella* Originating from Produce, Animal, and Clinical Sources†. *J. Food Prot.* **2005**, *68*, 906–912. [CrossRef]

23. Basler, C.; Nguyen, T.-A.; Anderson, T.C.; Hancock, T.; Behravesh, C.B. Outbreaks of Human Salmonella Infections Associated with Live Poultry, United States, 1990–2014. *Emerg. Infect. Dis.* **2016**, *22*, 1705–1711. [CrossRef] [PubMed]
24. Crecencio, R.B.; Brisola, M.C.; Bitner, D.; Frigo, A.; Rampazzo, L.; Borges, K.A.; Furian, T.Q.; Salle, C.T.P.; Moraes, H.L.S.; Faria, G.A.; et al. Antimicrobial susceptibility, biofilm formation and genetic profiles of *Escherichia coli* isolated from retail chicken meat. *Infect. Genet. Evol.* **2020**, *84*, 104355. [CrossRef] [PubMed]
25. Huang, K.; Tian, Y.; Tan, J.; Salvi, D.; Karwe, M.; Nitin, N. Role of contaminated organic particles in cross-contamination of fresh produce during washing and sanitation. *Postharvest Biol. Technol.* **2020**, *168*, 111283. [CrossRef]
26. Lopez-Galvez, F.; Gil, M.I.; Truchado, P.; Selma, M.V.; Allende, A. Cross-contamination of fresh-cut lettuce after a short-term exposure during pre-washing cannot be controlled after subsequent washing with chlorine dioxide or sodium hypochlorite. *Food Microbiol.* **2010**, *27*, 199–204. [CrossRef] [PubMed]
27. Dantas, S.T.; Rossi, B.F.; Bonsaglia, E.C.; Castilho, I.G.; Hernandes, R.T.; Fernandes, A.; Rall, V.L. Cross-contamination and biofilm formation by *Salmonella enterica* serovar Enteritidis on various cutting boards. *Foodborne Pathog. Dis.* **2018**, *15*, 81–85. [CrossRef] [PubMed]
28. Chia, T.; Goulter, R.; McMeekin, T.; Dykes, G.; Fegan, N. Attachment of different *Salmonella* serovars to materials commonly used in a poultry processing plant. *Food Microbiol.* **2009**, *26*, 853–859. [CrossRef]
29. Zhao, P.; Zhao, T.; Doyle, M.; Rubino, J.; Meng, J. Development of a model for evaluation of microbial cross-contamination in the kitchen. *J. Food Prot.* **1998**, *61*, 960–963. [CrossRef] [PubMed]
30. Carrasco, E.; Morales-Rueda, A.; García-Gimeno, R.M. Cross-contamination and recontamination by *Salmonella* in foods: A review. *Food Res. Int.* **2012**, *45*, 545–556. [CrossRef]
31. Moore, C.M.; Sheldon, B.W.; Jaykus, L.A. Transfer of *Salmonella* and Campylobacter from stainless steel to romaine lettuce. *J. Food Prot.* **2003**, *66*, 2231–2236. [CrossRef]
32. Gkana, E.; Lianou, A.; Nychas, G.-J.E. Transfer of *Salmonella enterica* Serovar Typhimurium from beef to tomato through kitchen equipment and the efficacy of intermediate decontamination procedures. *J. Food Prot.* **2016**, *79*, 1252–1258. [CrossRef]
33. Rateni, G.; Dario, P.; Cavallo, F. Smartphone-based food diagnostic technologies: A review. *Sensors* **2017**, *17*, 1453. [CrossRef]
34. Jayan, H.; Pu, H.; Sun, D.W. Recent development in rapid detection techniques for microorganism activities in food matrices using bio-recognition: A review. *Trends Food Sci. Technol.* **2020**, *95*, 233–246. [CrossRef]
35. González-Rivas, F.; Ripolles-Avila, C.; Fontecha-Umaña, F.; Ríos-Castillo, A.G.; Rodríguez-Jerez, J.J. Biofilms in the spotlight: Detection, quantification, and removal methods. *Compr. Rev. Food Sci. Food Saf.* **2018**, *17*, 1261–1276. [CrossRef]
36. Azeredo, J.; Azevedo, N.F.; Briandet, R.; Cerca, N.; Coenye, T.; Costa, A.R.; Desvaux, M.; Di Bonaventura, G.; Hébraud, M.; Jaglic, Z. Critical review on biofilm methods. *Crit. Rev. Microbiol.* **2017**, *43*, 313–351. [CrossRef]
37. Liu, Y.; Pu, H.; Sun, D.W. Hyperspectral imaging technique for evaluating food quality and safety during various processes: A review of recent applications. *Trends Food Sci. Technol.* **2017**, *69*, 25–35. [CrossRef]
38. Feng, C.-H.; Makino, Y.; Oshita, S.; Martín, J.F.G. Hyperspectral imaging and multispectral imaging as the novel techniques for detecting defects in raw and processed meat products: Current state-of-the-art research advances. *Food Control* **2018**, *84*, 165–176. [CrossRef]
39. Wakholi, C.; Kandpal, L.M.; Lee, H.; Bae, H.; Park, E.; Kim, M.S.; Mo, C.; Lee, W.H.; Cho, B.K. Rapid assessment of corn seed viability using short wave infrared line-scan hyperspectral imaging and chemometrics. *Sens. Actuators B Chem.* **2018**, *255*, 498–507. [CrossRef]
40. Qin, J.; Chao, K.; Kim, M.S.; Lu, R.; Burks, T.F. Hyperspectral and multispectral imaging for evaluating food safety and quality. *J. Food Eng.* **2013**, *118*, 157–171. [CrossRef]
41. Ye, X.; Iino, K.; Zhang, S. Monitoring of bacterial contamination on chicken meat surface using a novel narrowband spectral index derived from hyperspectral imagery data. *Meat Sci.* **2016**, *122*, 25–31. [CrossRef] [PubMed]
42. Quintelas, C.; Mesquita, D.P.; Lopes, J.A.; Ferreira, E.C.; Sousa, C. Near-infrared spectroscopy for the detection and quantification of bacterial contaminations in pharmaceutical products. *Int. J. Pharm.* **2015**, *492*, 199–206. [CrossRef] [PubMed]
43. Chen, C.Q.; Jiang, Q.Q.; Zhang, Z.C.; Shi, P.F.; Xu, Y.; Liu, B.; Xi, J.; Chang, S.Z. Hyperspectral Inversion of Petroleum hydrocarbon Contents in Soil Based on Continuum Removal and Wavelet Packet Decomposition. *Sustainability* **2020**, *12*, 4218. [CrossRef]
44. Pu, Y.Y.; Feng, Y.Z.; Sun, D.W. Recent progress of hyperspectral imaging on quality and safety inspection of fruits and vegetables: A review. *Compr. Rev. Food Sci. Food Saf.* **2015**, *14*, 176–188. [CrossRef]
45. Liu, Y.; Sun, D.W.; Cheng, J.H.; Han, Z. Hyperspectral imaging sensing of changes in moisture content and color of beef during microwave heating process. *Food Anal. Methods* **2018**, *11*, 2472–2484. [CrossRef]
46. Wu, D.; Sun, D.W. Potential of time series-hyperspectral imaging (TS-HSI) for non-invasive determination of microbial spoilage of salmon flesh. *Talanta* **2013**, *111*, 39–46. [CrossRef] [PubMed]
47. ElMasry, G.; Wang, N.; Vigneault, C. Detecting chilling injury in Red Delicious apple using hyperspectral imaging and neural networks. *Postharvest Biol. Technol.* **2009**, *52*, 1–8. [CrossRef]
48. Huang, Z.; Turner, B.J.; Dury, S.J.; Wallis, I.R.; Foley, W.J. Estimating foliage nitrogen concentration from HYMAP data using continuum removal analysis. *Remote Sens. Environ.* **2004**, *93*, 18–29. [CrossRef]
49. Malenovský, Z.; Homolová, L.; Zurita-Milla, R.; Lukeš, P.; Kaplan, V.; Hanuš, J.; Gastellu-Etchegorry, J.-P.; Schaepman, M.E. Retrieval of spruce leaf chlorophyll content from airborne image data using continuum removal and radiative transfer. *Remote Sens. Environ.* **2013**, *131*, 85–102. [CrossRef]

50. Bonah, E.; Huang, X.; Aheto, J.H.; Osae, R. Application of Hyperspectral Imaging as a Nondestructive Technique for Foodborne Pathogen Detection and Characterization. *Foodborne Pathog. Dis.* **2019**, *16*, 712–722. [CrossRef]
51. Foca, G.; Ferrari, C.; Ulrici, A.; Sciutto, G.; Prati, S.; Morandi, S.; Brasca, M.; Lavermicocca, P.; Lanteri, S.; Oliveri, P. The potential of spectral and hyperspectral-imaging techniques for bacterial detection in food: A case study on lactic acid bacteria. *Talanta* **2016**, *153*, 111–119. [CrossRef]
52. Jun, W.; Kim, M.S.; Cho, B.K.; Millner, P.D.; Chao, K.; Chan, D.E. Microbial biofilm detection on food contact surfaces by macro-scale fluorescence imaging. *J. Food Eng.* **2010**, *99*, 314–322. [CrossRef]
53. Siripatrawan, U.; Makino, Y.; Kawagoe, Y.; Oshita, S. Rapid detection of *Escherichia coli* contamination in packaged fresh spinach using hyperspectral imaging. *Talanta* **2011**, *85*, 276–281. [CrossRef]
54. Zhu, F.; Yao, H.; Hruska, Z.; Kincaid, R.; Brown, R.L.; Bhatnagar, D.; Cleveland, T.E. Integration of fluorescence and reflectance visible near-infrared (VNIR) hyperspectral images for detection of aflatoxins in corn kernels. *Am. Soc. Agric. Biol. Eng.* **2016**, *59*, 785–794. [CrossRef]
55. Kim, M.S.; Chen, Y.R.; Mehl, P.M. Hyperspectral reflectance and fluorescence imaging system for food quality and safety. *Trans. ASAE* **2001**, *44*, 721–729. [CrossRef]
56. Safavian, S.R.; Landgrebe, D. A survey of decision tree classifier methodology. *IEEE Trans. Syst. Man Cybern.* **1991**, *21*, 660–674. [CrossRef]
57. Fix, E.; Hodges, J.L. Discriminatory Analysis, Nonparametric Discrimination: Consistency Properties. *Int. Stat. Rev.* **1951**, *57*, 238–247. [CrossRef]
58. Everitt, B.S.; Landau, S.; Leese, M.; Stahl, D. Miscellaneous clustering methods. In *Clustering Methods*, 5th ed.; John Wiley & Sons, Ltd.: Chichester, UK, 2011; pp. 215–255. [CrossRef]
59. Kim, H.C.; Kim, D.J.; Bang, S.Y. Face recognition using LDA mixture model. *Pattern Recognit. Lett.* **2003**, *24*, 2815–2821. [CrossRef]
60. Ye, J. Least squares linear discriminant analysis. In Proceedings of the 24th International Conference on Machine Learning (ICML '07), New York, NY, USA, 20–24 June 2007; pp. 1087–1093. [CrossRef]
61. Balakrishnama, S.; Ganapathiraju, A. Linear discriminant analysis-a brief tutorial. *Inst. Signal Inf. Process.* **1998**, *18*, 1–8.
62. Jang, K.J.; Jung, S.Y.; Go, I.H.; Jeong, S.H. Discrimination model for cultivation origin of paper mulberry bast fiber and Hanji based on NIR and MIR spectral data combined with PLS-DA. *Anal. Sci. Technol.* **2019**, *32*, 7–16. [CrossRef]

sensors

Article

Deep Spectral-Spatial Features of Near Infrared Hyperspectral Images for Pixel-Wise Classification of Food Products

Hongyan Zhu [1], Aoife Gowen [2], Hailin Feng [3], Keping Yu [4] and Jun-Li Xu [2,*]

1 College of Electronic Engineering, Guangxi Normal University, Guilin 541004, China; hyzhu@mailbox.gxnu.edu.cn

2 UCD School of Biosystems and Food Engineering, University College of Dublin (UCD), Belfield, Dublin 4, Ireland; aoife.gowen@ucd.ie

3 School of Information Engineering, Zhejiang Agricultural and Forestry University, Hangzhou 310000, China; hlfeng@zafu.edu.cn

4 Global Information and Telecommunication Institute, Waseda University, Shinjuku, Tokyo 169-8050, Japan; keping.yu@aoni.waseda.jp

* Correspondence: junli.xu@ucd.ie

Received: 14 August 2020; Accepted: 14 September 2020; Published: 17 September 2020

Abstract: Hyperspectral imaging (HSI) emerges as a non-destructive and rapid analytical tool for assessing food quality, safety, and authenticity. This work aims to investigate the potential of combining the spectral and spatial features of HSI data with the aid of deep learning approach for the pixel-wise classification of food products. We applied two strategies for extracting spatial-spectral features: (1) directly applying three-dimensional convolution neural network (3-D CNN) model; (2) first performing principal component analysis (PCA) and then developing 2-D CNN model from the first few PCs. These two methods were compared in terms of efficiency and accuracy, exemplified through two case studies, i.e., classification of four sweet products and differentiation between white stripe ("myocommata") and red muscle ("myotome") pixels on salmon fillets. Results showed that combining spectral-spatial features significantly enhanced the overall accuracy for sweet dataset, compared to partial least square discriminant analysis (PLSDA) and support vector machine (SVM). Results also demonstrated that spectral pre-processing techniques prior to CNN model development can enhance the classification performance. This work will open the door for more research in the area of practical applications in food industry.

Keywords: hyperspectral; spatial-spectral features; classification; principal component analysis; convolutional neural network

1. Introduction

Hyperspectral imaging (HSI) was originally developed in the early 1970 for remote sensing applications [1]. The invention of the first charge-coupled device (CCD) detector played a crucial role in pushing this technology forward. In recent years, the technology has been reported to have applications in many diverse fields such as forensic science [2], pharmaceutical research [3], agriculture [4], and food science [5]. HSI goes beyond traditional imaging techniques by integrating spectral and spatial information from an object [6]. Therefore, the merits of spectroscopy and computer vision are both reflected in hyperspectral imaging. Spectroscopy identifies the analyte of interest based on its spectral signature, and imaging transforms this information into distribution maps for spatial visualization.

A key step in the successful implementation of HSI applications is the development of calibration models. For a classification task on near infrared hyperspectral imaging dataset acquired from benchtop instruments, chemometrics is currently considered as a popular tool that has been used for many years. Partial least squares discriminant analysis (PLSDA) [7] is a supervised classification modelling method that uses the PLS algorithm to predict the belonging of a sample to a specific class. PLSDA is increasingly used in hyperspectral data analysis for classification problems due to its capability to deal with multicollinearity problem in near infrared (NIR) spectra, which occurs because of very high intercorrelation between absorbances [8,9]. Nevertheless, only spectral features were used as input for classification models in most cases.

Machine learning (ML) techniques have been introduced for HSI data classification [10], which have been collected in an extensive list of detailed reviews, such as Li, et al. [11,12]. The ML field has experienced a significant revolution thanks to the development of new deep learning (DL) models since the early 2000s [13], which is supported by advances in computer technology. These models have gained popularity in the development of HSI classifiers [13,14]. For instance, support vector machine (SVM) was applied to HSI data for strawberry ripeness evaluation achieving classification accuracy over 85%. Convolutional neural network (CNN), being the current state-of-the-art in deep learning [15], first achieved success in the field of image recognition and has become an extremely popular tool for remotely sensed HSI data classification [10]. More importantly, CNN models show flexibility to deal with HSI data by introducing a one-dimensional CNN for processing spectral inputs [16], two-dimensional CNN for single or multiple wavelength images [17], and three-dimensional CNN (3D-CNN) for an intelligent combination of spectral and spatial image data [18,19]. Although CNN models have been successfully implemented for remote sensing applications, they are not often applied to HSI data of food products. Earlier this year, Al-Sarayreh, et al. [20] reported that 3D-CNN model approach applied to HSI data significantly enhanced the overall accuracy of red meat classification.

In this context, the current work aims to investigate the advantages and disadvantages of applying deep learning approaches to near infrared HSI data. The main objective is to compare CNN based modelling strategies against traditional chemometric (i.e., PLSDA) and machine learning (i.e., SVM) methods for the pixel-wise classification tasks of food products. We will apply a hybrid framework which involves first performing principal component analysis (PCA) to highlight the major spectral variation and then building 2-D CNN model from the first few PCs using spatial features. PLSDA, SVM and 3-D CNN are also applied for classifying HSI image data. The performance will be evaluated and compared in terms of efficiency and accuracy, exemplified through two case studies, i.e., classification of four sweet products and differentiation between white stripe ("myocommata") and red muscle ("myotome") pixels on salmon fillets.

2. Materials and Methods

2.1. Sweets Dataset

Spectral images of the sweet samples are acquired in the reflectance mode by employing a laboratory-based pushbroom hyperspectral imaging system. The main components of this system are an imaging spectrograph (Specim N17E, Spectral Imaging Ltd., Oulu, Finland) and an InGaAs camera (InGAs 12-bit SU320MS-1.7RT Sensors Unlimited, Inc., Princeton, NJ, USA). This configuration captures an image of a line across the sample, spanning the 320-pixel width of the sensor, and the spectrograph produces a spectrum for each of these pixels across the other dimension of the array, accounting for a two-dimensional image. The wavelength interval is 7 nm in the spectral range of 943–1643 nm, leading to 101 spectral bands. Direct reflectance spectra are used for subsequent data analysis.

This dataset consists of NIR hyperspectral images of four different sweets with different shapes, colors, and nutritional compositions among classes. Specifically, the details of the selected four products are as follows: raspberry flavor mushroom in pink and white color with a mushroom shape, mint humbugs in brown and golden stripe with an ellipse shape, teeth and lips in pink and white color with a teeth-like appearance, and tub in brown with a cola bottle shape. These sweets, made and purchased from Tesco Ireland Ltd., are labelled as raspberry flavor mushroom (RFM), Mint, Teeth, and Tub, respectively. Sweet samples are chosen because the different spatial information among classes has great potential for improving classification performance.

Four hypercubes of each sample type are obtained with the mean image showing in Figure 1, together with the representative red/green/blue (RGB) images captured by a computer vision system as described in Xu and Sun [21]. As seen, the first three samples of each type are selected as the training set for model development, while the remaining one serves as the validation set, leading to 12 hypercubes consisting of 27,807 pixels for training and 4 hypercubes of 8947 pixels for validation purpose. In addition to this, the developed model is tested on two mixed images containing two samples of each material. The mean images of the mixture are shown in Figure S1 of Supplementary Materials.

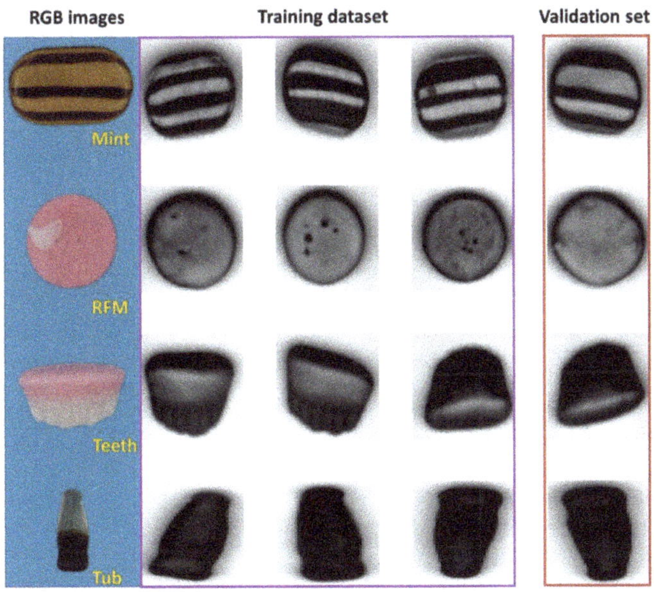

Figure 1. The RGB images of sweet samples and mean images at the spectral domain.

2.2. Salmon Dataset

NIR spectral images of farm-raised Atlantic salmon (*Salmon salar*) fillets are also collected in the reflectance configuration. The core components of the system included: an imaging spectrograph (ImSpector, N17E, Spectral Imaging Ltd., Oulu, Finland) collecting spectral images in a wavelength range of 900–1700 nm, a high performance camera with C-mount lens (Xeva 992, Xenics Infrared Solutions, Leuven, Belgium), two tungsten-halogen illuminating lamps (V-light, Lowel Light Inc., New York, NY, USA), a translation stage operated by a stepper motor (GPL-DZTSA-1000-X, Zolix Instrument Co., Beijing, China), and a computer installed with a data acquisition software (SpectralCube, Spectral Imaging Ltd., Oulu, Finland). Each fillet was individually placed on the moving table and then was scanned line by line at a speed of 2.7 cm/s adjusted to provide the same vertical and horizontal resolution (0.58 mm/pixel).

Salmon is valued as a fat-rich fish with a large proportion of lipids congregated in the white stripe of connective tissue ("myocommata"), segmenting the red-colored muscle ("myotome") tissue in vertical blocks and presenting a zebra-like appearance [22]. Previous study has demonstrated that the proportion of myocommata in a salmon fillet correlated well with its fat content [23]. In this sense, it is interesting to classify the white stripe from the red muscle because the proportions of the white strip in one fillet might contain some valuable information about the fat content and/or lipid oxidation [24].

The salmon dataset is of interest because there is spatial information on the salmon surface, yet it is unsure if this spatial information could help classification.

Overall, six salmon fillets are used with the mean images shown in Figure 2. The first three samples are included in the training set to develop classifiers, while the fourth sample is used as the validation set and the remaining two images are considered as the test set. These salmon fillets are obtained from three different batches. As seen from Figure 2, they are cut from different positions of fish. It can be discerned that samples are in different sizes and shapes. Some pixels with strong signals are also observed, which poses some challenges for the pixel classification. There could be due to the specular reflection of the illumination source at the salmon surface to produce regions with high-intensity values in the hyperspectral images. These regions act like a mirror and lead to saturation of CCD because of the white stripe or the existence of scales.

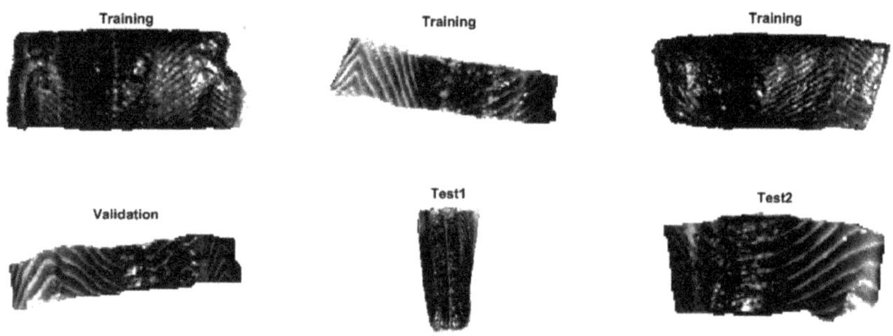

Figure 2. Mean images of salmon fillets at the spectral domain.

2.3. Background Removal

All data analysis is carried out using MATLAB (release R2019a, The MathWorks, Inc., Natick, MA, USA) incorporating functions from Deep Learning Toolbox, Statistics, and Machine Learning Toolbox, and additional functions written in-house.

Sweet samples are placed on a white tile for imaging. Background removal is carried out by subtracting a low-reflectance band from a high-reflectance band followed by a simple thresholding. In this context, the reflectance image at 1496 nm is subtracted from that of 957 nm to enhance the contrast between the sample and white tile. Afterwards, a threshold of 0.22 is applied for background removal. A binary mask for background removal is subsequently generated with all background regions set to zero.

Salmon samples are directly placed on the moving table for image acquisition. The background is removed from the salmon flesh image in the same manner. Bands 944 nm and 1450 nm are used followed by a thresholding value of 0.2. However, the selected bands and thresholding values might change depending on the segmentation result of individual hyperspectral image.

2.4. Spectral Pre-Processing

In the field of hyperspectral imaging, the most common practice is the adaptation of different pre-processing techniques [25]. Spectral preprocessing algorithms are mathematically used to improve spectral data. It aims to correct undesired effects such as random noise, length variation of light path, and light scattering resulting from variable physical sample properties or instrumental effects. This step is generally performed prior to multivariate modelling so as to reduce, eliminate, or standardize the impact on the spectra and to greatly enhance the robustness of the calibration model [26]. In this work, three spectral pre-processing methods are attempted comparatively, namely, standard normal variate (SNV), first derivative, and external parameter orthogonalization (EPO [27]). SNV is a mathematical transformation method of spectra, which is used to remove slope variation and correct light scattering effects. As one of the normalization methods, SNV is performed by first calculating the standard deviation and then normalizing by this value, thus giving the sample a unit standard deviation. First derivative using Savitzky-Golay (SG) method can reduce additive effects [28]. EPO decomposes a spectrum into two components: a useful component that has a direct relationship with the response variable, and a parasitic component that is from an external influence [29]. By removing the parasitic component through orthogonal transformation of spectra, the calibrated spectral model can be less sensitive to the external influence.

2.5. PLSDA and SVM Modelling

Partial least squares discriminant analysis (PLSDA) [30] and SVM are used to build classification models. It is common practice to "unfold" hypercubes such that the three-dimensional information is presented in two dimensions. Unfolding simply refers to rearranging spectra from a hypercube with three dimensions ((1) rows, (2) columns, and (3) wavelengths) to a matrix with two dimensions ((1) rows × columns against (2) wavelengths). Non-background pixels are extracted from each hypercube by unfolding and concatenated to make a two-dimensional matrix (X, i.e., a matrix where the rows represent observations and columns represent spectral features). PLSDA and SVM models are developed from X and Y (i.e., a matrix where the rows represent observations and columns represent the true classes). It is significant to select the proper number of latent variables (LVs). Inclusion of too few or too many LVs may lead to, respectively, under or over-fitting of the data and subsequently lead to poor future model performance [31]. In this work, venetian blinds cross-validation is applied to determine the optimal number of LVs, which is performed by checking the evolution of the accuracy with the number of LVs.

Multiclass support vector machine (SVM) with the error correcting output codes (ECOC) is also implemented for comparison. The SVM is a binary classifier which can be extended by fusing several of its kind into a multiclass classifier [32]. In this work, SVM decisions are fused using the ECOC approach, adopted from the digital communication theory [33].

2.6. PCA-CNN Modelling

This method starts with employing PCA on the global dataset to seek for the spectral variance among different sample types. To do this, it is necessary to unfold all image cubes in the training set along the spatial axis and takes all of the pixel spectra from each hypercube (omitting the background) and then concatenates them to make a two-dimensional matrix on which PCA is performed. PCA decomposes the original data matrix into scores and loadings. Each loading is a vector which provides information on the relative importance, or the weighting, of specific wavelengths relative to each other. The first PC describes the largest variance in the dataset and each following PC describes progressively less of the variance. Therefore, instead of using all loading vectors, we can opt to just use some of the earlier loading vectors to represent the original dataset. For samples in validation and test sets, the individual hypercube is first unfolded and then projected along the PC loadings by matrix multiplication producing PC scores matrices which are subsequently re-folded to form score images.

Score images from the first few PCs are used as the input for 2-D CNN model. For a pixel-based classification of hypercube $I(x, y, \lambda)$, where x and y are the width and the height of the image and λ denotes the number of spectral bands, it aims at predicting the label of each pixel within the image. Initially, the original hypercube $I(x, y, \lambda)$ is reduced to score images with the size of $x \times y \times d$ where d refers to the number of selected PCs. The next step is to extract a $k \times k \times d$ patch for each pixel, where k denotes the window size of the patch. In specific, each patch (i.e., the spatial context) is constructed surrounding a pixel, the center point of the patch. For the pixels that reside near the edge of the image, the patch includes some pixels belonging to the sample while the others belonging to the background.

In this work, the structure of the 2-D CNN consists of an input layer, a convolution (Conv) layer, a rectified linear unit (ReLU) layer, a pooling (POOL) layer, a dropout layer, a fully connected (FC) layer, a softmax layer, and an output layer. The convolutional layer convolves the input data by applying sliding convolutional filters and outputs the convolved features [34], that is, the feature maps. Each convolutional kernel outputs a feature map corresponding to a type of extracted features. Traditional convolution moves from left to right and from top to bottom with a step of 1. Strided convolution has a larger and user-defined step size for traversing the input. All feature maps are stitched and merged by the first fully connected layer to summarize all local features. The number of neural nodes in the fully connected layer changes with the convolution kernel size, the sampling kernel size, and the number of feature maps. For a classification task, it is a common practice to place a softmax layer after the last FC layer. The softmax function is used to compute the probability that each input data pattern belongs to a certain class. The kernel size, the number of feature maps, and the spatial size (i.e., the window size of the patch) are critical parameters in CNN model. These parameters were optimized based on a systematic way of tuning one parameter and fixing it followed by the same procedure for others.

2.7. Three-Dimensional CNN Modelling

As the image formed by hyperspectral bands may have some correlations, e.g., close spectral bands may account for similar images, it is desirable to take into account spectral correlations. Although the 2-D CNN model enables to use the spatial context, it is applied without consideration of spectral correlations. To address this issue, a 3-D CNN model is proposed to extract high-level spectral-spatial features from the original 3-D hyperspectral images. A patch ($k \times k \times \lambda$) for each pixel is extracted from hypercube and used as the input. The operational details of the 3-D CNN model are quite similar to those of the 2-D CNN model. Different from 2-D CNN, the convolution operator of this model is 3-D, whereas the first two dimensions are applied to capture the spatial context and the third dimension captures the spectral context. In addition to a Conv layer, a BN layer, a ReLU layer, a dropout layer, a FC layer, a softmax layer are included in the designed network structure.

2.8. Assessment of Classification Models

Essentially, the performance of a classifier is assessed by the data set classification accuracy index, i.e., % correct classification rate (%CCR). The ground truth for sweet samples is directly obtained by labelling after removing background (see Section 2.3). For salmon samples, a local thresholding strategy is applied on PC2 score images to obtain ground truth. Firstly, individual score image is divided into several sub-images and then an optimal threshold value is manually selected for each sub-image. Confusion matrix, also known as an error matrix, is used to evaluate the quality of the output of the classifier for validation and test sets. The elements in the diagonal are the elements correctly classified, while the elements out of the diagonal are misclassified. We also compute the percentages of all the examples belonging to each class that are correctly and incorrectly classified and show them on the far right of the confusion matrix. These metrics are often called the recall, also known as sensitivity (or true positive rate (TPR)) and false negative rate (FNR), respectively. The row at the bottom of the confusion matrix shows the percentages of all the examples predicted to belong to each class that are correctly and incorrectly classified. These metrics are often known as the precision (or positive predictive value (PPV)) and false discovery rate (FDR), respectively. In detail, they are calculated as below:

$$TPR = \frac{TP}{TP + FN} \tag{1}$$

$$FNR = \frac{FN}{FN + TP} \tag{2}$$

$$PPV = \frac{TP}{TP + FP} \tag{3}$$

$$FDR = \frac{FP}{FP + TN} \tag{4}$$

In these equations, TP, TN, FP and FN respectively refer to true positive, true negative, false positive and false negative. Desirable classification performance is characterized with higher CCR,

TPR, *PPV*, and lower *FNR* and *FDR*. Apart from these, classification and misclassification maps are also displayed to visualize where are the correctly and incorrectly classified pixels.

3. Results

3.1. Results of Sweet Dataset

3.1.1. Spectral Pre-Processing

Figure 3 shows the spectra averaged from one hypercube in the training set and the outcome of pre-treatments. RFM presents the highest reflectance across the whole spectral region (see Figure 3A), while Tub is the lowest. Discrimination among sweet types is highly possible owning to the observable difference in spectral profiles. SNV pre-processed results are displayed in Figure 3B. The combination of SNV and first derivative (window size of 11 and third order polynomial degree) is also applied, as shown in Figure 3C. It is noticed that the spectral difference among sweet samples is highlighted over 1400–1500 nm, which is related to water band due to hydrogen bonding [35].

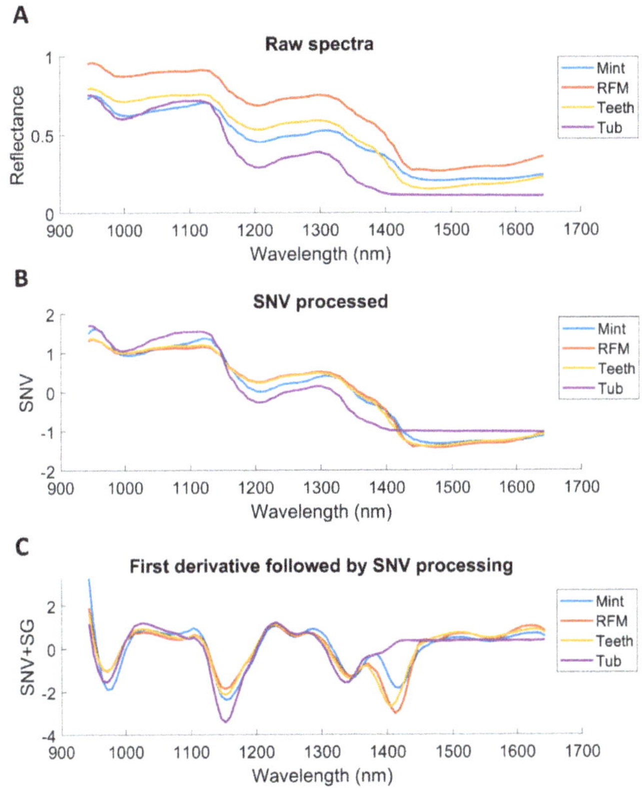

Figure 3. Influence of different spectral pre-processing methods on the mean spectra of one hypercube in the training set. (**A**) Raw spectra; (**B**) Pre-processed spectra with standard normal variate (SNV); (**C**) First derivative spectra (Saviztky-Golay with a window size of 11 and third order polynomial degree) followed by SNV.

3.1.2. PLSDA and SVM Modelling

Figure S2 shows the evolution of accuracy (%) with the number of LVs. In general, CCR (%) increases rapidly at the first few LVs and then remains constant, i.e., adding more variables will not improve accuracy. To avoid both underfitting and overfitting, LVs that contributed most to the enhancement of accuracy were selected, as per the outcome shown in Figure S2. The classification model performance of PLSDA and SVM, in terms of CCR (%) calculated on the validation set and prediction images, are shown in Table 1. In general, the accuracy was found to be higher than 99% for validation and test sets, suggesting the classifiers can generalize well on unknown samples. Apparently, pre-treatments enable the enhancement of the model's performance, as evidenced by the increased accuracy of test sets. It is also noticed that SVM outperformed PLSDA under the condition of using the same spectral pre-treatment.

Table 1. Model performance of sweet samples for validation and test images in terms of % correct classification rate (%CCR).

Model	Pre-Treatment	Time (min)	Training	Validation	Test1	Test2
PLSDA-I	-	<1	97.02	99.42	99.55	98.37
PLSDA-II	SNV	<1	99.37	99.41	99.58	99.16
PLSDA-III	SG+SNV	<1	99.44	99.33	99.69	99.52
SVM-I	-	<1	99.59	99.61	99.62	99.29
SVM-II	SNV	<1	99.99	99.97	99.83	99.69
SVM-III	SG+SNV	<1	99.99	99.97	99.80	99.70
PCA-CNN-I	-	2	99.95	99.01	99.39	99.00
PCA-CNN-II	SNV	2	100	98.79	100	100
PCA-CNN-III	SG+SNV	2	100	100	100	100
3D-CNN-I	-	14	97.72	98.64	98.05	97.94
3D-CNN-II	SNV	13	100	100	100	100
3D-CNN-III	SG+SNV	13	100	100	100	100

Note: PLSDA: partial least squares discriminant analysis; SVM: support vector machine; PCA: principal component analysis; CNN: convolutional neural network; SNV: standard normal variate; SG: First derivative using Saviztky-Golay.

Confusion matrices for test sets were obtained for PLSDA-III (pre-processed with SNV combined with first derivative) and SVM-II (pre-processed with SNV) and displayed in Figure 4. As illustrated, classification of Teeth pixels had the lowest TPR (i.e., sensitivity), suggesting that the true Teeth pixels are less likely to be recognized. Indeed, 10 pixels of Teeth were wrongly classified as Mint and 26 pixels as RFM in Test 1 image (Figure 4A), while 40 pixels of Teeth were incorrectly identified as RFM for Test 2 (Figure 4B), in consistent with Figure 3C where Teeth and RFM present close spectral profiles spanning the entire spectral region. The same observation can be found from SVM modelling result, in which 32 Teeth pixels were wrongly classified as RFM for Test 1 image (Figure 4C) and 51 pixels misclassified as RFM for Test 2 image (Figure 4D).

In order to produce classification maps, the mixture image was first unfolded with background pixels removed using masking to form a two-dimensional matrix on which the developed classifier could be applied. Finally, the resultant matrix with the predicted class assigned to each pixel needed to be refolded to generate classification maps, as shown in Figure 5. It is observed that most misclassification pixels are distributed along the edge of each object. It is also seen that SVM modelling produced less misclassified pixels.

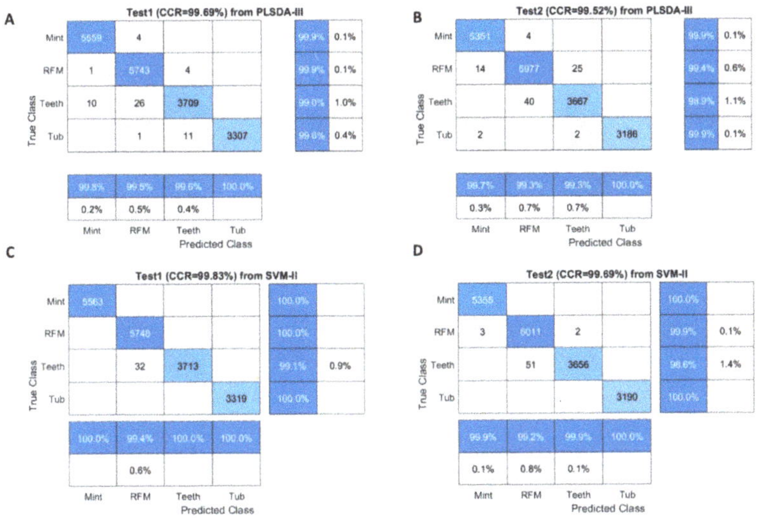

Figure 4. Confusion matrices for Test 1 image (**A**) and Test 2 image (**B**) obtained from PLSDA-III model; for Test 1 image (**C**) and Test 2 image (**D**) obtained from SVM-II model.

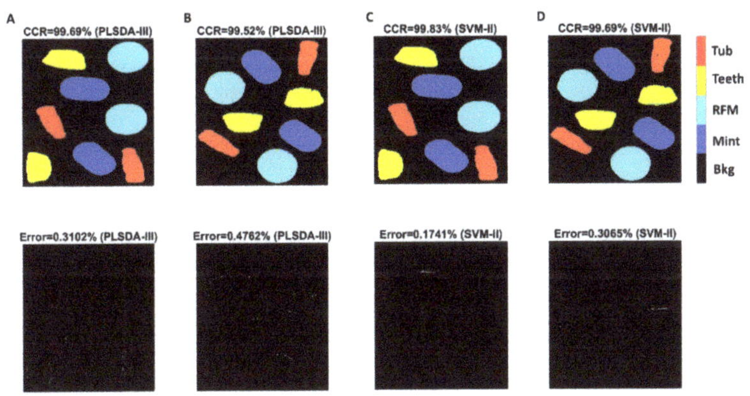

Figure 5. Classification and misclassification maps for Test 1 image (**A**) and Test 2 image (**B**) obtained from PLSDA-III model; for Test 1 image (**C**) and Test 2 image (**D**) obtained from SVM-II model.

3.1.3. PCA-CNN Modelling

First derivative (Saviztky-Golay with a window size of 11 and third order polynomial degree) followed by SNV pre-treated spectra were used to build PCA, with the first three PCs displayed in Figure 6. The pixels belonging to Tub can be easily separated from Teeth and RFM on PC1. The loading plot indicates that the band around 1410 nm mainly contributed to this separation, in agreement with the spectral profiles of Figure 3C. PC1 score images are shown in Figure S3 where different spatial patterns among classes are clearly perceivable, suggesting the suitability of using 2-D CNN modelling subsequently. Figure S4 showing the PC2 score image demonstrates that Mint pixels have higher values (red color) compared to others (blue color), implying the potential of separating Mint from other classes.

The original hypercube with 101 spectral variables was transformed to score images with the first 10 PCs (explaining 97% of variance) selected, followed by patch extraction with the window size of 17. For the training of 2-D CNN, the learning rate was set to 0.01, and the epoch of training was set to 100, the mini-batch size was set to 1024. Convolution layer was implemented with 20 feature maps (filter size = 5 × 5, stride = [2 2]), while the height and the width of the pool size are set to 2 with a stride of 2.

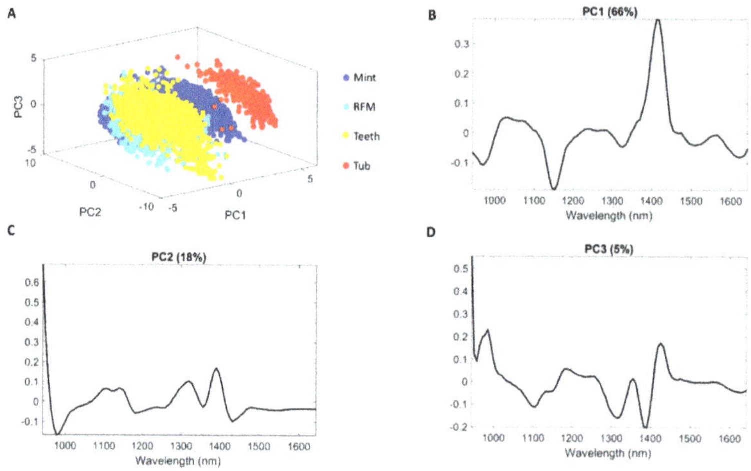

Figure 6. Principal component analysis score plots (**A**) and loading for PC1 (**B**), PC2 (**C**) and PC3 (**D**).

The accuracy and loss for training and validation sets were plotted against the number of iterations, as shown in Figure S5. The accuracy for training and validation both increase and then remain flat. The loss on the training set decreased rapidly for the first 100 iterations, suggesting that the network was learning fast to classify sweet samples. The loss of the validation set also decreased fast and stayed roughly within the small range as the training loss, implying that this model generalizes reasonably well to unseen data.

PCA-CNN model performances are also illustrated in Table 1. As shown, spectral pre-processing enabled the improvement of the model's performance, for instance, from the accuracy of 99.00% using raw spectra to accuracy of 100% using SNV pre-treated spectra. Compared to PLSDA and SVM modelling, PCA-CNN-III (pre-processed with first derivative followed by SNV) facilitated better predictive ability, providing 100% accuracy in validation and test sets. In terms of efficiency, PCA-CNN-III required two minutes for processing, which is acceptable compared to SVM and PLSDA.

3.1.4. Three-Dimensional CNN Modelling

The same pre-processing procedures (SNV, SG+SNV) were carried out in order to compare with raw spectra, after which a patch with the size of 7 × 7 × 101 was extracted from hypercube and used as the input. The same learning rate, and the number of epoch (see Section 3.1.3) were utilized in 3-D CNN model training. Convolution layer was performed with 3-D convolution operator (10 feature maps, filter size = 3 × 3 × 10, stride = [1 1 1]) with the training progress shown in Figure S6. Compared to PCA-CNN model (Figure S5), similar curve shapes (i.e., accuracy and loss) are noticed. Likewise, accuracy first soars and then remains stable after 400 iterations, while loss declines fast at the beginning and keeps flat, indicating that the model was not under or over-fitted.

The model performance of 3-D CNN is also displayed in Table 1. Again, we can observe that spectral pre-processing techniques greatly enhance classification performance. The accuracy of 100% was obtained from pre-processed spectra for training, validation, and both test sets. However, 3D-CNN-I model built from raw spectra delivered the worst performance with the lowest accuracy (in terms of validation and test sets) compared to other models. This suggests that spectral pre-treatment plays an important role in improving the effectiveness of 3-D CNN model developed from near-infrared HSI dataset.

The advantage of 3-D CNN is that it can exploit the spatial and spectral context simultaneously. In essence, PCA can exploit the spectral features, and then 2-D CNN can exploit the spatial context; therefore, the PCA-CNN method also enables the extraction of joint spectral-spatial information from each hypercube. Sweet samples suggest that PCA-CNN and 3-D CNN models after spectral preprocessing, i.e., PCA-CNN-III, 3D-CNN-II, and 3D-CNN-III from Table 1, delivered the best predictive ability with 100% accuracy for validation and test sets. Nevertheless, 3-D CNN required much longer time for the training process, i.e., 2 min of training 2-D CNN model versus 13 min of training 3-D CNN model using the same computer. Indeed, 3-D CNN model brings complexity into the classifier, increasing the number of parameters that each neural model needs to adjust during the training phase.

3.2. Results of Salmon Dataset

3.2.1. Spectral Pre-Processing

The mean spectra of white stripe and red muscle were computed from the first image of the training set. As seen from Figure 7A, higher reflectance is evidenced in white stripe owing to its bright white appearance. It is also observed that white stripe and red muscle present the different band shapes at 1210 nm and 1450 nm, which can be assigned to second overtone of CH_2 bond [36] and OH bond [35], respectively. SNV pre-treatment (Figure 7B) is seen to reduce some interfering variability. Using the mean spectrum of each fillet as the interference, the spectral difference between red muscle and white stripe is clear to observe after employing EPO (Figure 7C).

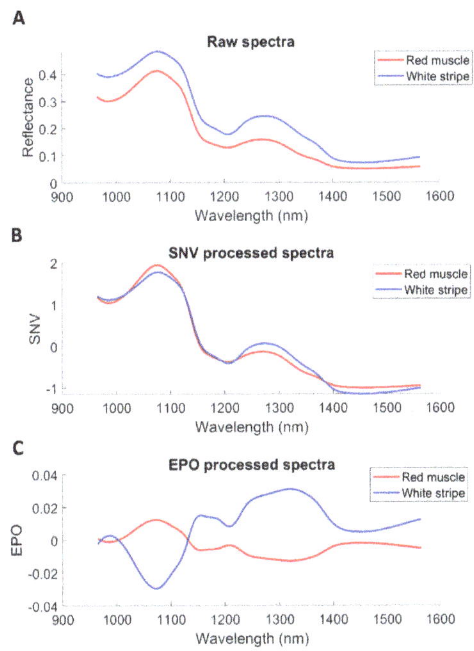

Figure 7. Influence of different spectral pre-processing methods on the mean spectra of one hypercube in the training set. (**A**) Raw spectra; (**B**) Pre-processed spectra with SNV; (**C**) Pre-processed spectra with EPO.

3.2.2. PLSDA and SVM Modelling

The selection of LVs for PLSDA modelling is shown in Figure S7. Table 2 summaries the classification performance of models built for salmon samples. It is noted that the prediction performance varies from sample to sample. For instance, the PLSDA model developed from raw spectra (i.e., PLSDA-I) presented the accuracy of 89.06% for the validation sample, while the inferior predictive ability is witnessed in test sets with the accuracy of 81.02 % for Test 2 fillet. Overall, pre-processing attempts to enhance model performance compared to using raw spectra, which is expected due to the reduction of variance that is irrelevant to classification. SVM outperformed PLSDA under the same pre-treatment condition. The best model performance was found using SVM classifier (SVM-III) built from EPO pre-treated spectra.

Table 2. Model performance of salmon samples for validation and test images in terms of % correct classification rate (%CCR).

Model	Pre-Treatment	Time (min)	Training	Validation	Test1	Test2
PLSDA-I	-	<1	92.89	89.06	82.54	81.02
PLSDA-II	SNV	<1	92.48	93.10	83.75	87.97
PLSDA-III	EPO	<1	90.37	87.81	89.00	85.85
SVM-I	-	<1	94.13	93.72	83.27	87.37
SVM-II	SNV	<1	94.96	87.74	83.81	80.85
SVM-III	EPO	<1	93.02	93.43	94.32	95.22
PCA-CNN-I	-	8	94.51	92.97	84.76	87.85
PCA-CNN-II	SNV	7	94.46	92.86	84.32	85.76
PCA-CNN-II	EPO	7	95.02	93.63	93.96	93.29
3D-CNN-I	-	43	92.76	92.17	81.99	85.44
3D-CNN-II	SNV	43	96.21	93.16	84.87	87.71
3D-CNN-III	EPO	43	93.25	92.19	90.79	91.27

Note: PLSDA: partial least squares discriminant analysis; SVM: support vector machine; PCA: principal component analysis; CNN: convolutional neural network; SNV: standard normal variate; SG: First derivative using Saviztky-Golay.

Confusion matrices for validation and test sets were computed from SVM-III (model built from EPO pre-treated data) due to its better performance in general and displayed in Figure 8. For test sets, it was found that the classification of red muscle pixels has high sensitivity (over 99%), implying the strong ability to correctly identify red muscle pixels on the salmon surface. On the other hand, the sensitivity of identifying white stripe was relatively low, meaning that it is more likely to wrongly classify actual white stripe pixels into the red muscle category.

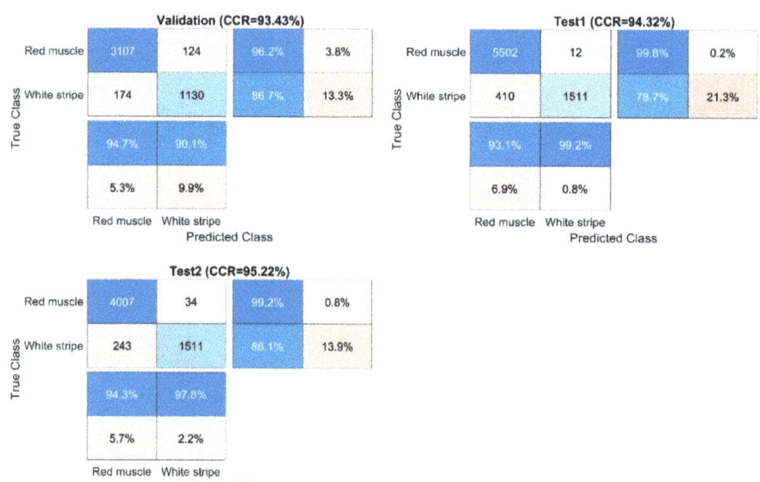

Figure 8. Confusion matrices for validation and test sets obtained by the SVM-III model built from EPO pre-treated spectra.

Prediction maps and misclassification maps were subsequently developed from SVM-III and shown in Figure 9. Meanwhile, the ground truth images are displayed in Figure S8. Visually, the misclassified pixels are distributed along the white stripe. It is consistent with the confusion matrix (Figure 8) where the sensitivity of red muscle is much higher than that of white stripe.

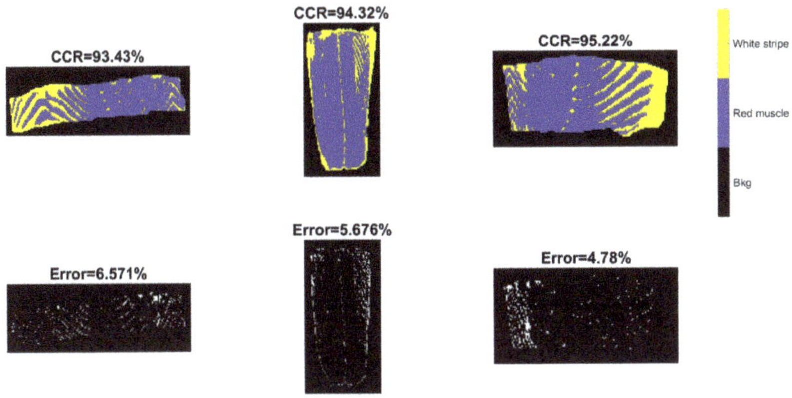

Figure 9. Classification and misclassification maps for validation (the **left** image) and test sets (the **middle** and **right** images) obtained by the support vector machine (SVM) model built from EPO pre-treated spectra (SVM-III of Table 2).

3.2.3. PCA-CNN Modelling

PCA is the key to the proposed method which aims to extract spectral and spatial information from hyperspectral images. To visualize PCA results, scores and loadings obtained from EPO pre-processed spectra are presented in serval figures. Figure 10 shows the scatter plots of score values from the first four PCs. It can be seen that PC1 and PC2 express the major difference between these two classes, facilitating the separation into two clusters on the PC1-PC2 scatter plot. In the case of spectral loadings (see Figure S9), it is found that the separation mostly relies on the influence of the band over 1000–1100 nm, 1250–1350 nm, and 1400–1500 nm. Additionally, PC1 and PC2 score images are illustrated in Figures S10 and S11 of Supplementary Materials, respectively. PC1 score images exhibit some noisy pixels corresponding to the pixels with strong signals in the reflectance images of Figure 2. White stripe pixels presenting the blue colors are slightly distinguishable from red muscle pixels in PC2; however, there is an area of red pixels occurred in individual salmon fillet mostly due to the higher thickness. The surface of salmon fillet was usually not flat, with varying thickness from region to region, making it a challenging task for pixel classification.

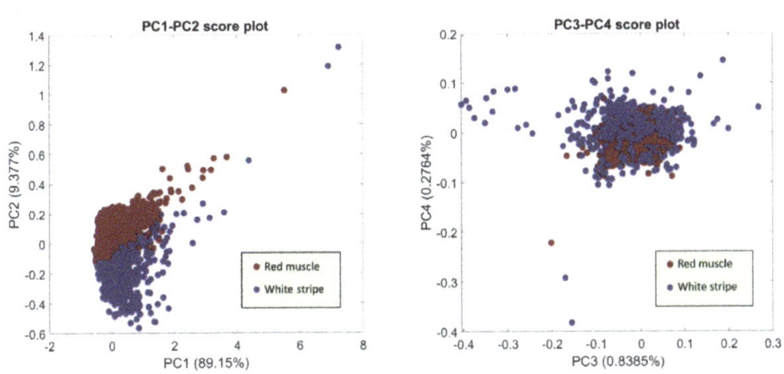

Figure 10. PCA scatter plots from the first four PCs. Explained variance of each PC is indicated in bracket.

PCA-CNN models were respectively developed from raw, SNV, and EPO pre-processed spectra, as the results shown in Table 2. In all cases, the window size were set to 7 and the first 5 PCs (explaining 99% of variance for raw spectra, 98% of variance for SNV, and 99% for EPO pre-processed spectra) were used for 2-D CNN modelling, meaning that the classification of pixels was based on the patch of $7 \times 7 \times 5$. Figure S12 shows training process for the model built from SNV pre-treated spectra. The accuracy for training and validation both increase and then remain flat, while the loss decreases rapidly at the beginning and keeps stable at the late stage. EPO pre-treatment outperforms SNV with higher accuracy for test sets. EPO pre-treatment improved the accuracy of Test 1 image from 84.76%, using raw spectra to 93.96%. Figure 11 shows confusion matrices for validation and test sets calculated from the PCA-CNN-III (pre-processed with EPO) model. In addition, classification and misclassification maps are exhibited in Figure 12 for visualization purpose.

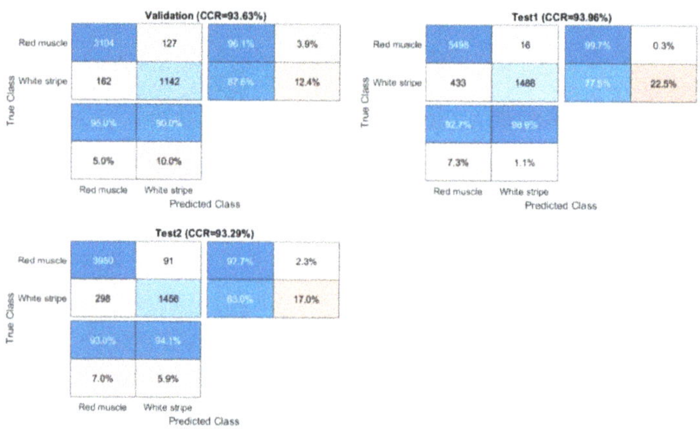

Figure 11. Confusion matrices for validation and test sets obtained by the PCA-CNN model built from EPO pre-treated spectra.

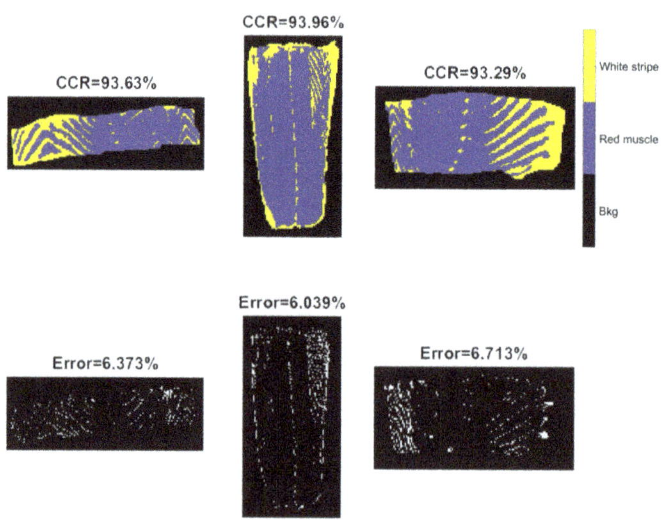

Figure 12. Classification and misclassification maps for validation (the **left** image) and test sets (the **middle** and **right** images) obtained by the PCA-CNN model built from EPO pre-treated spectra.

3.2.4. Three-Dimensional CNN Modelling

3-D CNN models were developed from raw, SNV and EPO pre-treated spectra, with the training progress of 3D-CNN-II (SNV pre-processed spectra) shown in Figure S13. Compared to the 2-D CNN model (Figure S12), similar curve shapes (i.e., accuracy and loss) are noticed. Originally, the training set was characterized as 5-dimensional dataset ($7 \times 7 \times 180 \times 1 \times 21222$ where 7 represents the window size of each patch, 180 denotes to the number of spectral variables, and 21,222 refers to the number of observations). However, CNN training produced a "CPU out of memory" error message. As a result, we extracted every fourth observation and formed a reduced training set with the size of $7 \times 7 \times 180 \times 1 \times 5306$. Again, spectral pre-treatments improved the model performance, which was more obvious for the test sets. According to Table 2, the 3D-CNN-III model leads to the accuracy of 92.19%, 90.79%, and 91.27%, respectively, for validation of the Test 1, and Test 2 images, which are inferior to that of the PCA-CNN-III model. Confusion matrices for validation and test sets are computed and illustrated in Figure 13. There is little distinguishable difference in classification maps obtained from PCA-CNN-III (Figure 12) and 3D-CNN-III (Figure 14), although less misclassified pixels are found for PCA-CNN-III on closer inspection.

For classification of white stripe pixels from red muscle, the best performance was achieved by using SVM developed from EPO pre-treated spectra (SVM-III), followed by PCA-CNN-III. This is probably because the spatial information was unable to make much contribution for this classification task. Deep learning strategies (PCA-CNN and 3-D CNN) show better predictive ability than PLSDA with overall higher accuracy. In terms of running time, 3-D CNN is the most time-consuming, while SVM and PLSDA are the fastest. In spite of using a reduced training set equivalent to 25% of original data, 3-D CNN (43 min) is still much slower than 2-D CNN training (7 min) given the same programming environment.

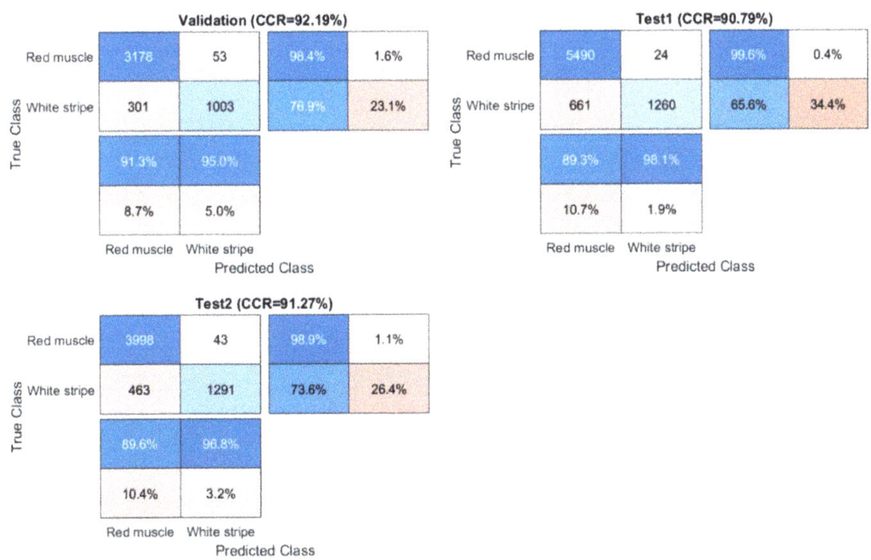

Figure 13. Confusion matrices for validation and test sets obtained by the 3-D CNN model built from EPO pre-treated spectra.

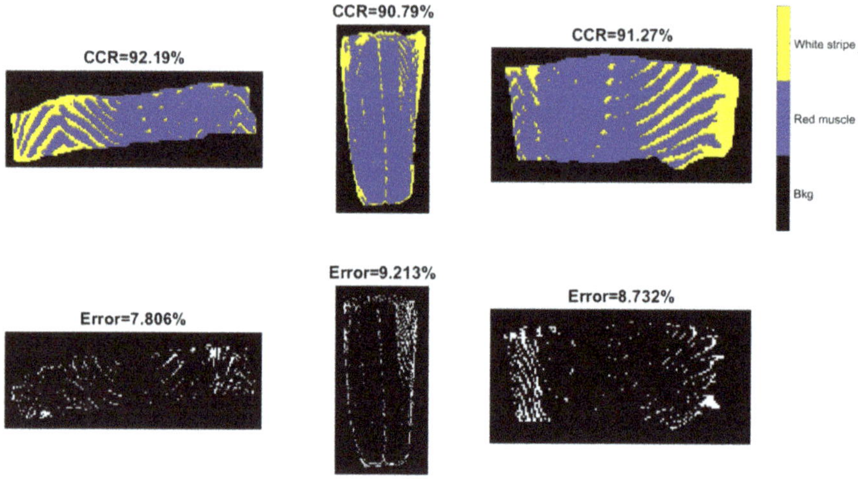

Figure 14. Classification and misclassification maps for validation (the **left** image) and test sets (the **middle** and **right** images) obtained by the 3-D CNN model built from EPO pre-treated spectra.

4. Discussion

This work intends to compare different supervised classifiers for NIR hyperspectral imaging data acquired from benchtop instruments. Using PLSDA, the relevant sources of data variability are modelled by LVs, which are a linear combination of the original variables, and, consequently, it allows graphical visualization and understanding of the different data patterns and relations by LV scores and loadings. Theoretically, PLSDA combines dimensionality reduction and discriminant analysis into one algorithm and is especially applicable to modelling high dimensional data. Therefore, it has demonstrated great success in modelling hyperspectral imaging datasets for diverse purposes. In this work, however, PLSDA presented the inferior modelling performance in both example datasets.

For sweet samples, there are distinctive spatial patterns among classes, which could potentially contribute to the classification. PLSDA and SVM focus exclusively on the spectral domain despite the inherent spatial-spectral duality of the hyperspectral dataset. In other words, the hyperspectral data are considered not as an image but as an unordered listing of spectral vectors where the spatial coordinates can be shuffled arbitrarily without affecting classification modeling results [37]. As we can observe from classification and misclassification maps of sweet samples, PLSDA and SVM classifiers exhibit random noise in pixel-based classification (significantly less in the CNN-based methods), because they ignore spatial-contextual information when providing a pixel prediction. Therefore, pixel-based CNN models outperform traditional chemometric technique (i.e., PLSDA) and machine learning (SVM) in terms of every aspect of classification performance for sweet samples, e.g., higher accuracy, sensitivity, and precision. On the other hand, the spectral difference is the main source for classification between white strip and red muscle classes of salmon samples. Therefore, inclusion of spatial information by applying CNN based strategies (i.e., PCA-CNN and 3-D CNN) cannot necessarily enhance model performance.

PCA-CNN and 3-D CNN both enable to use the conjunction of spatial and spectral information, therefore achieving better classification results compared to PLSDA. PCA is based on the fact that neighboring bands of hyperspectral images are highly correlated and often convey almost the same information about the object. In this sense, PCA facilitates to transform the original data so to remove the correlation among the bands. Technically, the first few PC score images may contain most of the information contained in the entire hyperspectral image data; hence, classifications using the most significant PCA bands yield the same class patterns as when entire hyperspectral data sets are used. Our two hyperspectral datasets (i.e., sweet and salmon) suggested that similar predictive capability is witnessed between these two CNN-based strategies. However, in terms of the runtime of modelling training, the PCA-CNN classifier requires much less time than 3-D CNN partly due to the computational complexity of the 3-D convolution layer. Moreover, it is observed that the number of parameters that each model needs to adjust during the training phase, being the PLSDA model with fewest parameters and the 3-D CNN the one with the most parameters to fit.

5. Conclusions

In this work, PLSDA, SVM, PCA-CNN, and 3-D CNN models for pixel classification were developed and compared in terms of accuracy and efficiency. The results from sweet dataset strongly support the fact that joint spectral and spatial features are more useful than focusing only on spectral features, making the CNN-based modelling ideal for the extraction of highly discriminative features for classification purposes. Nevertheless, salmon dataset demonstrated that SVM model outperformed CNN based methods because spatial information is less important for this classification task. PCA-CNN and 3-D CNN delivered similar classification results, yet the run-time to implement PCA-CNN is much faster than 3-D CNN, suggesting that the use of the PCA approach prior to hyperspectral image classification is beneficial and effective. It significantly reduces the amount of data to be handled and achieves practically acceptable and accurate classification results that are comparable with those obtained using the entire hyperspectral image data. This work also demonstrated the significance of applying spectral pre-processing techniques to complex HSI scenes before classification. Although CNN modelling is powerful in feature extraction, spectral pre-processing techniques manage to remove or reduce some unwanted variance and therefore enhance the classification performance. The proposed CNN based modelling framework from this work could be adopted for solving similar classification problems in food and agriculture applications.

Supplementary Materials: The following are available online at http://www.mdpi.com/1424-8220/20/18/5322/s1, Figure S1: Mean images of the test set with mixed sweet samples at the spectral domain, Figure S2: The evolution of accuracy with the number of LVs developed from (A) raw spectra; (B) SNV pre-treated spectra; (C) first derivative spectra followed by SNV pre-treatment, Figure S3: PC1 score images for the training and validation sets, Figure S4: PC2 score images for the training and validation sets, Figure S5: Training progress of PCA-CNN-III with accuracy (top) and loss (bottom) plotted against iteration, Figure S6: Training progress of 3-D CNN with accuracy (top) and loss (bottom) plotted against iteration, Figure S7: The evolution of accuracy with the number of LVs developed from (A) raw spectra; (B) SNV pre-treated spectra; (C) EPO pre-processed spectra, Figure S8: Ground truth for the validation and test sets, Figure S9: PCA loadings for the first four PCs, Figure S10: PC1 score images of all salmon fillets, Figure S11: PC2 score images of all salmon fillets, Figure S12: Training progress of PCA-CNN-II with accuracy (top) and loss (bottom) plotted against iteration, Figure S13: Training progress of 3D-CNN-II with accuracy (top) and loss (bottom) plotted against iteration.

Author Contributions: H.Z. contributes to data analysis, manuscript writing, revising and proofreading; A.G., H.F. and K.Y. contribute to conceiving and designing the analysis, manuscript review and proofreading; J.-L.X. contributes to conceptualization, data collection, performing data analysis, part of manuscript writing, review and proofreading. All authors have read and agreed to the published version of the manuscript.

Funding: Funding for this research was provided by the Talents Project of Guangxi Normal University (DC2000002490).

Conflicts of Interest: The authors declare no conflict of interest.

References

1. Goetz, A.F.; Vane, G.; Solomon, J.E.; Rock, B.N. Imaging spectrometry for earth remote sensing. *Science* **1985**, *228*, 1147–1153. [CrossRef] [PubMed]
2. Reed, G.; Savage, K.; Edwards, D.; Daeid, N.N. Hyperspectral imaging of gel pen inks: An emerging tool in document analysis. *Sci. Justice* **2014**, *54*, 71–80. [CrossRef] [PubMed]
3. Alexandrino, G.L.; Amigo, J.M.; Khorasani, M.R.; Rantanen, J.; Friderichsen, A.V.; Poppi, R.J. Unveiling multiple solid-state transitions in pharmaceutical solid dosage forms using multi-series hyperspectral imaging and different curve resolution approaches. *Chemom. Intell. Lab.* **2017**, *161*, 136–146. [CrossRef]

4. Mahesh, S.; Jayas, D.; Paliwal, J.; White, N. Hyperspectral imaging to classify and monitor quality of agricultural materials. *J. Stored Prod. Res.* **2015**, *61*, 17–26. [CrossRef]
5. Xu, J.L.; Sun, D.W. Identification of freezer burn on frozen salmon surface using hyperspectral imaging and computer vision combined with machine learning algorithm. *Int. J. Refrig.* **2017**, *74*, 149–162. [CrossRef]
6. Gowen, A.A.; Feng, Y.; Gaston, E.; Valdramidis, V. Recent applications of hyperspectral imaging in microbiology. *Talanta* **2015**, *137*, 43–54. [CrossRef]
7. Barker, M.; Rayens, W. Partial least squares for discrimination. *J. Chemom.* **2003**, *17*, 166–173. [CrossRef]
8. Garrido-Novell, C.; Pérez-Marin, D.; Amigo, J.M.; Fernández-Novales, J.; Guerrero, J.E.; Garrido-Varo, A. Grading and color evolution of apples using RGB and hyperspectral imaging vision cameras. *J. Food Eng.* **2012**, *113*, 281–288. [CrossRef]
9. Ivorra, E.; Girón, J.; Sánchez, A.J.; Verdú, S.; Barat, J.M.; Grau, R. Detection of expired vacuum-packed smoked salmon based on PLS-DA method using hyperspectral images. *J. Food Eng.* **2013**, *117*, 342–349. [CrossRef]
10. Paoletti, M.; Haut, J.; Plaza, J.; Plaza, A. Deep learning classifiers for hyperspectral imaging: A review. *ISPRS J. Photogramm.* **2019**, *158*, 279–317. [CrossRef]
11. Li, S.; Song, W.; Fang, L.; Chen, Y.; Ghamisi, P.; Benediktsson, J.A. Deep learning for hyperspectral image classification: An overview. *IEEE Trans. Geosci. Remote* **2019**, *57*, 6690–6709. [CrossRef]
12. Audebert, N.; Le Saux, B.; Lefèvre, S. Deep learning for classification of hyperspectral data: A comparative review. *IEEE Geosci. Remote Sens. Mag.* **2019**, *7*, 159–173. [CrossRef]
13. Schmidhuber, J. Deep learning in neural networks: An overview. *Neural Networks* **2015**, *61*, 85–117. [CrossRef]
14. Zhu, X.X.; Tuia, D.; Mou, L.; Xia, G.S.; Zhang, L.; Xu, F.; Fraundorfer, F. Deep learning in remote sensing: A comprehensive review and list of resources. *IEEE Geosci. Remote Sens. Mag.* **2017**, *5*, 8–36. [CrossRef]
15. Guo, A.J.; Zhu, F. A CNN-based spatial feature fusion algorithm for hyperspectral imagery classification. *IEEE Trans. Geosci. Remote* **2019**, *57*, 7170–7181. [CrossRef]
16. Hu, W.; Huang, Y.; Wei, L.; Zhang, F.; Li, H. Deep convolutional neural networks for hyperspectral image classification. *J. Sensors* **2015**, *2015*, 12. [CrossRef]
17. Chen, Y.; Jiang, H.; Li, C.; Jia, X.; Ghamisi, P. Deep feature extraction and classification of hyperspectral images based on convolutional neural networks. *IEEE Trans. Geosci. Remote* **2016**, *54*, 6232–6251. [CrossRef]
18. Hamida, A.B.; Benoit, A.; Lambert, P.; Amar, C.B. 3-D deep learning approach for remote sensing image classification. *IEEE Trans. Geosci. Remote* **2018**, *56*, 4420–4434. [CrossRef]
19. Li, Y.; Zhang, H.; Shen, Q. Spectral–spatial classification of hyperspectral imagery with 3D convolutional neural network. *Remote Sens.* **2017**, *9*, 67. [CrossRef]
20. Al-Sarayreh, M.; Reis, M.; Yan, W.Q.; Klette, R. Potential of deep learning and snapshot hyperspectral imaging for classification of species in meat. *Food Control* **2020**, *117*, 107332. [CrossRef]
21. Xu, J.L.; Sun, D.W. Computer vision detection of salmon muscle gaping using convolutional neural network features. *Food Anal. Methods* **2018**, *11*, 34–47. [CrossRef]
22. Borderías, A.J.; Gómez-Guillén, M.C.; Hurtado, O.; Montero, P. Use of image analysis to determine fat and connective tissue in salmon muscle. *Eur. Food Res. Technol.* **1999**, *209*, 104–107. [CrossRef]
23. Segtnan, V.H.; Høy, M.; Lundby, F.; Narum, B.; Wold, J.P. Fat distribution analysis in salmon fillets using non-contact near infrared interactance imaging: A sampling and calibration strategy. *J. Near Infrared Spec.* **2009**, *17*, 247–253. [CrossRef]
24. Xu, J.L.; Riccioli, C.; Sun, D.W. Efficient integration of particle analysis in hyperspectral imaging for rapid assessment of oxidative degradation in salmon fillet. *J. Food Eng.* **2016**, *169*, 259–271. [CrossRef]
25. Rinnan, Å.; van den Berg, F.; Engelsen, S.B. Review of the most common pre-processing techniques for near-infrared spectra. *TrAC Trend Anal. Chem.* **2009**, *28*, 1201–1222. [CrossRef]
26. Reich, G. Near-infrared spectroscopy and imaging: Basic principles and pharmaceutical applications. *Adv. Drug Deliv. Rev.* **2005**, *57*, 1109–1143. [CrossRef] [PubMed]
27. Roger, J.M.; Chauchard, F.; Bellon-Maurel, V. EPO–PLS external parameter orthogonalisation of PLS application to temperature-independent measurement of sugar content of intact fruits. *Chemom. Intell. Lab.* **2003**, *66*, 191–204. [CrossRef]
28. Martens, H.; Næs, T. Pretreatment and linearization. In *Multivariate Calibration*; John Wiley & Sons Ltd.: New York, NY, USA, 2011; pp. 314–356.

29. Wijewardane, N.K.; Ge, Y.; Morgan, C.L. Moisture insensitive prediction of soil properties from VNIR reflectance spectra based on external parameter orthogonalization. *Geoderma* **2016**, *267*, 92–101. [CrossRef]
30. Brereton, R.G.; Lloyd, G.R. Partial least squares discriminant analysis for chemometrics and metabolomics: How scores, loadings, and weights differ according to two common algorithms. *J. Chemom.* **2018**, *32*, e3028. [CrossRef]
31. Gowen, A.; Downey, G.; Esquerre, C.; O'Donnell, C. Preventing over-fitting in PLS calibration models of near-infrared (NIR) spectroscopy data using regression coefficients. *J. Chemom.* **2011**, *25*, 375–381. [CrossRef]
32. Übeyli, E.D. ECG beats classification using multiclass support vector machines with error correcting output codes. *Digit. Signal Process.* **2007**, *17*, 675–684. [CrossRef]
33. Dietterich, T.G.; Bakiri, G. Solving multiclass learning problems via error-correcting output codes. *J. Artif. Intell. Res.* **1994**, *2*, 263–286. [CrossRef]
34. Chu, J.L.; Krzyżak, A. Analysis of Feature Maps Selection in Supervised Learning Using Convolutional Neural Networks. In *Advances in Artificial Intelligence. Canadian AI 2014. Lecture Notes in Computer Science*; Sokolova, M., van Beek, P., Eds.; Springer: Cham, Switzerland, 2014; Volume 8436. [CrossRef]
35. Suh, E.J.; Woo, Y.A.; Kim, H.J. Determination of water content in skin by using a FT near infrared spectrometer. *Arch. Pharm. Res.* **2005**, *28*, 458–462. [CrossRef]
36. González-Martın, I.; González-Pérez, C.; Hernández-Méndez, J.; Alvarez-Garcıa, N. Determination of fatty acids in the subcutaneous fat of Iberian breed swine by near infrared spectroscopy (NIRS) with a fibre-optic probe. *Meat Sci.* **2003**, *65*, 713–719. [CrossRef]
37. Xu, J.L.; Gowen, A.A. Spatial-spectral analysis method using texture features combined with PCA for information extraction in hyperspectral images. *J. Chemom.* **2020**, *34*, e3132. [CrossRef]

sensors MDPI

Article

Study on the Optimization of Hyperspectral Characteristic Bands Combined with Monitoring and Visualization of Pepper Leaf SPAD Value

Ziran Yuan [1,2], Yin Ye [1,2,*], Lifei Wei [3], Xin Yang [1,2] and Can Huang [3]

1 Soil and Fertilizer Research Institute, Anhui Academy of Agricultural Sciences, Hefei 230031, China; yuanziran11@163.com (Z.Y.); yangxin20210928@163.com (X.Y.)
2 Key Laboratory of Nutrient Cycling and Resources Environment of Anhui Province, Anhui Academy of Agricultural Sciences, Hefei 230031, China
3 Faculty of Resources and Environmental Science, Hubei University, Wuhan 430062, China; weilifei2508@hubu.edu.cn (L.W.); 201711110811067@stu.hubu.edu.cn (C.H.)
* Correspondence: yeyin1218@163.com

Abstract: Chlorophyll content is an important indicator of plant photosynthesis, which directly affects the growth and yield of crops. Using hyperspectral imaging technology to quickly and non-destructively estimate the soil plant analysis development (SPAD) value of pepper leaf and its distribution inversion is of great significance for agricultural monitoring and precise fertilization during pepper growth. In this study, 150 samples of pepper leaves with different leaf positions were selected, and the hyperspectral image data and SPAD value were collected for the sampled leaves. The correlation coefficient, stability competitive adaptive reweighted sampling (sCARS), and iteratively retaining informative variables (IRIV) methods were used to screen characteristic bands. These were combined with partial least-squares regression (PLSR), extreme gradient boosting (XGBoost), random forest regression (RFR), and gradient boosting decision tree (GBDT) to build regression models. The developed model was then used to build the inversion map of pepper leaf chlorophyll distribution. The research results show that: (1) The IRIV-XGBoost model demonstrates the most comprehensive performance in the modeling and inversion stages, and its R^2_{cv}, $RMSE_{cv}$, and MAE_{cv} are 0.81, 2.76, and 2.30, respectively; (2) The IRIV-XGBoost model was used to calculate the SPAD value of each pixel of pepper leaves, and to subsequently invert the chlorophyll distribution map of pepper leaves at different leaf positions, which can provide support for the intuitive monitoring of crop growth and lay the foundation for the development of hyperspectral field dynamic monitoring sensors.

Keywords: pepper leaf; SPAD value; hyperspectral inversion; characteristic waveband selection

Citation: Yuan, Z.; Ye, Y.; Wei, L.; Yang, X.; Huang, C. Study on the Optimization of Hyperspectral Characteristic Bands Combined with Monitoring and Visualization of Pepper Leaf SPAD Value. *Sensors* **2022**, *22*, 183. https://doi.org/10.3390/s22010183

Academic Editor: Mercedes Del Río Celestino

Received: 22 November 2021
Accepted: 22 December 2021
Published: 28 December 2021

1. Introduction

Chlorophyll content is one of the most important indicators of the health status of crops and is significant for guiding crop fertilization and field management in different crop growth periods [1]. SPAD values can be directly used as relative values to characterize chlorophyll content. A portable chlorophyll meter is usually used to measure the SPAD value of plant leaves to directly characterize the relative plant chlorophyll content. However, leaves need to be repeatedly inserted during the process, which makes large-scale chlorophyll detection using this method difficult. Research shows that the SPAD value can be used to accurately derive hyperspectral remote sensing data in a non-destructive and pollution-free manner at a low price. In recent years, hyperspectral remote sensing has become a powerful tool for chlorophyll content estimation. Because it is rapid, non-destructive, and capable of detecting chlorophyll over large areas, it is of great significance for crop growth monitoring, precise fertilization, and yield evaluation [2,3].

Hyperspectral imaging technology combines the advantages of both spectrum and image. It has a high resolution and multi-band capabilities. Further, it integrates an atlas, combining traditional imaging and spectral technologies [4]. Changes in plant chlorophyll content lead to changes in the plant reflectance spectrum characteristics [5]. The use of hyperspectral technology to obtain plant growth parameters provides a theoretical basis for measuring chlorophyll, which makes it possible to monitor the growth of crops across a large area [6]. Traditional chlorophyll determination methods mainly rely on chemical experiments, are labor intensive, consume a lot of material resources, and require sample destruction. Although portable chlorophyll meters can measure chlorophyll content in real time, they require manual and repeated measurements, which limits their application in the monitoring of large areas. Furthermore, portable devices can only provide information about the chlorophyll content at a certain point of the leaf, which is not sufficient to obtain an accurate whole-leaf chlorophyll concentration [7].

Hyperspectral technology can not only quantitatively predict the chlorophyll content of the plant leaf but also perform inversion research and image presentation on the distribution of the leaf's chlorophyll content. Zhao et al. used this technology in combination with vegetation index analysis to develop a method that uses hyperspectral imaging technology to obtain five different images in real time to facilitate measurements of leaf water status, relative water content, and equivalent water thickness in tomato varieties [8]. Daughtry and Wu et al. analyzed the accuracy of more than 10 spectral indices, such as MCARI and OSAVI, to estimate the chlorophyll concentration in maize leaves [9,10]. Yu et al. collected samples of leaves, roots, and stems of pepper plants and determined the nitrogen content using a random frog algorithm combined with the partial least-squares method to establish the nitrogen content growth model of the pepper plant [11]. Their results show that hyperspectral imaging is a very promising technology and has great potential for determining the spatial distribution of nitrogen content in pepper plants.

However, there are no studies that use hyperspectral imaging to examine the differences in spatial distribution of SPAD values in leaves located at different positions on pepper plants. Therefore, to ascertain the response of pepper plants' leaf chlorophyll spatial distribution during the growth process, this study adopted hyperspectral imaging technology to develop a method for diagnosing the SPAD value and mapping the spatial distribution of chlorophyll in leaves located at different positions. Four algorithms were used to screen the sensitive wavelengths of pepper leaf chlorophyll diagnosis. These were combined with four regression models to establish a SPAD value diagnostic model. This lays a foundation for the dynamic response of chlorophyll during the growth season of pepper plants.

2. Materials and Methods

2.1. Sample Collection

The study area was located in Wuhu Dehong Ecological Agriculture Co., Ltd. (118°12′ E, 31°26′ N), Shuangba Village, Shenxiang Town, Jiujiang District, Wuhu City, China. It has a subtropical temperate monsoon climate, with sufficient sunlight and rainfall. The experimental variety was Wanjiao 177, the planting time was 20 July 2020, and the collection time was 7 September 2020. The pepper samples were collected at the seedling stage. The fertilization level was selected according to the local conventional fertilization level. The pepper leaves are arranged in descending order according to the leaf growth sequence and are divided into upper, middle, and lower leaves. The upper leaves were the smallest in size, while the lower leaves were the largest. The size of the middle leaves was in between the sizes of the upper and lower leaves (Figure 1). The leaves of the pepper plants were artificially plucked. Fifty leaves were randomly collected from three leaf positions of different pepper plants. Hence, 150 leaf samples in total were placed in a sealed bag to keep the leaves fresh, and taken back to the laboratory immediately to obtain hyperspectral image data.

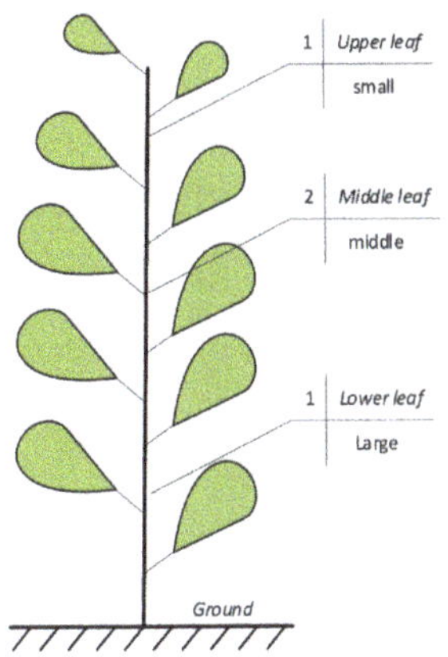

Figure 1. Pepper plant leaf position.

2.2. Chlorophyll Determination

The SPAD-502 Plus chlorophyll meter (Konica Minoita, Tokyo, Japan) was used to measure chlorophyll content. SPAD values can be directly used as relative values to characterize chlorophyll content [12–14]. The chlorophyll meter has the following characteristics: measurement area: 2×3 mm^2; measurement accuracy: ± 1.0 SPAD unit; and measurement range: -9.9–199.9 SPAD unit. Each leaf was divided into six plots (as shown in Figure 2). Three measurements were recorded for each plot, and the average value was taken as the final result of the chlorophyll content of the leaves.

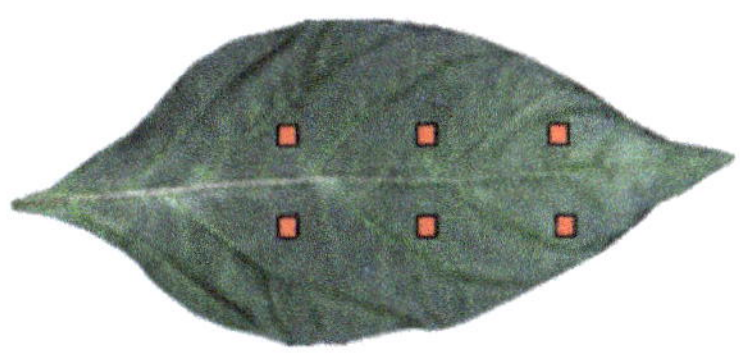

Figure 2. Sampling area of pepper leaves.

The formula used to calculate the SPAD values is shown in Equation (1):

$$\text{SPAD} = K \cdot \lg\left(\frac{IRt/IR0}{Rt/R0}\right) \tag{1}$$

where K is a constant; IRt is the incident 940 nm infrared light intensity passing through the blade; $IR0$ is the emitted infrared light intensity; Rt is the incident 650 nm red light intensity passing through the blade; and $R0$ is the emitted red light intensity.

2.3. Hyperspectral Data Collection

After completing the chlorophyll measurements, the leaves were cleaned with ultrapure water, and the excess surface water was removed using an absorbent paper. Figure 3 shows a schematic diagram of the hyperspectral imaging system used in this study (Wuxi Dualix Spectral Image Technology Co., Ltd. (formerly Sichuan Dualix Spectral Image Technology Co., Ltd.), Wuxi, China, Model: GaiaSorter). The imaging system mainly includes a tungsten halogen lamp as the light source, a hyperspectral camera, an electronically controlled mobile platform, a server and computer control, and other parts.

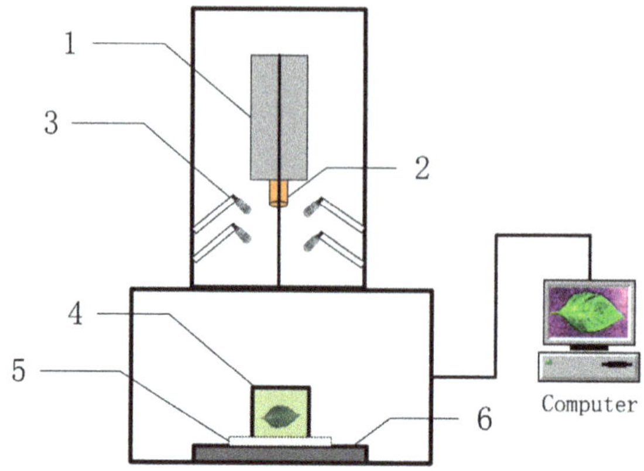

Figure 3. Schematic diagram of the GaiaSorter hyperspectral imaging system. 1. Hyperspectral imager, 2. imaging lens, 3. halogen lamp, 4. sample table, 5. correction whiteboard, and 6. electric translation table.

The height between the hyperspectral camera and the displacement platform was 60 cm, and the height between the halogen tungsten light source and the displacement platform was 40 cm. The wavelength range was 400–1000 nm, and the spectral resolution was 3.6 nm. Experiments were performed in a dark box to perform image correction on the collected spectral images. The image correction formula is given in Equation (2).

$$R_{ref} = \frac{DN_{raw} - DN_{dark}}{DN_{white} - DN_{dark}} \tag{2}$$

where R_{ref} is the corrected image, DN_{raw} is the original image, DN_{white} is the whiteboard image, and DN_{dark} is the blackboard corrected image.

2.4. Spectral Extraction

ENVI 5.3 was used to read the hyperspectral image data of pepper leaves and select six representative rectangular regions of interest (avoiding leaf veins) in the image (Figure 2) as the original spectrum of the sample. A weighted average spectrum was also obtained, which was used as the original spectral data (Figure 4).

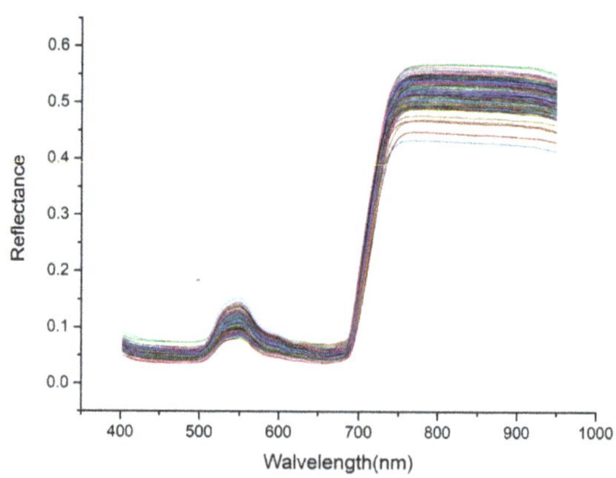

Figure 4. Original spectral curve.

2.5. Research Methods

2.5.1. Correlation Coefficient Method

Spearman's correlation coefficient, which is an index that measures the association between two sets of variables, was used to describe the relationship between the spectral characteristics and SPAD value of pepper leaves [15,16]. We used a monotonic equation to evaluate the correlation between the two statistical variables. The formula used is shown in Equation (3):

$$\rho = \frac{\sum_{i=1}^{N}(x_i - \overline{x})(y_i - \overline{y})}{\sqrt{\sum_{i=1}^{N}(x_i - \overline{x})^2 \sum_{i=1}^{N}(y_i - \overline{y})^2}} \tag{3}$$

where ρ represents the correlation coefficient, x_i is the reflectance of the ith band, y_i is the SPAD value of the ith leaf sample, $\overline{x}$ is the average reflectance, and $\overline{y}$ is the average SPAD value of the pepper leaves.

2.5.2. Stability Competitive Adaptive Reweighted Sampling (sCARS)

sCARS is an advanced wavelength selection method that gradually removes unimportant variable information to achieve the purpose of collecting informative variables [17,18]. The algorithm defines the critical wavelength as the wavelength with the largest absolute value of the regression coefficient in a multivariate linear model (such as PLSR). sCARS can be summarized as follows:

1. Select N wavelength subsets from N Monte Carlo sampling [19] runs in an iterative and competitive manner. In each sampling process, a fixed proportion of samples is randomly selected to establish a calibration model.
2. Perform a two-step process to select characteristic wavelengths: use an exponential decrease function [17] for wavelength selection and use adaptive reweighted sampling to achieve competitive wavelength selection.
3. Use cross-validation [20] to select the subset with the smallest cross-validation root mean square error (RMSECV).

2.5.3. Iteratively Retaining Informative Variables

Iteratively retaining informative variables (IRIV) is a feature variable selection algorithm based on the binary matrix shift filter (BMSF) [21]. Each row of the matrix (containing random combination of the variables) separately establishes partial least-squares models and uses RMSECV to evaluate the effect of different random variable combination models [22,23]. Based on the model cluster analysis method, the average value of RMSECV with and without the variable is calculated for each wavelength, and the difference between the two, known as the difference of mean values (DMEAN), is obtained. The non-parametric test method, Mann–Whitney U test, is used for hypothesis testing [22,24]. Each iteration generates different DMEAN and p values. Both the strongly and weakly informative wavelength variables are retained. After multiple iterations, the uninformative wavelength variables and interfering wavelength variables are eliminated, and finally, reverse elimination is performed to obtain the optimal characteristic wavelength variable.

Step 1: The raw data of m samples of p variables are formed into a matrix A containing only the numbers 0 and 1, where the number 1 represents a variable used for modeling, and the number 0 means that the variable was not used for the modeling. The RMSECV value obtained by five-fold cross-validation was used as the evaluation standard, and the vector of $m \times 1$ size was recorded as $RMSECV_0$. substitute 1 in the ith column (i = 1, 2, ..., p) of matrix A for 0, and 0 for 1 to obtain matrix B. The partial least squares (PLS) model is also established in each row of matrix B, and the vector of $m \times 1$ size is recorded as $RMSECV_i$.

Step 2: Define φ_0 and φ_i to evaluate the importance of each variable as follows:

$$\varphi_{0k} = \begin{cases} k^{th}RMSECV_0 & A_{ki} = 1 \\ k^{th}RMSECV_i & B_{ki} = 1 \end{cases} ; \varphi_{ik} = \begin{cases} k^{th}RMSECV_0 & A_{ki} = 0 \\ k^{th}RMSECV_i & B_{ki} = 0 \end{cases} \tag{4}$$

where k^{th} represents the kth line in the vector, and the $k^{th}RMSECV_0$ and $k^{th}RMSECV_i$ represent the values of the kth row in the vectors $RMSECV_0$ and $RMSECV_i$, respectively. The mean values of φ_0 and φ_i are denoted as $M_{i,in}$ and $M_{i,out}$, respectively, and the two mean values are subtracted to obtain $DMEAN_i$. If $DMEAN_i < 0$, it is a strongly informative variable or a weakly informative variable; if $DMEAN_i > 0$, it is an uninformative variable or an interfering variable.

$$DMEAN_i = M_{i,in} - M_{i,out} \tag{5}$$

$p = 0.05$ was defined as the threshold for the Mann–Whitney U test [21], where the p value, denoted as p_i, is computed by the Mann–Whitney U test with the distribution of φ_0 and φ_i. The smaller the p_i value, the more significant the difference between the two distributions. Finally, the variables were divided into the four categories (strongly informative variables, weakly informative variables, uninformative variables, and interfering variables).

Step 3: Strongly informative variables and weakly informative variables are retained for each iteration, and uninformative variables and interfering variables are eliminated, so that a new subset of variables is generated. Return to step 1 for the next iteration until there are only strong and weak informative variables left. The defined variable types are listed in Table 1.

Table 1. Variable classification rules.

Wavelength Variable Type	Classification Rules
Strongly informative	$DMEAN_i < 0,\ P_i < 0.05$
Weakly informative	$DMEAN_i < 0,\ P_i > 0.05$
Uninformative	$DMEAN_i > 0,\ P_i > 0.05$
Interfering	$DMEAN_i > 0,\ P_i < 0.05$

Step 4: The backward elimination of the reserved variables is undertaken as follows:

(a) Denote t as the number of reserved variables.

(b) For all the reserved variables, obtain the RMSECV value with five-fold cross-validation using PLS, which is denoted as θ^t.

(c) Leave out the ith variable and apply five-fold cross-validation to the remaining $t-1$ variables to obtain the RMSECV value θ_{-i}. Conduct this for all variables $i = 1, 2, \ldots, t$.

(d) If $\min\{\theta_{-i}, 1 \le i \le t\} > \theta^t$, step (g) is performed.

(e) When excluding the ith variable with the minimum RMSECV value, remove the ith variable and change t to be $t-1$.

(f) Repeat steps (a)–(e).

(g) The remaining variables are the final informative variables.

2.5.4. Partial Least-Squares Regression

Partial least-squares regression (PLSR) is a spectral analysis method that includes multiple linear regression, canonical correlation analysis, and principal factor analysis. The main objective of PLSR is to establish a linear model of independent variables, particularly in cases where two groups containing a large number of highly linearly correlated variables are analyzed. PLSR is also used when the number of samples is less than the number of variables to avoid overfitting [25–27]. The principle of PLSR is as follows. First, extract the mutually independent components $(x_1, x_{2,\ldots,}x_m)$ from the independent variable $T_h(h = 1, 2, \ldots)$. The extracted principal components carry as many original components as possible. Then, extract the independent components $(y_{1,}y_2, \ldots, y_m)$ from the independent variable $U_h = (h = 1, 2, \ldots)$. The covariance between T_h and U_h must be maximized, and the regression equation between the extracted components and the dependent variable is established through the multiple regression method. The basic model of the PLSR is:

$$X = T_h P^T + E \tag{6}$$

$$Y = U_h Q^T + F \tag{7}$$

where P and Q are $m \times h$ orthogonal load matrices, and E and F are error terms, which are random variables that follow a normal distribution.

2.5.5. Extreme Gradient Boosting (XGBoost)

XGBoost is a distributed gradient boosting algorithm based on classification and regression trees. XGBoost is popular in the fields of machine learning and data mining and has excellent judgment and recognition capabilities. The basic principle is to weigh the results of multiple decision trees (weak classifiers) as the final output (strong classifier) [28]. XGBoost achieves good control of model complexity by adding rule items to the objective function, thereby solving the problem of collinearity between the variables to a certain extent and preventing overfitting of the model. In the XGBoost model, the second-order Taylor series is used for the cost function, and the first-order and second-order derivatives are used to approximate the optimization of the objective function closer to the actual value, thereby improving the prediction accuracy [29,30].

2.5.6. Random Forest Regression (RFR)

RFR is an integrated statistical learning classification and regression algorithm that combines multiple decision trees to produce similar predictions for different features of the same phenomenon [31]. The output is the average of all the decision tree results in a random forest, assuming that the training set is independently extracted from the distribution of random vectors. The prediction result of the model is the mean of the k regression trees.

2.5.7. Gradient Boosting Decision Tree (GBDT)

GBDT is a comprehensive algorithm with a strong learning strategy. Although the original purpose was to solve the classification problem, it has been successfully applied in the field of regression [32,33].

$$F_m(x) = F_{m-1}(x) + h_m(x) \tag{8}$$

Here, $h_m(x)$ represents the basic functions of the weak learners. In GBDT, the basic function h_m is a small regression tree of fixed size, and the GBDT model $F_m(x)$ can be regarded as the sum of m small regression trees. A new tree is generated for each iteration, m. A simple tree is determined by the deviation between the experimental measurements and all previous model (i.e., gradient) predictions. Then, the regression tree is incorporated into the GBDT model.2.5.8. Software

CA, sCARS, and IRIV were programmed in MATLAB Version 2017b. SPXY and the regression models (PLSR, XGBoost, RFR, GBDT) were written in Python/Jupyter Notebook. The machine learning algorithms in the scikit-learn packages were also used.

2.6. Accuracy Evaluation

A 10-fold cross-validation was used to evaluate the accuracy of the model. The original dataset was randomly divided into 10 subsets with approximately equal sample sizes. Nine of them were combined as the training set in turn, and the one remaining set was used as the test set. In each test, the evaluation index, such as the correct rate, was calculated, and the generalization ability of the model was evaluated by taking the average value of the evaluation index after k tests. The parameters of determination coefficients (R^2_{cv}), root mean square error ($RMSE_{cv}$), and mean absolute error (MAE_{cv}) generated by 10- fold cross validation were used to measure the accuracy of the models. The closer R^2_{cv} is to 1, the better the stability of the model and the higher the degree of fit. The $RMSE_{cv}$ and MAE_{cv}

were used to test the predictive ability of the model. The smaller the $RMSE_{cv}$ and MAE_{cv}, the better the predictive ability.

$$R^2 = \left(\frac{\sum_{i=1}^{n} (x_i - \overline{x})(y_i - \overline{y})}{\sqrt{\sum_{i=1}^{n} (x_i - \overline{x})^2} \sqrt{\sum_{i=1}^{n} (y_i - \overline{y})^2}} \right) \quad (9)$$

$$RMSE = \sqrt{\frac{\sum_{i-1}^{n} (y_i - \hat{y}_i)^2}{n}} \quad (10)$$

$$MAE = \frac{1}{m} \sum_{i=1}^{n} |y_i - \hat{y}_i| \quad (11)$$

where n is the number of samples, y_i is the measured value, $\hat{y}_i$ is the predicted value, and $\overline{y}$ is the average of the measured values.

2.7. Technical Roadmap

In this study, 150 samples of pepper leaves with different leaf positions were selected as the research object, and the hyperspectral image data and chlorophyll content of the pepper leaves were obtained. The technical roadmap is illustrated in Figure 5. The hyperspectral images were first white-calibrated, and then the original spectral data were obtained through the region of interest. The CA, sCARS, and IRIV methods were used, respectively. The IRIV screens the characteristic bands and uses PLSR, XGBoost, RFR, and GBDT to construct regression models. A 10-fold cross-validation was used as the accuracy evaluation index to filter out uninformative variables. The optimal algorithm reuses the constructed model to establish the inversion map of pepper leaf chlorophyll distribution, which lays the foundation for exploring the dynamic response of pepper chlorophyll during the growth period.

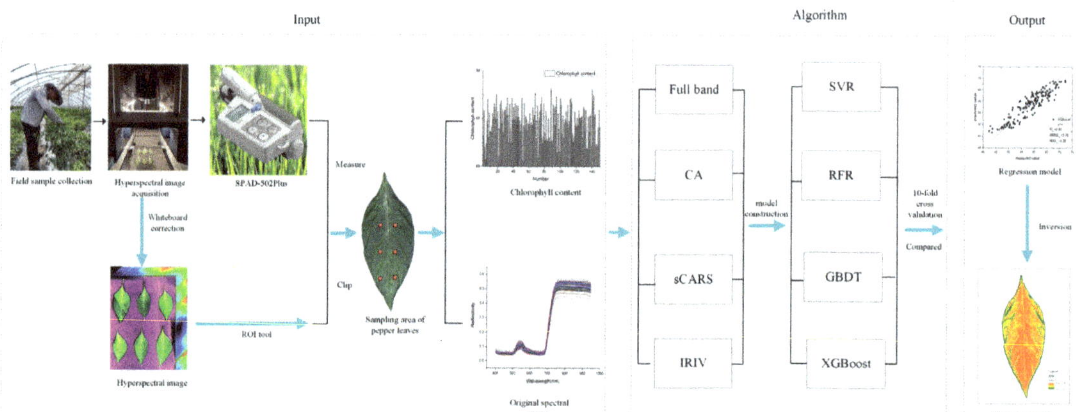

Figure 5. Technical Roadmap.

3. Results

3.1. Selection of Characteristic Band Based on CA Algorithm

Spearman's correlation analysis was performed between the original spectral reflectance of the whole wave band (400–1000 nm) and the SPAD values of pepper leaves. The spectral reflectance of each band was correlated with the SPAD value and a correlation curve was drawn. As shown in Figure 6, the overall correlation was relatively high, and the volatility was relatively large. In visible light (533–560 nm), the correlation is highly negative. After 697 nm, the correlation tends to be stable and continues to increase. Through the significance level test of p=0.01, the band with a correlation greater than 0.65, was finally extracted as the sensitive band. This significant band range was 403–475 nm, with a total of 76 bands, accounting for 43.18% of the overall variable. They are 533.3 nm, 536.7 nm, 540 nm, 543.4 nm, 546.7 nm, 550.1 nm, 553.4 nm, 556.8 nm, 560.1 nm, 697.1 nm, 700.6 nm, 704.1 nm, 707.6 nm, 711.1 nm, 767.6 nm, and 771.1–990.4 nm.

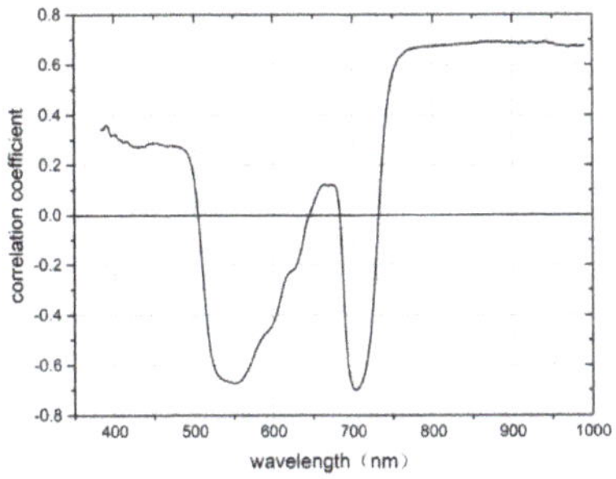

Figure 6. Correlation of SPAD values and spectral reflectance.

3.2. Selection of Characteristic Band Based on SCARS Algorithm

Using the original spectrum as the input spectrum, the specific calculation process of the sCARS algorithm is shown in Figure 7. Figure 7a shows that as the number of sCARS iterations increases, the number of wavelengths retained gradually decreases. The decrease speed is from fast to slow, indicating that sCARS has two stages, "rough selection" and "selection" in the process of screening characteristic bands. Figure 7b shows the change in trend of 10-fold cross-validation, which has a trend from large to small and then to large. When the operation reaches 459 times, the value is the smallest, which means that at 459 times, the wavelength that affects the SPAD value modeling of the pepper leaf is eliminated. The smallest is the best selection of the band subset, and a total of 46 bands were selected, accounting for 26.14% of the overall variable. They are 386.6 nm, 392.9 nm, 402.5 nm, 415.4 nm, 431.5 nm, 526.7 nm, 530.0 nm, 590.5 nm, 593.9 nm, 597.3 nm, 600.7 nm, 610.9 nm, 614.3 nm, 617.7 nm, 624.6 nm, 641.7 nm, 645.1 nm, 676.2 nm, 679.7 nm, 683.2 nm, 693.6 nm, 711.1 nm, 718.1 nm, 732.2 nm, 832.1 nm, 850.2 nm, 853.8 nm, 868.4 nm, 872.0 nm, 875.7 nm, 879.3 nm, 890.3 nm, 894.0 nm, 916.0 nm, 919.7 nm, 923.4 nm, 927.1 nm, 930.8 nm, 938.2 nm, 945.6 nm, 953.0 nm, 960.5 nm, 971.7 nm, 979.2 nm, 982.9 nm, and 986.7 nm.

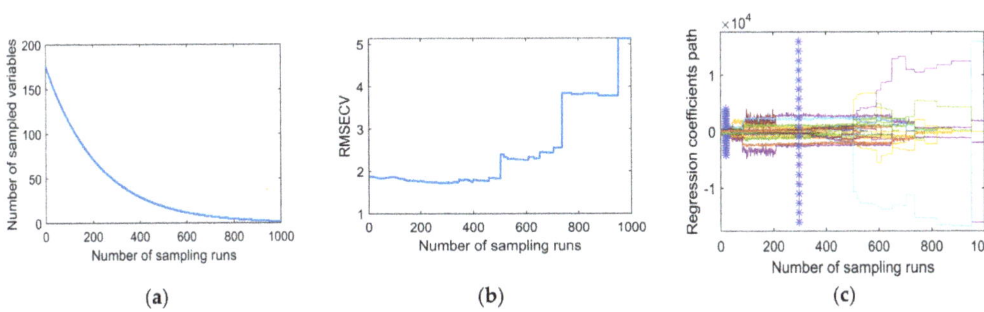

Figure 7. Characteristic variable selection process of sCARS algorithm. (**a**) Changes in the number of waveband variables. (**b**) Validation of RMSECV. (**c**) Path of variable regression coefficients.

3.3. Selection of Characteristic Band Based on IRIV Algorithm

The purpose of the IRIV algorithm is to eliminate irrelevant variables and retain variables related to the SPAD value of pepper leaves. The algorithm uses a 5-fold cross-validation method to establish a partial least-squares model to select the characteristic variables. The IRIV algorithm has gone through seven rounds. As shown in Figure 8, the number of iteration variables in the first three rounds decreased rapidly, from 176 to 48, and then the rate of variable reduction slowed down. After the 6th iteration, the uninformative variables and interfering variables are completely eliminated. In general, only variables with a large amount of information are selected as the best set of variables. Although they have significant positive effects, they are not always optimal because the positive effects of weakly informative variables are ignored. Thus, weakly informative variables are retained at this stage. Therefore, IRIV is used to search for important variables through multiple iterative loops until there are no uninformative or interfering variables, and the optimal characteristic wavelength variable is obtained through reverse elimination. A total of 26 bands were selected, accounting for 14.77% of the overall variables. They are 477.1 nm, 490.3 nm, 510.1 nm, 526.7 nm, 597.3 nm, 600.7 nm, 610.9 nm, 614.3 nm, 617.7 nm, 624.6 nm, 628 nm, 638.3 nm, 648.6 nm, 676.2 nm, 725.1 nm, 728.7 nm, 839.3 nm, 853.8 nm, 861.1 nm, 868.4 nm, 875.7 nm, 879.3 nm, 894 nm, 916 nm, 945.6 nm, and 979.2 nm.

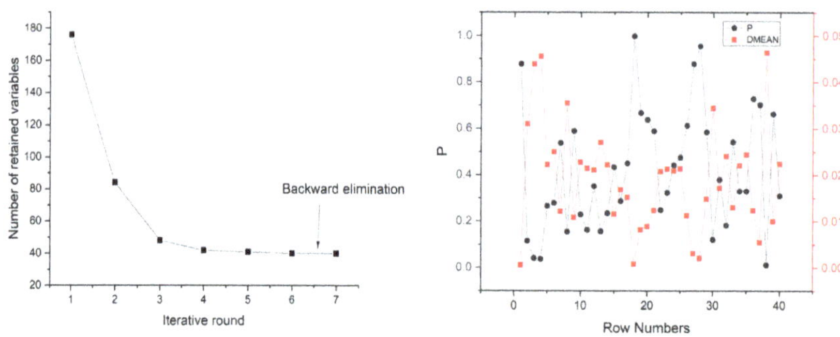

(a) (b)

Figure 8. IRIV algorithm selection process: (**a**) The change in the number of retained informative variables in each round; (**b**) Changes in P value and DMEAN in the sixth round.

3.4. Screening Results

As shown in Figure 9, the order of the three methods used to simplify the model capacity is as follows: IRIV > sCARS > CA. The CA, sCARS, and IRIV algorithms selected 76, 46, and 26 characteristic variables for modeling, accounting for only 43.18%, 26.14%, and 14.77% of the entire band, respectively. The sensitive wavelengths of pepper leaf SPAD value were concentrated between 415.4–431.5 nm, 526.7–676.2 nm, and 839.3–979.2 nm, indicating that these three bands are closely related to pepper leaf SPAD value, as shown in Figure 9, where the blue line part is the same part of the band selected by the three feature selection methods, and they are 853.8 nm, 868.4 nm, 875.7 nm, 879.3 nm, 916 nm, 945.6 nm, and 979.2 nm. This may be related to the plant nutritional status. When the nutritional status is good, the content of chlorophyll in leaves is high, there are more cell layers, and the gap between mesophyll and cells is thick, which can further increase the spectral reflectance. Finally, the higher the SPAD value, the higher the reflectance, and the same correlation is also high, which provides a reliable mathematical basis for the chlorophyll diagnosis model of pepper leaves [34].

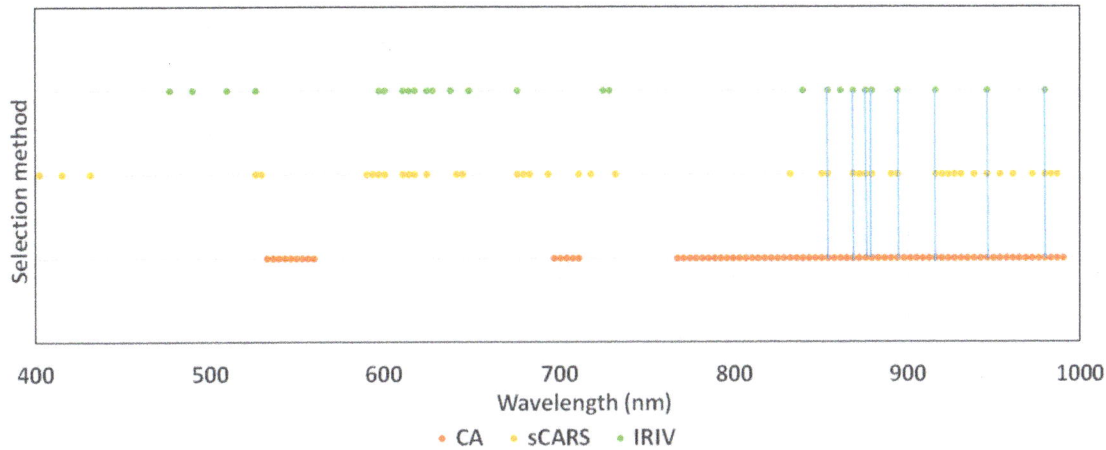

Figure 9. Comparison chart of optimal variable distribution.

3.5. Optimal Algorithm Selection

3.5.1. Accuracy Comparison of Different Methods

A comprehensive comparison of the model prediction results established by different variable selection methods can be seen in Table 2. According to the 10-fold cross-validation discriminant results, the model based on the characteristic variables of the IRIV algorithm achieves the highest accuracy, and the modeling accuracy of each model is relatively high. R^2_{cv} is above 0.8, and the accuracy of the four models constructed by it is much greater than that of the other three methods. It can be seen that the IRIV method is an effective variable selection method and is better than the full band, CA, and sCARS methods. This also shows that the IRIV algorithm is an effective means of improving the accuracy of model prediction and can improve modeling efficiency. In addition, a comparison of the four modeling methods indicates that the characteristic variable modeling of the PLSR algorithm achieves the highest accuracy. However, in terms of the overall accuracy, there is not much difference among the four models.

Table 2. Comparison of accuracy of different methods.

Selection Method	Number of Bands	Modeling Algorithm	R^2_{cv}	$RMSE_{cv}$	MAE_{cv}
Full bands	176	PLSR	0.52	2.57	2.11
		XGBoost	0.48	2.80	2.28
		RFR	0.42	2.95	2.83
		GBDT	0.50	2.76	2.19
CA	76	PLSR	0.48	2.59	2.1
		XGBoost	0.29	3.00	2.39
		RFR	0.41	2.95	2.4
		GBDT	0.44	2.84	2.23
sCARS	46	PLSR	0.55	2.59	2.13
		XGBoost	0.54	2.68	2.17
		RFR	0.43	2.92	2.32
		GBDT	0.53	2.74	2.17
IRIV	26	PLSR	0.84	2.46	2.02
		XGBoost	0.81	2.76	2.30
		RFR	0.80	2.85	2.28
		GBDT	0.80	2.82	2.22

3.5.2. Model Construction Based on the Bands selected by the IRIV Algorithm

Figure 10 shows a scatter plot of the four estimation models under IRIV feature variable screening. From the fitting effect, the results of the four modeling methods were evenly distributed on both sides of the 1:1 straight line. This shows that selecting effective feature variables from the full band spectral data and using these feature variables to build a prediction model can not only greatly simplify the model and reduce the amount of model calculations, but also improve the prediction ability and robustness of the built model. It also shows that the model constructed using this method can be used in the actual monitoring of the SPAD value of pepper leaves.

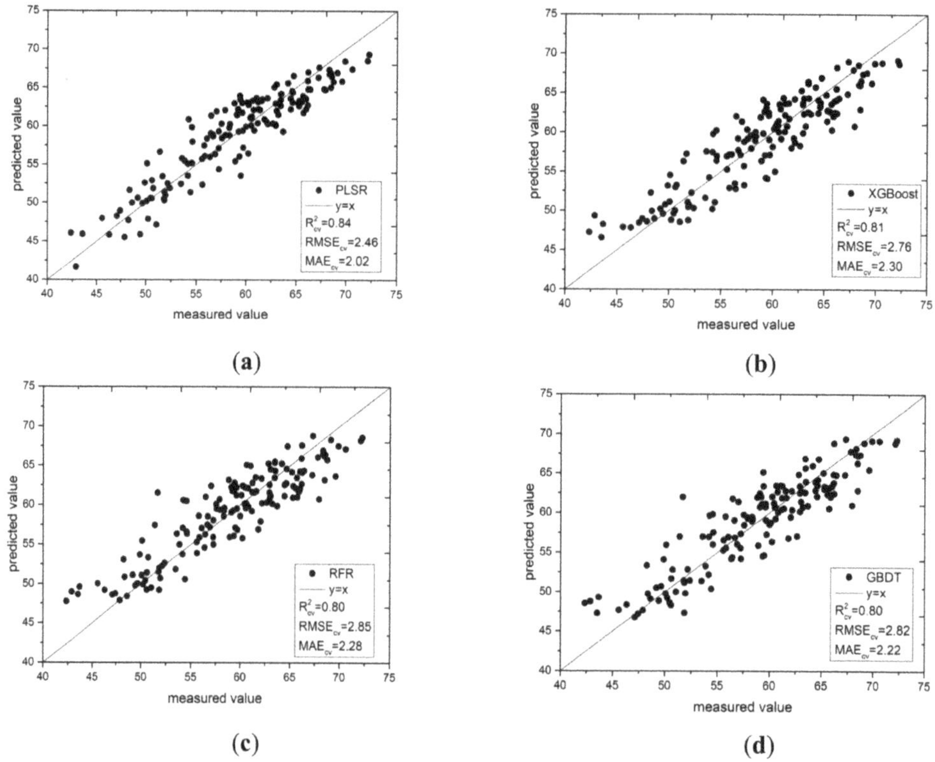

Figure 10. Scatter plot of measured and predicted values of the four models: (**a**) PLSR; (**b**) XGBoost; (**c**) RFR; and (**d**) GBDT.

3.6. *Chlorophyll Distribution of Pepper Leaves*

Using the IRIV-XGBoost model, we estimated the SPAD value of each pixel of pepper leaves, and then drew the chlorophyll distribution map of the pepper leaves. Each SPAD value corresponds to a specific color in the color table. The specific steps are as follows:

Step 1: Hyperspectral images of pepper leaves were obtained under 26 characteristic wavelengths selected by the IRIV algorithm.

Step 2: The reflectivity of each pixel in the characteristic wavelength image was extracted.

Step 3: The SPAD value of each pixel was calculated, and a gray distribution map was obtained.

Step 4: The gray distribution map was used to draw the SPAD distribution map of the pepper leaves at different leaf positions.

As shown in Figures 11–13, different colors (green, yellow, and red) and color depth represent the SPAD value of pepper leaves at different concentrations. Overall, leaf chlorophyll spreads around the central vein. In the lower leaf, the overall color was evenly distributed, and the yellow and red were darker, while the middle leaf and upper leaf chlorophyll were lighter in yellow and red. The distribution of SPAD value in different leaf positions can be seen intuitively: lower > middle > upper, which is consistent with the actual measurements regarding the distribution and changes in the trend of pepper leaf SPAD values, as well as with the growth law of the pepper seedling stage.

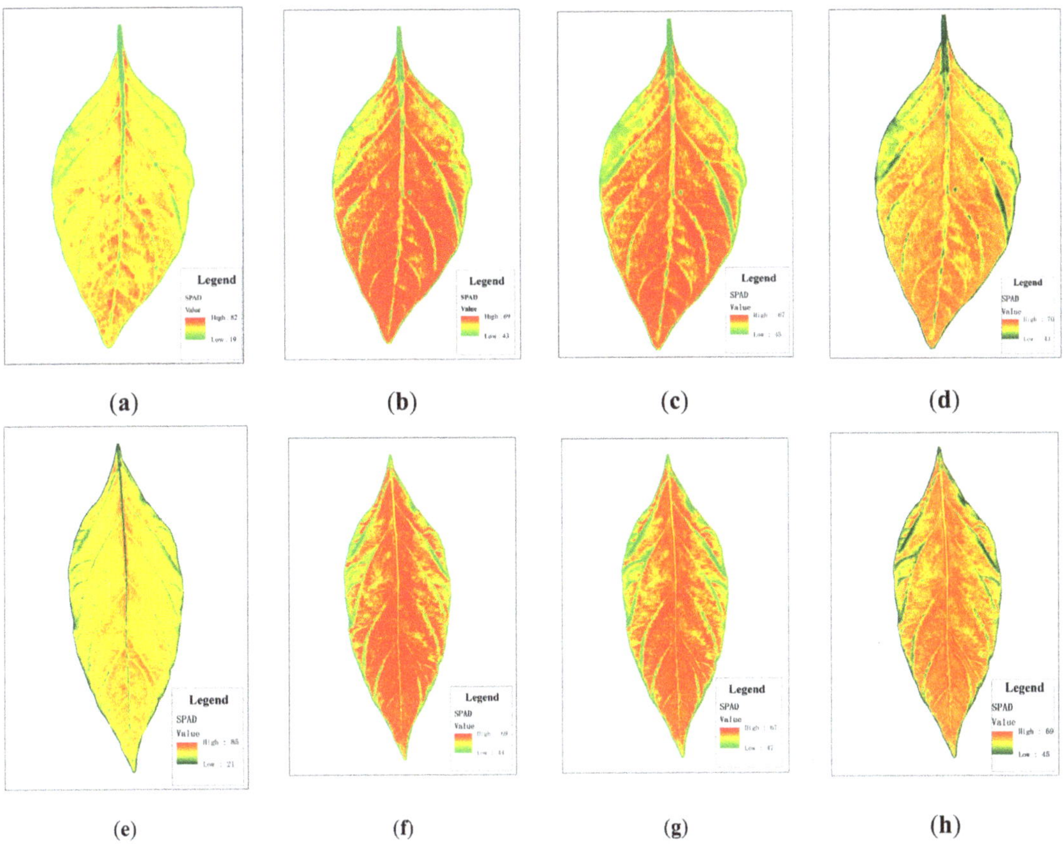

Figure 11. Distribution of SPAD value in the lower leaf in different models: (**a**,**e**—PLSR), (**b**,**f**—XGBoost), (**c**,**g**—RFR), (**d**,**h**—GBDT).

As shown in Table 3, the statistical information of the inversion graph constructed by the three nonlinear models of XGBoost, RFR, and GBDT is relatively close to the true value, while the statistical results of the linear model PLSR show a maximum value of 82 and a minimum value of 2. This is inconsistent with the actual situation. In terms of overall performance, IRIV-XGBoost performed the best.

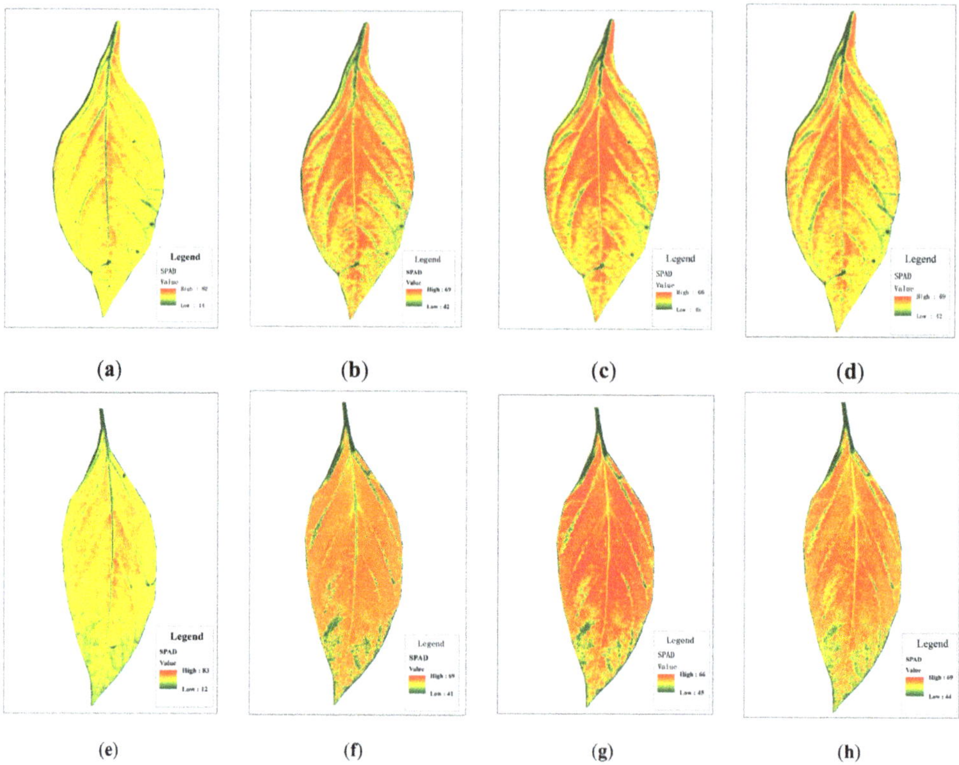

Figure 12. Distribution of SPAD value in the middle leaf in different models: (**a**,**e**—PLSR), (**b**,**f**—XGBoost), (**c**,**g**—RFR), (**d**,**h**—GBDT).

Table 3. Statistical information of chlorophyll inversion map of pepper leaves under different models and different leaf positions.

Leaf Position	Measured Value	Model Method	Min Value	Max Value
Lower leaf	66.0	PLSR	19	82
		XGBoost	43	69
		RFR	46	67
		GBDT	43	70
	69.0	PLSR	21	85
		XGBoost	44	69
		RFR	47	67
		GBDT	45	69
Middle leaf	61.0	PLSR	14	82
		XGBoost	42	69
		RFR	45	66
		GBDT	42	69
	60.6	PLSR	12	83
		XGBoost	41	69
		RFR	45	66
		GBDT	44	69
Upper leaf	48.3	PLSR	3	73
		XGBoost	42	67
		RFR	45	65
		GBDT	42	68
	50.5	PLSR	2	72
		XGBoost	42	67
		RFR	45	64
		GBDT	42	68

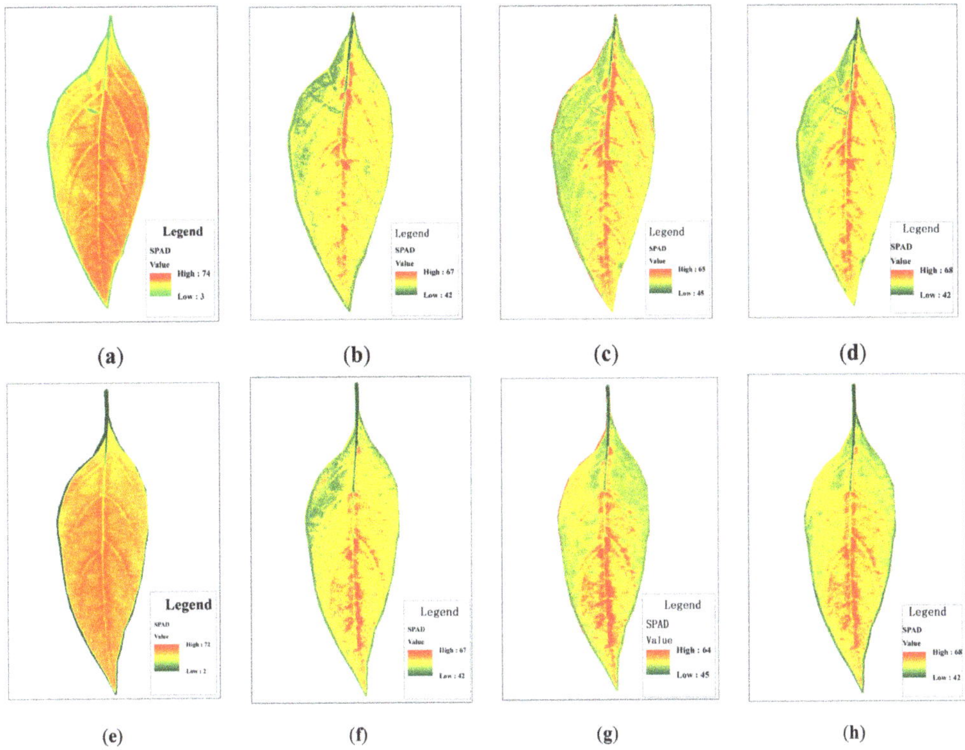

Figure 13. Distribution of SPAD value in the upper leaf in different models: (**a**,**e**—PLSR), (**b**,**f**—XGBoost), (**c**,**g**—RFR), (**d**,**h**—GBDT).

3.7. Statistical Summary Based on the IRIV-XGBoost Algorithm

The SPAD inversion images of pepper leaves obtained by the IRIV-XGBoost algorithm were separately counted. From the mean and standard deviation of each pixel of the inversion image (Figure 14), most of the predicted values are consistent with the measured values, and the predicted and measured values have strong correlation. This shows that the use of hyperspectral imaging technology to construct the SPAD distribution map of pepper leaves is effective, realizes the rapid and accurate acquisition of the SPAD of pepper leaves at a small area scale, and provides a theoretical basis for later crop growth and the development of new equipment.

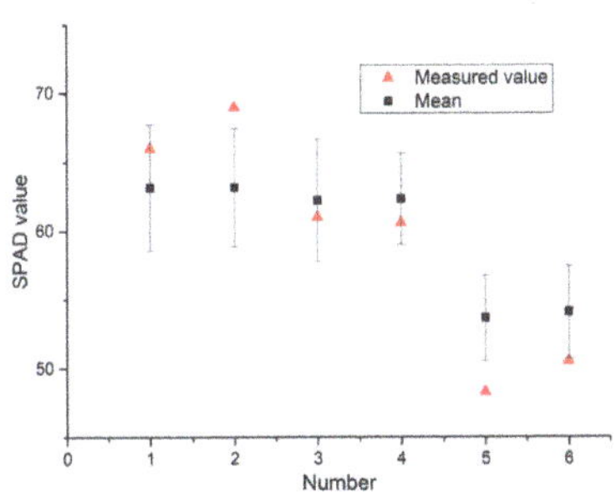

Figure 14. Predicted SPAD value and measured value with the standard deviation as error bars (No.1,2—Figure 11b,f, No.3,4—Figure 12b,f, No.5,6—Figure 13b,f).

4. Discussion

CA, sCARS, and the IRIV algorithm respectively select 76, 46, and 26 characteristic variables for modeling. The results show that prior to modeling, screening the characteristic variables of the original spectrum not only ensures the accuracy of the model, but also greatly reduces the complexity of the model. There are two reasons for this phenomenon: (1) A large number of spectral bands in hyperspectral data provides us with rich spectral information. At the same time, it also leads to redundant information and increases the complexity of data processing, which increases the calculation deviation of statistical parameters. The extraction of characteristic parameters can effectively reduce the dimension of hyperspectral data so as to achieve the effect of optimizing the model [35,36]. (2) The IRIV strategy considers the synergetic effect among variables through random combination. By means of this, only strongly informative and weakly informative variables are retained in each round. This is due to their positive effect under the condition of random combinations among variables. When compared with two outstanding variable selection methods, the outstanding performance of IRIV indicates that it is a good alternative to variable selection in multivariate calibration [22].

The three nonlinear models, XGBoost, RFR, and GBDT, obtained similar results in the hyperspectral imaging inversion stage. All three achieved good results and conformed to the measured value distribution and growth law of pepper. However, the PLSR does not match the actual situation in the inversion stage. Although the accuracy of PLSR in the modeling stage was slightly higher than that of the other three models, it performed poorly in the inversion stage. This is because the PLSR model is a linear model, and it has certain limitations when dealing with high-dimensional data. PLSR can solve the problems of multiple variables and multiple correlations between variables, but it will lose part of the effectiveness after the principal component transformation of the data. Therefore, PLSR is weak in solving nonlinear problems [37,38], and the three nonlinear models of XGBoost, RFR, and GBDT can better solve the complex nonlinear relationship between hyperspectral images and SPAD value. The model has good anti-noise ability, high model accuracy, and good robustness [39].

As shown in Figures 11–13, the SPAD value of pepper leaves exhibited a stepped distribution. The farther away from the center of the plant, the lower the chlorophyll index value. The lower leaves contained higher SPAD value than the upper leaves. The reasons for this analysis may be as follows: (1) Chlorophyll is a light-absorbing substance and an important nutritional indicator. Plant nutrients are transported from the stem upward through the center of the plant to the edge of the leaf, so the SPAD value in the center of the plant is slightly higher than in the edge of the leaf. (2) Since the collected pepper plants are in the seedling stage and the lower leaves are still in the vigorous growth period, they contain more mesophyll, and the leaf functions characterized by chlorophyll are growing vigorously. The leaves are only formed during the seedling stage, and they are in a vigorous growth period. Respiration was strong. Although the stomatal conductance is high, many internal structures are imperfect, so the SPAD value is relatively low. As the leaf age increased, the leaf structure became complete, and the SPAD value gradually increased.

5. Conclusions

Hyperspectral data for pepper leaves located at different positions on the plant were collected to analyze the differences in the SPAD value distribution and the dynamic characteristics of the growth period of the pepper plants. The average spectra of the SPAD value measurement positions of pepper leaves were extracted, and CA, sCARS, and IRIV were used to screen feature bands. These methods were combined with PLSR, XGBoost, RFR, and GBDT to construct regression models, and the distribution of SPAD value in pepper leaves at different leaf positions was drawn. The main conclusions of this study are as follows:

(1) A comprehensive comparison of the full band, CA, sCARS, and IRIV variable screening feature bands was undertaken to construct a variety of SPAD value estimation

models and the model capabilities were tested through 10-fold cross-validation. The estimation capabilities of the different models were quite different. The IRIV algorithm achieved the highest accuracy, above 0.8, which greatly reduces the complexity of the model while ensuring the accuracy of the model.

(2) Four modeling methods were compared: PLSR, XGBoost, RFR, and GBDT. The accuracy of PLSR in the modeling stage is slightly higher than that of the other three models, but it performs poorly in the inversion stage. XGBoost is better suited to solve the complex nonlinear relationship between hyperspectral images and SPAD value. The model has good anti-noise ability, high model accuracy, and good robustness.

(3) The IRIV-XGBoost model was used to calculate the SPAD value of each pixel of pepper leaves and then invert the chlorophyll distribution map of pepper leaves at different leaf positions, which can reflect the dynamic response of pepper leaf chlorophyll in plants in the seedling stage and finally realize the non-destructive detection of pepper leaf content for different leaves and the visual expression of chlorophyll distribution. This result is consistent with the distribution and change trend of the SPAD value of pepper leaves when measured, and it is also in line with the growth law of pepper seedling stage. In future, the dynamics of different growth periods need to be tested and verified in the field to lay a foundation for the overall dynamic diagnosis of pepper canopy.

Author Contributions: Z.Y. and Y.Y. were responsible for the overall design of the study and contributed to the proofreading of the manuscript. Z.Y. performed the experiments, analyzed and interpreted the data, wrote the manuscript, and helped with the proofreading of the manuscript. L.W. contributed to the design of the study and proofreading of the manuscript. X.Y. and C.H. analyzed and interpreted the data. All authors have read and agreed to the published version of the manuscript.

Funding: This research was funded by the National Natural Science Foundation of China (31901154), the research team project of Anhui Academy of Agricultural Sciences (2021YL0085), and the Provincial Science and Technology Innovation Park Demonstration Project (201907d06020011).

Institutional Review Board Statement: Not applicable.

Informed Consent Statement: Not applicable.

Data Availability Statement: Not applicable.

Conflicts of Interest: The authors declare no conflict of interest.

References

1. Dordas, C. Nitrogen Nutrition Index and Leaf Chlorophyll Concentration and its relationship with Nitrogen Use Efficiency in Barley (*Hordeum vulgare* L.). *J. Plant Nutr.* **2017**, *40*, 1190–1203.
2. Ata-Ul-Karim, S.T.; Cao, Q.; Zhu, Y.; Tang, L.; Rehmani, M.I.A.; Cao, W. Non-destructive Assessment of Plant Nitrogen Parameters Using Leaf Chlorophyll Measurements in Rice. *Front. Plant Sci.* **2016**, *7*, 1829.
3. Liu, B.; Shen, W.; Yue, Y.M.; Li, R.; Tong, Q.; Zhang, B. Combining spatial and spectral information to estimate chlorophyll contents of crop leaves with a field imaging spectroscopy system. *Precis. Agric.* **2017**, *18*, 491–506.
4. Manley, M. Near-infrared spectroscopy and hyperspectral imaging: Non-destructive analysis of biological materials. *Chem. Soc. Rev.* **2014**, *43*, 8200.
5. Sun, H.; Zheng, T.; Liu, N.; Cheng, M.; Li, M.; Zhang, Q. Vertical distribution of chlorophyll in potato plants based on hyperspectral imaging. *Trans. Chin. Soc. Agric. Eng.* **2018**, *34*, 149–156.
6. Liu, D.; Sun, D.W.; Zeng, X.A. Recent Advances in Wavelength Selection Techniques for Hyperspectral Image Processing in the Food Industry. *Food Bioprocess Technol.* **2014**, *7*, 307–323.
7. Ji-Yong, S.; Xiao-Bo, Z.; Jie-Wen, Z.; Kai-Liang, W. Nondestructive diagnostics of nitrogen deficiency by cucumber leaf chlorophyll distribution map based on near infrared hyperspectral imaging. *Sci. Horticult.* **2012**, *138*, 190–197.
8. Zhao, T.; Nakano, A.; Iwaski, Y.; Umeda, H. Application of Hyperspectral Imaging for Assessment of Tomato Leaf Water Status in Plant Factories. *Appl. Sci.* **2020**, *10*, 4665.
9. Daughtry, C.; Walthall, C.L.; Kim, M.S.; Colstoun, E.; Iii, M.M. Estimating Corn Leaf Chlorophyll Concentration from Leaf and Canopy Reflectance. *Remote Sens. Environ.* **2000**, *74*, 229–239.
10. Wu, C.; Zheng, N.; Tang, Q.; Huang, W. Estimating chlorophyll content from hyperspectral vegetation indices: Modeling and validation. *Agric. For. Meteorol.* **2008**, *148*, 1230–1241.

11. Yu, K.Q.; Zhao, Y.R.; Li, X.L.; Shao, Y.N.; Fei, L.; Yong, H. Hyperspectral Imaging for Mapping of Total Nitrogen Spatial Distribution in Pepper Plant. *PLoS ONE* **2014**, *9*, e116205.
12. Liu, Y.; Fan, M. Application Feasibility of SPAD-502 in Diagnosis of Potato Nitrogen Nutrient Status. *Chin. Potato J.* **2012**, *26*, 45–48.
13. Coste, S.; Baraloto, C.; Leroy, C.; Marcon, É.; Renaud, A.; Richardson, A.D.; Roggy, J.C.; Schimann, H.; Uddling, J.; Hérault, B. Assessing foliar chlorophyll contents with the SPAD-502 chlorophyll meter: A calibration test with thirteen tree species of tropical rainforest in French Guiana. *Ann. For. Sci.* **2010**, *67*, 607.
14. Nau, J.; Prokopová, J.; Ebíek, J.; Pundová, M. SPAD chlorophyll meter reading can be pronouncedly affected by chloroplast movement. *Photosynth. Res.* **2010**, *105*, 265–271.
15. Wang, D.; Aixia, S.U.; Liu, W. Trend Analysis of Vegetation Cover Changes Based on Spearman Rank Correlation Coefficient. *J. Appl. Sci.* **2019**, *37*, 519–528.
16. Amarkhil, Q.; Elwakil, E.; Hubbard, B. A Meta-Analysis of Critical Causes of Project Delay Using Spearman's Rank and Relative Importance Index Integrated Approach. *Can. J. Civil. Eng.* **2020**, *48*, 1498–1507.
17. Li, H.; Liang, Y.; Xu, Q.; Cao, D. Key wavelengths screening using competitive adaptive reweighted sampling method for multivariate calibration. *Anal. Chim. Acta* **2009**, *648*, 77–84.
18. Liu, G.H.; Xia, R.S.; Jiang, H.; Mei, C.-L.; Huanng, Y.-H. A Wavelength Selection Approach of Near Infrared Spectra Based on SCARS Strategy and Its Application. *Spectrosc. Spectr. Anal.* **2014**, *34*, 2094–2097.
19. Holmes, E.W.; Kahn, S.E.; Molnar, P.A.; Bermes, E.W. Verification of reference ranges by using a Monte Carlo sampling technique. *Clin. Chem.* **2020**, *40*, 2216–2222.
20. Guo, L.N.; Fan, G.S. Support Vector Machines for Surface Soil Density Prediction based on Grid Search and Cross Validation. *Chin. J. Soil Sci.* **2018**, *49*, 512–518.
21. Zhang, H.; Wang, Y.; Liu, C.; Jiang, H. Influence of surfactant CTAB on the electrochemical performance of manganese dioxide used as supercapacitor electrode material. *J. Alloys Compd.* **2012**, *517*, 8.
22. Yun, Y.H.; Wang, W.; Tan, M.; Liang, Y.; Li, H.; Cao, D.; Lu, H.; Xu, Q. A strategy that iteratively retains informative variables for selecting optimal variable subset in multivariate calibration. *Anal. Chim. Acta* **2014**, *807*, 36–43.
23. Zhang, H.; Wang, H.; Dai, Z.; Chen, M.S.; Yuan, Z. Improving accuracy for cancer classification with a new algorithm for genes selection. *BMC Bioinform.* **2012**, *13*, 298.
24. Mann, H.B.; Whitney, D.R. On a Test of Whether one of Two Random Variables is Stochastically Larger than the Other. *Ann. Math. Stat.* **1947**, *18*, 50–60.
25. Nawar, S.; Buddenbaum, H.; Hill, J.; Kozak, J. Modeling and Mapping of Soil Salinity with Reflectance Spectroscopy and Landsat Data Using Two Quantitative Methods (PLSR and MARS). *Remote Sens.* **2014**, *6*, 10813–10834. [CrossRef]
26. Wang, S.; Chen, Y.; Wang, M.; Zhao, Y.; Li, J. SPA-Based Methods for the Quantitative Estimation of the Soil Salt Content in Saline-Alkali Land from Field Spectroscopy Data: A Case Study from the Yellow River Irrigation Regions. *Remote Sens.* **2019**, *11*, 967.
27. Weng, Y.; Qi, H.; Fang, H. PLSR-Based Hyperspectral Remote Sensing Retrieval of Soil Salinity of Chaka-Gonghe Basin in Qinghai Province. *Acta Pedol. Sin.* **2010**, *47*, 1255–1263.
28. Chen, T.; Guestrin, C. XGBoost: A Scalable Tree Boosting System. In Proceedings of the 22nd ACM SIGKDD International Conference on Knowledge Discovery and Data Mining, San Francisco, CA, USA, 13–17 August 2016; pp. 785–794.
29. Jia, Y.; Jin, S.; Savi, P.; Gao, Y.; Tang, J.; Chen, Y.; Li, W. GNSS-R Soil Moisture Retrieval Based on a XGboost Machine Learning Aided Method: Performance and Validation. *Remote Sens.* **2019**, *11*, 1655.
30. Loggenberg, K.; Strever, A.; Greyling, B.; Poona, N. Modelling Water Stress in a Shiraz Vineyard Using Hyperspectral Imaging and Machine Learning. *Remote Sens.* **2018**, *10*, 202.
31. Tong, Y.; Lu, K.; Yang, Y.; Li, J.; Qian, J. Can Natural Language Processing Help Differentiate Inflammatory Intestinal Diseases in China? Models Applying Random Forest and Convolutional Neural Network Approaches. *BMC* **2020**, *20*, 248.
32. Liu, L.; Ji, M.; Dong, Y.; Zhang, R.; Buchroithner, M. Quantitative Retrieval of Organic Soil Properties from Visible Near-Infrared Shortwave Infrared (Vis-NIR-SWIR) Spectroscopy Using Fractal-Based Feature Extraction. *Remote Sens.* **2016**, *8*, 1035.
33. Jin, X.; Zhu, X.; Li, S.; Wang, W.; Qi, H. Predicting Soil Available Phosphorus by Hyperspectral Regression Method Based on Gradient Boosting Decision Tree. *Laser Optoelectron. Prog.* **2019**, 56.
34. Zhao, Z.H.; Zhang, C.Z.; Zhang, J.B.; Wu, Y.J.; Zhang, H.M.; Lu, C.Y. The Hyperspectral Inversion for Estimating Maize Chlorophyll Contents Based on Regression Analysis. *Chin. Agric. Sci. Bull.* **2021**, *35*, 7–16.
35. Sarathjith, M.C.; Das, B.S.; Wani, S.P.; Sahrawat, K.L. Variable indicators for optimum wavelength selection in diffuse reflectance spectroscopy of soils. *Geoderma* **2016**, *267*, 1–9.
36. Xu, S.; Zhao, Y.; Wang, M.; Shi, X. Determination of rice root density from Vis–NIR spectroscopy by support vector machine regression and spectral variable selection techniques. *Catena* **2017**, *157*, 12–23.
37. Pinto, C.F.; Parab, J.S.; Sequeira, M.D.; Naik, G.M. Development of Altera NIOS II Soft-core system to predict total Hemoglobin using Multivariate Analysis. *J. Phys. Conf. Ser.* **2021**, *1921*, 12039.

38. Wei, L.; Yuan, Z.; Yu, M.; Huang, C.; Cao, L. Estimation of Arsenic Content in Soil Based on Laboratory and Field Reflectance Spectroscopy. *Sensors* **2019**, *19*, 3904.
39. Raissi, M.; Karniadakis, G.E. Hidden Physics Models: Machine Learning of Nonlinear Partial Differential Equations. *J. Comput. Phys.* **2018**, *357*, 125–141.

sensors

MDPI

Article

The Use of a Micro Near Infrared Portable Instrument to Predict Bioactive Compounds in a Wild Harvested Fruit—Kakadu Plum (*Terminalia ferdinandiana*)

Eshetu Bobasa [1,2], Anh Dao T. Phan [1,2], Michael Netzel [1,2], Heather E. Smyth [1,2], Yasmina Sultanbawa [1,2] and Daniel Cozzolino [1,2,*]

[1] ARC Industrial Transformation Training Centre for Uniquely Australian Foods, Queensland Alliance for Agriculture and Food Innovation, The University of Queensland, 39 Kessels Rd, Coopers Plains, Brisbane, QLD 4108, Australia; e.bobasa@uq.edu.au (E.B.); a.phan1@uq.edu.au (A.D.T.P.); m.netzel@uq.edu.au (M.N.); h.smyth@uq.edu.au (H.E.S.); y.sultanbawa@uq.edu.au (Y.S.)

[2] Centre for Nutrition and Food Sciences, Queensland Alliance for Agriculture and Food Innovation, The University of Queensland, St. Lucia, Brisbane, QLD 4072, Australia

* Correspondence: d.cozzolino@uq.edu.au

Citation: Bobasa, E.; Phan, A.D.T.; Netzel, M.; Smyth, H.E.; Sultanbawa, Y.; Cozzolino, D. The Use of a Micro Near Infrared Portable Instrument to Predict Bioactive Compounds in a Wild Harvested Fruit—Kakadu Plum (*Terminalia ferdinandiana*). *Sensors* **2021**, *21*, 1413. https://doi.org/10.3390/s21041413

Academic Editors: Mercedes Del Río Celestino and Rafael Font Villa

Received: 19 January 2021
Accepted: 13 February 2021
Published: 18 February 2021

Publisher's Note: MDPI stays neutral with regard to jurisdictional claims in published maps and institutional affiliations.

Abstract: Kakadu plum (KP; *Terminalia ferdinandiana* Exell, Combretaceae) is an emergent indigenous fruit originating from Northern Australia, with valuable health and nutritional characteristics and properties (e.g., high levels of vitamin C and ellagic acid). In recent years, the utilization of handheld NIR instruments has allowed for the in situ quantification of a wide range of bioactive compounds in fruit and vegetables. The objective of this study was to evaluate the ability of a handheld NIR spectrophotometer to measure vitamin C and ellagic acid in wild harvested KP fruit samples. Whole and pureed fruit samples were collected from two locations in the Kimberley region (Western Australia, Australia) and were analysed using both reference and NIR methods. The standard error in cross validation (SECV) and the residual predictive deviation (RPD) values were 1.81% dry matter (DM) with an RPD of 2.1, and 3.8 mg g^{-1} DM with an RPD of 1.9 for the prediction of vitamin C and ellagic acid, respectively, in whole KP fruit. The SECV and RPD values were 1.73% DM with an RPD of 2.2, and 5.6 mg g^{-1} DM with an RPD of 1.3 for the prediction of vitamin C and ellagic acid, respectively, in powdered KP samples. The results of this study demonstrated the ability of a handheld NIR instrument to predict vitamin C and ellagic acid in whole and pureed KP fruit samples. Although the RPD values obtained were not considered adequate to quantify these bioactive compounds (e.g., analytical quantification), this technique can be used as a rapid tool to screen vitamin C in KP fruit samples for high and low quality vitamin C.

Keywords: near infrared; vitamin C; ellagic acid; wild harvest; Kakadu plum

1. Introduction

Kakadu plum (KP; *Terminalia ferdinandiana* Exell, Combretaceae) is an emerging indigenous fruit originating from Northern Australia, with valuable health and nutritional characteristics and properties such as high levels of vitamin C, ellagic acid, and other polyphenolic compounds [1–6]. Kakadu plum is the most common name for this fruit, and it is found from the Kimberley (Western Australia) to Darwin (Northern Territory) and Queensland regions [1–6]. Commercial harvesting of KP fruit started in the late 1990s. While the vast majority of production is from wild harvested fruit [1–6], some commercial orchards can be found in Australia. Like many wild-harvested native foods, weather conditions, including drought, bushfires, and cyclones, might have an impact on the volume of fruit available, so production is highly variable from year to year [1–6]. The main harvest time is January, although some trees have multiple flowerings and can produce fruit up until July, depending on the region. The production of this fruit is estimated to average

15–17 tonnes per annum [1–6]. Although the KP is commercialised as whole fruit, it can be processed as a pureed or dehydrated powder (e.g., freeze dried) [7]. The dehydrated powder is used as a functional food ingredient in order to add value to a wide range of different food products (e.g., yogurts and ice creams), a common practice in the food industry [1–7]. It is well recognised that the health benefits of native plants are attributed to the content of antioxidant compounds such as natural ascorbic acid (vitamin C) and polyphenols, including gallic and ellagic acids [4,5]. These antioxidants have become very important in human health and nutrition, motivating the rapidly expanding search for plant sources containing these compounds in the wild (e.g., native plants). Kakadu plant materials (e.g., fruit and leaves) have high quantities of ellagic acid, together with the bioactive forms of vitamin C (ascorbic acid), making this plant very attractive as a source of natural antioxidants [4–6].

In recent years, applications based on the use of vibrational spectroscopy (near infrared, mid infrared, and Raman) have been utilised to quantify and monitor the composition and nutritional value in a wide range of plant and fruit materials [8–13]. In particular, the use of near infrared (NIR) spectroscopy has demonstrated that it can be a versatile tool to analyse different types of samples and conditions [8–13]. These recent developments in portable and handheld instrumentation have opened a new window for utilising these types of instruments to analyse and monitor the composition of fruit and vegetables [8–14]. In this context, the utilization of handheld instrumentation is allowing the quantification of antioxidants and bioactive compounds in native or wild harvest fruit samples like KP fruit.

Therefore, the objective of this study was to evaluate the ability of a handheld NIR instrument combined with chemometrics to measure vitamin C and ellagic acid concentrations in KP fruit samples.

2. Materials and Methods

2.1. Samples

Kakadu plum fruit samples were wild harvested in January 2020 from two different locations in the Kimberley region (Western Australia, Australia). Ten KP trees from each site were randomly selected for harvesting (approximately 50–100 fruit per tree). The samples were stored and transported to the laboratory under refrigerated conditions and then immediately stored at −80 °C for further analysis. The frozen fruit samples were thawed at room temperature (20 °C) before the NIR and reference analyses. After NIR scanning, the fruit samples were blended into a puree using a mortar and pestle. Consequently, the obtained pureed samples were analysed as a puree using the same NIR spectrophotometer as described in the section below (infrared spectroscopic measurements). Following the NIR analysis, the pureed samples were lyophilized (Lindner and May Ltd., Windsor, QLD, Australia) and finely ground using a Retsch MM301 cryomill (Retsch GmbH, Haan, Germany) in order to provide a uniform powder for the determination of vitamin C and ellagic acid. After all of the fruit samples underwent NIR scanning, representative samples were selected using principal component analysis in order to be utilised for further reference analysis and calibration development.

2.2. Infrared Spectroscopic Measurements

The NIR spectra of either whole (n = 60) or pureed (n = 60) KP fruit samples were collected using a portable NIR spectrophotometer (Micro-NIR 1700, Viavi, Milpitas, CA, USA) operating in a 950–1600 nm wavelength range, with a spectral resolution of 10 nm with no moving parts (Viavi Solutions, 2015, Milipitas, CA, USA). The NIR instrument was connected through a USB interface to a notebook computer running proprietary software (MicroNIR Prov 3.1, Viavi, Milpitas, CA, USA) for the acquisition of the diffuse reflectance spectra of the samples (Viavi Solutions, 2015, Milipitas, CA, USA). The controlling parameters for the spectral data acquisition were set at 50 min integration time and with an average of 50 scans (MicroNIR Prov 3.1, Viavi, Milpitas, CA, USA). The reference spectra for the absorbance/reflectance calculations were collected using Spectralon® after the consecutive scanning of 10 samples.

2.3. Determination of Ellagic Acid

The extraction and analysis of ellagic acid (EA) were conducted according to the method previously reported by Williams and collaborators, with some modifications [5,6]. Briefly, 100 mg of powdered samples were extracted with 80% methanol containing 0.01N HCl using a vortex, followed by sonication for 10 min. The free EA released in the supernatant (referred to as extract A) was collected after being centrifuged (3220× *g*, 5 min at 20 °C; Eppendorf Centrifuge 5810 R, Hamburg Germany), whereas the residues were extensively extracted with absolute methanol in order to completely release the remaining free EA (extract B).

In order to measure the EA existing under bound form (e.g., ellagitannins), hydrolysis was conducted following the method reported by Williams and collaborators [5,6]. The obtained extract A was added into a 5 mL Reacti-Therm vial (Fisher Scientific, Bellefonte, PA, USA) and subjected to overnight hydrolysis at 90 °C using 2N HCl. The EA released after hydrolysis was dissolved in methanol (referred to as extract C) before the UPLC-PDA analysis.

EA in three different extracts was analysed using a Waters AcquityTM UPLC-PDA System (Waters, Milford, MA, USA). The compound was separated on a Waters BEH Shield RP C18 column (100 × 2.1 mm i.d; 1.7 µm) maintained at 35 °C. The mobile phases included 0.1% formic acid (FA) in Milli-Q water (A) and 0.1% FA in methanol (B). The flow rate was 0.3 mL/min, with the following gradient elution for B: 35% B isocratic conditions for 5 min, 50% B for 10 min, and 100% B for 15 min. The contents of free EA (extracts A and B) and total free and bound EA (extracts B and C) were quantified using an external calibration curve of ellagic acid acquired at 254 nm [5,6].

2.4. Determination of Vitamin C

The extraction and analysis of vitamin C in the powder samples were conducted following a method previously described elsewhere [15]. Briefly, 100 mg of powdered KP samples were extracted with 3% meta-phosphoric acid containing 8% acetic acid and 1 mL ethylenediaminetetraacetic acid (EDTA). The reduction of dehydroascorbic acid (DHAA), which was also present in the extracts/samples, to ascorbic acid (L-AA) was performed [15,16]. The total vitamin C (L-AA + DHAA) was determined using a Waters UPLC-PDA system and a Waters HSS-T3 column (150 × 2.1 mm i.d; 1.8 µm; 25 °C), with water with 0.1% formic acid as the mobile phase (0.3 mL/min) under isocratic elution. Vitamin C was quantified using an external calibration curve of ascorbic acid acquired at 245 nm [15].

2.5. Data Analysis

The NIR spectra were pre-processed (second derivative, second order polynomial, 21 smoothing points) using The Unscrambler software (version 11, CAMO, Oslo, Norway) [17]. A principal component analysis was conducted using The Unscrambler software, after a second derivative with cross validation (full cross validation) [18]. Partial least squares regression (PLS) was used to relate the NIR spectra with the content of vitamin C and ellagic acid in the KP fruit samples analysed. To evaluate the performance of the PLS models, validations were performed on two different datasets. For the purpose of this study, the original dataset was split into two subsets of 70% (e.g., calibration) and 30% (e.g., validation), using the Kennard-Stone algorithm [19]. Thus, 40 uniformly distributed samples were selected and used in the calibration, while 20 samples were used for validation. By performing data partitioning, knowledge of the training dataset did not affect the test dataset, and the predictive power of the created model subsequently increased. Leave-one-out cross-validation was applied on the calibration set for internal validation, and the test set was used to externally validate the generated models. The coefficient of determination

(R^2), the standard error in cross validation (SECV), and the residual predictive deviation (RPD) were used to evaluate the calibration models developed [18,20–22].

3. Results and Discussion

Table 1 reports the descriptive statistics (e.g., average, standard deviation, range, and coefficient of variation) for the measurement of the dry matter, vitamin C, and ellagic acid content in the KP fruit samples used to develop the NIR calibrations. Table 2 shows the cross validation and validation statistics for the prediction of vitamin C and ellagic acid in the set of whole and pureed KP fruit samples analysed. As stated above, the SECV and the RPD (SD/SECV) were used to evaluate the ability of the PLS models developed to predict these parameters [18,20]. SECV is a quantitative measure of how precise the samples are predicted during validation where the bias is a systematic deviation of the predicted values from the true value due to a particular measurement method [18,20]. The SECV and RPD values were 1.81% dry matter (DM) with an RPD of 2.1, and 3.8 mg g^{-1} DM with an RPD of 1.9 for the prediction of vitamin C and ellagic acid, respectively, in the set of whole KP fruit samples. Using the set of pureed KP samples, the SECV and RPD values were 1.73% DM with an RPD of 2.2, and 5.6 mg g^{-1} DM with an RPD of 1.3 for the prediction of vitamin C and ellagic acid, respectively. According to other authors, an RPD value between 2 and 2.5 might indicate that rough quantitative predictions could be possible, while a value between 2.5 and 3 or above might be associated with good and excellent prediction accuracy [18,20–23]. The RPD values in this study were between 1.3 to 2.2 for the prediction of vitamin C and ellagic acid. Similar SECV values were reported by other authors using mid infrared spectroscopy to predict ellagic acid in coastal oak samples [24].

Table 1. Descriptive statistics for the measurement of vitamin C and ellagic acid in Kakadu plum fruit samples analysed using NIR spectroscopy.

	% DM	VIT C (% DM)	EA (mg g^{-1} DM)
Average	16.4	12.5	20.64
SD	1.2	3.81	7.7
Minimum	14.2	7.8	7.6
Maximum	18.7	19.3	31.5
CV (%)	7.3	30.4	37.4

CV—coefficient of variation (CV = SD/mean); DM—dry matter; EA—total ellagic acid; SD—standard deviation; VIT C—vitamin C.

Table 2. Cross validation and validation statistics for the prediction of ellagic acid and vitamin C in whole and pureed Kakadu plum sample analyses using near infrared reflectance spectroscopy.

		R^2_{CV}	SECV	Slope	Bias	LV	RPD_{CV}	r	SEP
Whole	VIT C (% DM)	0.55	1.81	0.53	0.029	8	2.1	0.85	2.0
	EA (mg g^{-1} DM)	0.57	3.8	0.61	−0.007	11	1.96	0.55	7.5
Puree	VIT C (% DM)	0.86	1.73	0.87	0.10	8	2.2	0.89	1.9
	EA (mg g^{-1} DM)	0.48	5.6	0.57	0.002	11	1.3	0.56	6.2

CV—cross validation; DM—dry matter; LV—number of optimal latent variables used to develop the models; VIT C—vitamin C; EA—total ellagic acid; R^2CV—coefficient of determination in cross validation; r—correlation coefficient in prediction; RPD—SD/SECV; SECV—standard error for cross validation; SEP—standard error of prediction.

R^2 indicates the percentage of variance present in the true component values, which will be reproduced in the prediction (18, 20–23). Depending on the R^2 values obtained during the calibration process, the NIR models can be classified as follows: possessing a low correlation ($0.26 < R^2 < 0.49$), models that can be used to discriminate between a low and high composition of samples ($0.50 < R^2 < 0.64$), models that can be used for a rough prediction of the composition ($0.65 < R^2 < 0.81$), possessing a good correlation ($0.82 < R^2$

< 0.90), and having excellent precision ($R^2 > 0.90$) [18,20–23]. The PLS calibration models developed using the pureed KP samples explained between 48% and 86% of the variation related to vitamin C and ellagic acid, while 55% to 57% of the variation was explained in the calibration models using the whole KP fruit samples. The observed differences in the PLS models were associated with sample presentation (whole vs. pureed fruit).

It has been reported that the NIR spectra are comprised of wide bands originating from overlapping absorptions corresponding to overtones and combinations of vibrational modes involving C-H, O-H, and N-H chemical bonds [8,25,26]. Although the water absorption bands related to the O-H bonds are predominant in the NIR spectra of fruit such as KP fruit, other molecules can be measured [8,25,26]. Carbohydrates, organic acids, proteins, and other minor compounds can exhibit wide absorption bands as a result of complex hydrogen bonding interactions with different molecules in the NIR wavelength range [8,24,25]. Therefore, the interpretation of the NIR spectra is not as straight forward as the interpretation of the MIR region [8,25,26].

In order to understand the basis of the NIR calibrations developed, the PLS loadings were analysed and interpreted for each of the sample presentations used to develop the calibrations for vitamin C and ellagic acid (e.g., whole or pureed fruit; Figures 1 and 2). The relationships between the wavelength and PLS latent variables/loadings imply that these wavelengths contribute to explaining the developed models [18,20–22]. Therefore, the value and direction (e.g., positive and negative) of the PLS loading indicated the contributions of individual wavelengths to the model [18,20–22]. It has been reported that when PLS models are developed for the same parameters, using different pre-processing or sampling presentation modes for the same sample, they can utilise different wavelengths or loadings. In this study, sample presentations (whole vs. pureed) were shown to have an effect by explaining the observed differences in the PLS calibrations and loadings. The loadings used by the PLS calibrations for the measurement of vitamin C and ellagic acid in the KP puree samples are shown in Figure 1. The loadings for vitamin C were observed at wavelengths of around 1137 nm (C-H combination, aromatic groups), 1217 nm ($C-H_2$), 1299 nm (first overtone of C-H combination), 1465 nm (N-H associated with secondary amines), and 1558 nm (O-H), whereas for ellagic acid, the most important wavelengths were observed at 1174 nm (C-H), 1310 nm (first overtone of C-H combination), 1410 nm (O-H bonds), and 1510 nm (N-H amide) [8,25–33]. The PLS loadings observed for the calibrations developed using the whole KP fruit samples are shown in Figure 2. The main loadings were observed at 1093 nm (C-H, aromatic groups), 1347 nm (C-H), 1465 nm (N-H), and 1570 nm (N-H) for vitamin C, while for ellagic acid, four wavelengths were observed to influence the models, at 1155 nm (C-H), 1242 nm (C-H), 1440 nm, and 1508 nm (C-H and N-H) [8,25–33]. It has been observed that the calibration models for the same parameters used similar wavelengths, and these might indicate that the sample presentation (whole vs. pureed fruit samples) might not have a greater effect on the information collected by the NIR instrument for the prediction of the bioactive compounds in the set of KP fruit samples analysed.

Figure 3 shows the scatter plot for the validation of the measurement of vitamin C and ellagic acid in the pureed samples. The influence of the region or origin of the samples was observed upon cross-validation models developed for vitamin C (bimodal distribution as a result of region). However, this trend was not observed for the prediction of ellagic acid in the KP fruit samples analysed. In addition, one and three outlier samples were observed in the prediction of vitamin C and ellagic acid, respectively. A detailed analysis of these outlier samples indicated that they corresponded to spectral outliers. These results are in agreement with those reported by other authors, who indicated that region might have an effect on the concentration of some of these bioactive compounds [1–3,7]. It is well known that vitamin C is an important parameter because of its important health and antioxidant properties, which have received a great deal of attention, thus necessitating the development of rapid analytical methods [26]. However, some authors have reported unsatisfactory results using short wavelengths in the NIR region or when the samples contain low concentrations of vitamin C (less than 10 g L^{-1}) [22,28–33]. Another reported issue might be related to the effect of moisture and its interference when determining the presence of compounds with low concentrations. Recently, Oliveira-Folador and collaborators [33] suggested that the high water content of the pulp of fruit (approximately 84%) contributes to the inherent complexity of NIR spectra. This might also be explained by the fact that the NIR spectral range used is highly sensitive to elements that modify light diffusion, such as physical structure and the presence and content of water in the sample [20,21]. The physical structure of the fruit has been reported to have a large effect on the acquisition of spectra, and this is strongly influenced by the light scattering phenomena, as reported by other authors in different types of fruit and vegetables [22,28–32]. Figure 4 shows the principal component score plot of the KP samples scanned as whole and pureed fruit. Two groups were observed related to the sample presentation used. Whole samples tended to scatter along principal component one (50% of the variation), while most of the pureed samples were clustered together. It is also important to highlight that the NIR spectrum of fresh materials is essentially composed of a large set of overtones and combination bands. This combination, together with the complex chemical composition of a typical fruit or vegetable, makes the near infrared spectrum highly complex [8,25]. Regardless of these issues, the NIR region used in this study showed a high applicability for the rapid screening of samples for high, medium, and low vitamin C and ellagic acid.

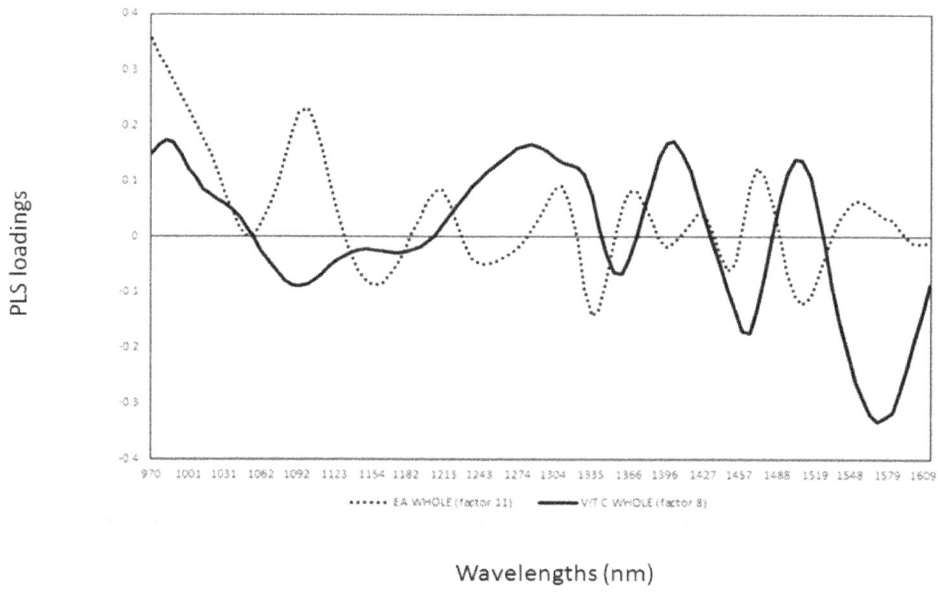

Figure 1. Partial least square loadings for the measurement of vitamin C and ellagic acid in whole Kakadu plum fruit samples.

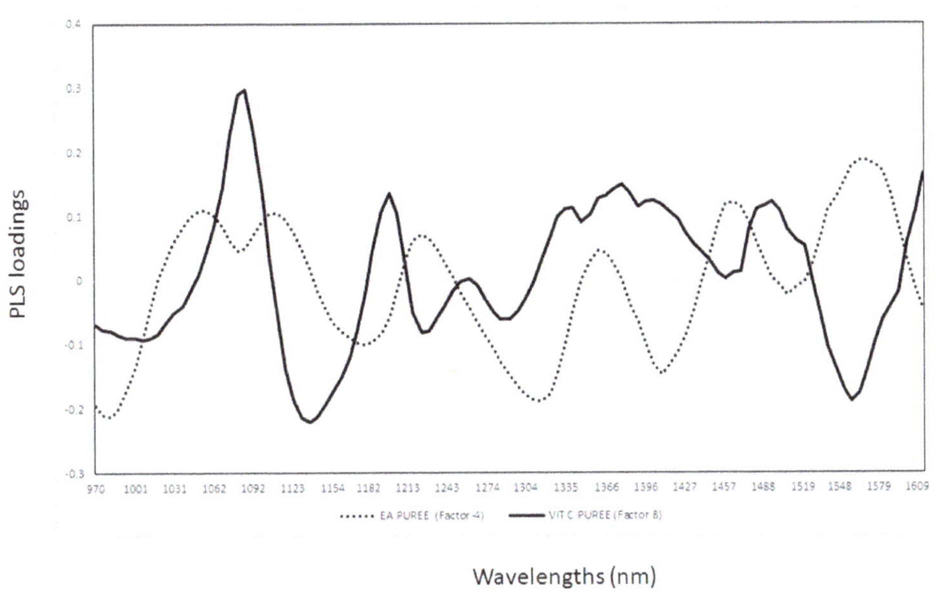

Figure 2. Partial least square loadings for the measurement of vitamin C and ellagic acid in pureed Kakadu plum fruit samples.

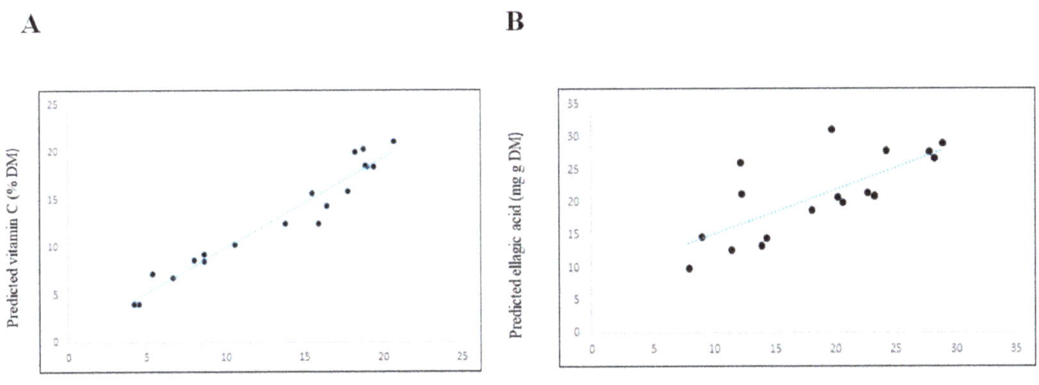

Figure 3. Scatter plot for the validation (n = 20) of the measurement of vitamin C (Panel (**A**)) and ellagic acid (Panel (**B**)) in the pureed Kakadu plum samples.

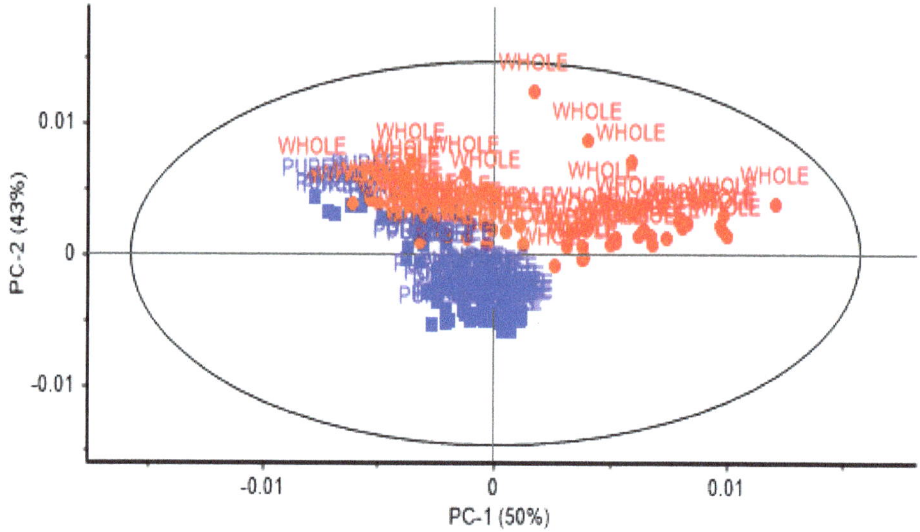

Figure 4. Principal component score plot of Kakadu plum samples analysed as pureed or as whole fruit using near infrared spectroscopy.

4. Conclusions

The results of this study showed the ability of a handheld NIR instrument to predict vitamin C and ellagic acid in both whole and pureed KP fruit samples. Although the RPD values obtained are not considered adequate to quantify these bioactive compounds, they can be used to quickly screen the fruit for high- and low-quality vitamin C. The handheld instrument used in this study can be an alternative for rapid and throughput screening of raw materials in remote areas, where it might not be appropriate to use other types of instruments to assess fruit quality (e.g., bioactive compounds). However, further studies are needed to optimize the prediction models for these bioactive compounds and to evaluate the effect of region/origin and harvest (years) in order to make the models more robust for routine applications.

Author Contributions: Conceptualization, Y.S. and D.C.; methodology, Y.S., M.N. and D.C.; software, A.D.T.P. and E.B.; validation, D.C.; formal analysis, A.D.T.P. and E.B.; investigation, E.B. and Y.S.; data curation, D.C.; writing—original draft preparation, D.C. and Y.S.; writing—review and editing, D.C., Y.S., M.N., H.E.S., A.D.T.P. and E.B.; supervision, Y.S.; funding acquisition, Y.S. All authors have read and agreed to the published version of the manuscript.

Funding: Funding support from the CRC for Developing Northern Australia Limited Project AT.2.1718031, for improving the efficiency of Kakadu plum value chains to grow a robust and sustainable industry, and from the Australian Research Council (ARC) Industrial Transformation Training Centre (ITTC) for Uniquely Australian Foods (grant number: IC180100045).

Acknowledgments: The authors acknowledge the Traditional Owners of the lands on which the *Terminalia ferdinandiana* fruit were harvested, and respect the knowledge and experience the traditional owners have regarding the care, harvesting, and use of these plants.

Conflicts of Interest: The authors declare no conflict of interest.

References

1. Netzel, M.; Netzel, G.; Tian, Q.; Schwartz, S.; Konczak, I. Native Australian fruits—A novel source of antioxidants for food. *Innov. Food Sci. Emerg. Technol.* **2007**, *8*, 339–346. [CrossRef]
2. Konczak, I.; Maillot, F.; Dalar, A. Phytochemical divergence in 45 accessions of *Terminalia ferdinandiana* (Kakadu plum). *Food Chem.* **2014**, *151*, 248–256. [CrossRef]
3. Konczak, I.; Zabaras, D.; Dunstan, M.; Aguas, P. Antioxidant capacity and hydrophilic phytochemicals in commercially grown native Australian fruits. *Food Chem.* **2010**, *123*, 1048–1054. [CrossRef]
4. Konczak, I.; Zabaras, D.; Dunstan, M.; Aguas, P.; Roulfe, P.; Pavan, A. *Health Benefits of Australian Native Foods: An Evaluation of Health-Enhancing Compounds*; RIRDC: Canberra, Australia, 2009.
5. Williams, D.J.; Edwards, D.; Pun, S.; Chaliha, M.; Sultanbawa, Y. Profiling ellagic acid content: The importance of form and ascorbic acid levels. *Food Res. Int.* **2014**, *66*, 100–106. [CrossRef]
6. Williams, D.J.; Edwards, D.; Pun, S.; Chaliha, M.; Burren, B.; Tinggi, U.; Sultanbawa, Y. Organic acids in Kakadu plum (*Terminalia ferdinandiana*): The good (ellagic), the bad (oxalic) and the uncertain (ascorbic). *Food Res. Int.* **2016**, *89*, 237–244. [CrossRef]
7. Sultanbawa, Y. Processing of native plant foods and ingredients. In *Australian Native Plants; Cultivation and Uses in the Health and Food Industries*; CRC Press: Boca Raton, FL, USA, 2016.
8. Nicolai, B.M.; Beullens, K.; Bobelyn, E.; Peirs, A.; Saeys, W.; Theron, K.I.; Lammertyn, J. Non-destructive measurement of fruit and vegetable quality by means of NIR spectroscopy: A review. *Postharvest Biol. Technol.* **2007**, *46*, 99–118. [CrossRef]
9. Sorak, D.; Herberholz, L.; Iwascek, S.; Altinpinar, S.; Pfeifer, F.; Siesler, H.W. New developments and applications of handheld Raman, mid-infrared, and near infrared spectrometers. *App. Spectrosc. Rev.* **2012**, *47*, 83–115. [CrossRef]
10. Bec, K.B.; Huck, C.W. Breakthrough Potential in Near-Infrared Spectroscopy: Spectra Simulation. A Review of Recent Developments. *Front. Chem.* **2019**. [CrossRef]
11. Saeys, W.; Trong, N.N.D.; Van Beers, R.; Nicolaï, B.M. Multivariate calibration of spectroscopic sensors for postharvest quality evaluation: A review. *Postharvest Biol. Technol.* **2019**, *158*, 110981. [CrossRef]
12. Cattaneo, T.; Stellari, A. Review: NIR Spectroscopy as a Suitable Tool for the Investigation of the Horticultural Field. *Agronomy* **2019**, *9*, 503. [CrossRef]
13. Beć, K.B.; Grabska, J.E.; Huck, C.W. Near-Infrared Spectroscopy in Bio-Applications. *Molecules* **2020**, *25*, 2948. [CrossRef]
14. Crocombe, R.A. Portable Spectroscopy. *Appl. Spectrosc.* **2018**, *72*, 1701–1751. [CrossRef] [PubMed]
15. Campos, F.M.; Ribeiro, S.M.R.; Della Lucia, C.M.; Pinheiro-Sant'Ana, H.M.; Stringheta, P.C. Optimization of methodology to analyze ascorbic and dehydroascorbic acid in vegetables. *Química Nova* **2009**, *32*, 87–91. [CrossRef]
16. Spinola, V.; Mendes, B.; Camara, J.S.; Castilho, P.C. An improved and fast UHPLC-PDA methodology for determination of L-ascorbic and dehydroascorbic acids in fruits and vegetables. Evaluation of degradation rate during storage. *Anal. Bioanal. Chem.* **2012**, *403*, 1049–1058. [CrossRef]
17. Savitzky, A.; Golay, M.J.E. Smoothing and Differentiation of Data by Simplified Least Squares Procedures. *Anal. Chem.* **1964**, *36*, 1627–1639. [CrossRef]
18. Bureau, S.; Cozzolino, D.; Clark, C. Contributions of Fourier-transform mid infrared (FT-MIR) spectroscopy to the study of fruit and vegetables: A review. *Postharvest Boil. Technol.* **2019**, *148*, 1–14. [CrossRef]
19. Kennard, R.W.; Stone, L.A. Computer Aided Design of Experiments. *Technometrics* **1969**, *11*, 137–148. [CrossRef]
20. Williams, P.; Dardenne, P.; Flinn, P. Tutorial: Items to be included in a report on a near infrared spectroscopy project. *J. Near Infrared Spectrosc.* **2017**, *25*, 85–90. [CrossRef]
21. Cozzolino, D. The sample, the spectra and the maths–the critical pillars in the development of robust and sound vibrational spectroscopy applications. *Molecules* **2020**, *25*, 3674. [CrossRef] [PubMed]
22. Williams, P.; Norris, K. *Near-Infrared Technology in the Agricultural and Food Industries*; American Association of Cereal Chemist: St. Paul, MN, USA, 2002.
23. Toledo-Martín, E.M.; García-García, M.C.; Font, R.; Moreno-Rojas, J.M.; Gómez, P.; Salinas-Navarro, M.; Del Río-Celestino, M. Application of visible/near-infrared reflectance spectroscopy for predicting internal and external quality in pepper. *J. Sci. Food Agric.* **2016**, *96*, 3114–3125. [CrossRef]
24. Conrad, A.O.; Rodriguez-Saona, L.E.; McPherson, B.A.; Wood, D.L.; Bonello, P. Identification of Quercus agrifolia (coast live oak) resistant to the invasive pathogen Phytophthora ramorum in native stands using Fourier transform infrared (FT-IR) spectroscopy. *Front. Plant Sci.* **2014**, *5*, 521. [CrossRef]
25. Murray, I.; Williams, P. Chemical Principals of Near-Infrared Technology. In *Near Infrared Technology in the Agricultural and Food Industries*; American Association of Cereal Chemists: St. Paul, MN, USA, 1987; pp. 143–167.
26. Workman, J., Jr.; Weyer, L. *Practical Guide to Interpretive Near-Infrared Spectroscopy*, 2nd ed.; Taylor & Francis Group: Abingdon, UK; CRC Press: Boca Raton, FL, USA, 2012.
27. Panicker, C.Y.; Varghese, H.T.; Philip, D. FT-IR, FT-Raman and SERS spectra of Vitamin C. *Spectrochim. Acta Part A* **2006**, *65*, 802–804. [CrossRef] [PubMed]
28. Kramchote, S.; Nakano, K.; Kanlayanarat, S.; Ohashi, S.; Takizawa, K.; Bai, G. Rapid determination of cabbage quality using visible and near-infrared spectroscopy. *LWT Food Sci. Technol.* **2014**, *59*, 695–700. [CrossRef]
29. Maniwara, P.; Nakano, K.; Boonyakiat, D.; Ohashi, S.; Hiroi, M.; Tohyama, T. The use of visible and near infrared spectroscopy for evaluating passion fruit postharvest quality. *J. Food Eng.* **2014**, *143*, 33–43. [CrossRef]

30. Pissard, A.; Fernaandez Pierna, J.A.; Baeten, V.; Sinnaeve, G.; Lognay, G.; Mouteau, A.; Dupont, P.; Rondia, A.; Lateur, M. Non-destructive measurement of vitamin C, total polyphenol and sugar content in apples using near-infrared spectroscopy. *J. Sci. Food Agric.* **2013**, *93*, 238–244. [CrossRef]
31. Rodrigues Borba, K.; Spricigo, P.C.; Aykas, D.P.; Mitsuyuki, M.C.; Colnago, L.A.; Ferreira, M.D. Non-invasive quantification of vitamin C, citric acid, and sugar in 'Valencia' oranges using infrared spectroscopies. *J. Food Sci. Technol.* **2020**. [CrossRef]
32. Tierno, R.; López, A.; Riga, P.; Arazuri, S.; Jarén, C.; Benedicto, L.; Ruiz de Galarreta, J.I. Phytochemicals determination and classification in purple and red fleshed potato tubers by analytical methods and near infrared spectroscopy. *J. Sci. Food Agric.* **2016**, *96*, 1888–1899. [CrossRef]
33. Oliveira-Foladora, G.; de Oliveira Bicudo, M.; Forville de Andrade, E.; Renard, C.M.G.; Bureau, S.; de Castilhos, F. Quality traits prediction of the passion fruit pulp using NIR and MIR Spectroscopy. *LWT Food Sci. Technol.* **2018**, *95*, 172–178. [CrossRef]

sensors

MDPI

Article

Rapid Estimation of Potato Quality Parameters by a Portable Near-Infrared Spectroscopy Device

Olga Escuredo *, Laura Meno, María Shantal Rodríguez-Flores and Maria Carmen Seijo

Department of Vegetal Biology and Soil Sciences, Faculty of Sciences, University of Vigo, 32004 Ourense, Spain; laura.meno@uvigo.es (L.M.); mariasharodriguez@uvigo.es (M.S.R.-F.); mcoello@uvigo.es (M.C.S.)
* Correspondence: oescuredo@uvigo.es

Abstract: The aim of the present work was to determine the main quality parameters on tuber potato using a portable near-infrared spectroscopy device (MicroNIR). Potato tubers protected by the Protected Geographical Indication (PGI "Patata de Galicia", Spain) were analyzed both using chemical methods of reference and also using the NIR methodology for the determination of important parameters for tuber commercialization, such as dry matter and reducing sugars. MicroNIR technology allows for the attainment/estimation of dry matter and reducing sugars in the warehouses by directly measuring the tubers without a chemical treatment and destruction of samples. The principal component analysis and modified partial least squares regression method were used to develop the NIR calibration model. The best determination coefficients obtained for dry matter and reducing sugars were of 0.72 and 0.55, respectively, and with acceptable standard errors of cross-validation. Near-infrared spectroscopy was established as an effective tool to obtain prediction equations of these potato quality parameters. At the same time, the efficiency of portable devices for taking instantaneous measurements of crucial quality parameters is useful for potato processors.

Keywords: NIR spectrometer; intact potato; dry matter; reducing sugars; chemometrics; MPLS

Citation: Escuredo, O.; Meno, L.; Rodríguez-Flores, M.S.; Seijo, M.C. Rapid Estimation of Potato Quality Parameters by a Portable Near-Infrared Spectroscopy Device. *Sensors* **2021**, *21*, 8222. https://doi.org/10.3390/s21248222

Academic Editors: Mercedes Del Río Celestino and Rafael Font Villa

Received: 12 November 2021
Accepted: 6 December 2021
Published: 9 December 2021

1. Introduction

The potato is a traditional crop and the base of human diet in many world regions. In the past two decades, potato production has experienced a greater growth compared with other tubers, due to its high yield and human nutritional value [1,2]. This increase is favored by the need to meet the increased demand for food due to the world population growth [2]. Undoubtedly, the potatoes have played an important role in food availability, and are currently still holding this role.

The potatoes are characterized as a good source of starch with unique characteristics compared with the basic starches of cereals. In addition to fresh consumption, tubers can be destined for the processing industry (such as chips, flakes, dry and frozen potatoes), for additional food ingredients, such as tuber seed for field cultivation, animal feed, and in the chemical industry as a source of starch and ethanol [2,3]. The potato in Spain is mainly intended for fresh consumption, industrial processing, and animal feed. The highest production of this tuber is found in the Northwest of Spain. Specifically, in Galicia, this traditional agricultural activity is an important source of economic income for families [4], in order for the production to be covered by the Protected Geographical Indication (PGI) Patata de Galicia [5]. Kennebec, Agria, and Fina de Carballo are the potato varieties, which are protected by the designation of origin comprising the largest extensions of this crop in Galicia.

The potato tubers are stored for several months after their harvest, in order to meet the market demands throughout the year. This increases their marketability and generates an additional benefit for growers, processors, and consumers [6,7]. The composition of the potato varies with the storage time and cultivar type [7–9]. Generally, the respiration and the evaporation rate increase during maturity and tuber storage. This produces an increase in weight loss and peroxidase enzymatic activity together with a decrease in starch and ascorbic acid content. As a consequence, reducing sugars are synthesized [8,10,11]. Therefore, the nutritional quality degradation of potato during their storage is induced, with changes in starch content, dry matter, and reducing sugars [7,8,12]. Reducing sugar is an important indicator for evaluating the quality of raw material in the processing of potato industry [13]. Excess levels of reducing sugars cause an unacceptable non-enzymatic brown color for fried products, due to the reaction between the reducing sugars and the α-amino groups of amino acids [6,13,14]. Therefore, high reducing sugars in tubers are not suitable for processing.

The industry has innovated and invested in improved post-harvest storage, preserving potato quality for the seed, fresh, and processing sectors [3,15]. The acceptance of the tuber in fresh markets depends mainly on its external appearance [9], but with dependence on the internal composition, which is determined by destructive analytical procedures. The amount of dry matter and reducing sugars are the main physicochemical parameters that determine the industrial yield, quality, and flavor of potato tubers. The most used conventional analytical techniques are colorimetric and titration measurement methods [13]. In addition, the operators themselves based on their professional experience select the potato tubers by hand on the processing belt. However, this operation is not enough to guarantee optimal quality and compliance with quality standards. Analytical methods commonly employed to determine the main compounds of potatoes do not seem to be suitable for in-line applications in the food industry since they require a large amount of time and are destructive. Therefore, ensuring the minimum level of quality of basic foods that is accepted by the consumer requires assessing its quality by swift and non-destructive techniques [6,16,17]. As a result, the importance of quantifying the dry matter and the sugars in-line during potato processing ensures optimal quality and discards unsuitable tubers for marketing.

In the last years, visible- and near-infrared (VIS-NIR) spectroscopy has contributed to providing non-destructive methods for the evaluation of the internal quality of fresh fruits and vegetables or cereals [7,17–20]. The advantages of NIRS are time saving, offering the ability to record many quality characteristics or ingredients with a single measurement. Some studies have been conducted to test the near-infrared spectroscopy measuring quality parameters of potatoes, such as sugars or dry matter content in laboratory [6,12,14,17,21–25]. According to some researchers, the results of these studies are difficult to compare since some are focused on whole tubers, unpeeled or peeled, in cross-sections or crushed in the form of puree [6,25]. However, these researchers have shown the potential of the NIR technology for the application in the potato industry. With the appearance of miniaturized or portable spectrometers, NIR spectral analysis became feasible directly in the field or during food processing [26,27]. The advance in the technological improvement of portable systems of NIR spectrometers is displacing the benchtop instruments, due to the advantages in the food industry [28]. In addition, this technique is favored by the increased availability of low-cost portable devices, which can be more easily implemented into the processing line. The estimation of potato quality parameters has not been applied with modern and portable systems of this type.

Currently, the food industry faces the challenge of the demand for high quality products with the possibility of monitoring much of the product in real time, but meeting with the requirements of food safety and traceability [29]. Spectroscopic sensors are optimal instruments for real time analysis of analytical techniques [27], with direct measurements in situ, which are very flexible and rugged, without the use of chemical reagents and waste [16,17,20,26,29]. The objective of this paper was to investigate the feasibility for measuring the main quality parameters of intact potatoes by means of a portable near-infrared (MicroNIR) spectroscopy device. The estimation of dry matter and reducing sugars content in potatoes was validated with NIR-spectra data and chemometrics.

2. Materials and Methods

2.1. Potato Samples

NIR recordings were performed directly on the tubers before chemical analysis in the laboratory. A superficial cleaning of the tubers was carried out to eliminate possible particles that could interfere with the spectral acquisition and obtain a representative sample of the whole tubers. The spectral and chemical measurements were carried out on six replicates (N = 534), resulting in an average value (N = 89) that was used for the subsequent chemometric treatments.

The sampled potatoes were grown during the crop seasons of 2019 and 2020 in A Limia region (Northwest of Spain). Two types of potato varieties were analyzed: Kennebec (N = 48) and Agria (N = 41). The choice of these types of potatoes is due to the different commercial destinations in this geographical region. Kennebec is intended for fresh consumption, and Agria for the potato processing industry. These potato cultivars are the ones with the highest production in the geographical area and are covered under the designation "PGI Patata de Galicia" recognized by the European Union.

2.2. Destructive Measurements of the Reference Quality Parameters

For reference analytical procedures, the tubers were gently washed to remove traces of soil adhering to the skin. Once the potatoes were dried, they were cut into four pieces for chemical analysis. Two alternate parts of the tuber were taken for the dry matter analysis by thermogravimetry, and the other two parts were used for the quantification of the reducing sugars content by a colorimetry method.

2.2.1. Dry Matter Content

A piece of 5 g of each fresh potato was weighed to obtain the fresh weight (FW). Then, the sample cubes were placed in a dryer at 60 °C for 24 h. After this time, the samples were weighed to obtain the dry weight (DW). The dry matter content expressed in percentage was calculated according to Equation (1), based on the weight before and after drying.

$$\text{Dry matter (\%)} = [(\text{FW} - \text{DW})/\text{FW}] \times 100 \quad (1)$$

2.2.2. Reducing Sugars Content

The potato pieces selected for the determination of reducing sugars were crushed to form a puree. The potato solutions were prepared with 50 g of each potato puree dissolved in 200 g of distilled water. Then, 5 mL of the potassium oxalate solution (5%), 5 mL of the zinc acetate solution (0.1 M), and 5 mL of the potassium ferrocyanide solution (10.6%) were added to each mashed potato mixture (potato solution) to remove the reducing materials that were not sugars. At the same time, a blank solution was prepared. Thereafter, the potato and blank solutions were filtered. The oxidation of the reduced sugars was carried out with the ferricyanide solution. For this, 500 μL of the filtrated sample was deposited in a test tube with 10 mL of ferricyanide for 15 min in a boiling bath. The more intense the yellow color of the oxidation-reduction reaction, the greater the amount of ferricyanide remained unreacted, and the sample contained less reducing sugars. Finally, the intensity of the oxidation-reduction reaction of the solutions was measured by spectrometry at 422 nm at room temperature. Glucose solutions (0.2–0.8 g/L) as a reference standard were used for the calibration curve (R^2 = 0.99). The reducing sugars content was expressed in g/100 g.

2.3. Near-Infrared Spectroscopy: Instrumentation and Spectral Data Acquisition

The NIR measurements of tuber samples were obtained using the portable MicroNIR Pro v2.5 equipment (MicroNIR 1700 ES, VIAVI, Santa Rosa, CA, USA) coupled to an instrument that is designed to measure the diffuse reflectance in the NIR region of the electromagnetic spectrum [30]. The portable MicroNIR system is easily handled and sized (45 mm diameter × 42 mm height; 60 g of weight), and it is equipped with a 128-pixel detector array [30]. The MicroNIR system employs a linear variable filter (LVF) as the dispersing element. The LVF is connected to a linear indium gallium arsenide (InGaAs array detector) into the equipment, which results in an extremely compact and rugged spectral engine with no moving parts [28]. Uncooled detectors of this type are often used since they offer good performance and cover the major part of the NIR spectral region [27,28]. The ultra-compact spectroscopic engine is coupled with a tungsten lamps diffuse illumination system.

NIR measurements were taken by the direct application with the MicroNIR spectrometer on tubers. Six replicate spectra were recorded for each sample and the average of the spectra was calculated (Figure 1). Spectra were recorded using the instrument acquisition software MicroNIR™ Pro v.2.2 (VIAVI, Santa Rosa, CA, USA) at intervals of 6 nm in the spectra in a range between 900–1700 nm. Spectral data were downloaded directly from the NIR equipment to a laptop connected through a USB port. However, this miniaturized spectrometer has the advantage of operating while it is connected by an USB interface to a tablet or wirelessly connected to a smartphone [28]. MicroNIR used a Spectralon® ceramic tile as a white reference (100% reflectance) of polytetrafluoroethylene (~99%). The obtained spectra were combined into the spectral matrix, where the diffuse reflectance signal of the NIR spectrum is expressed as reflectance (R), using the values of log (1/R) for the chemometric analyses.

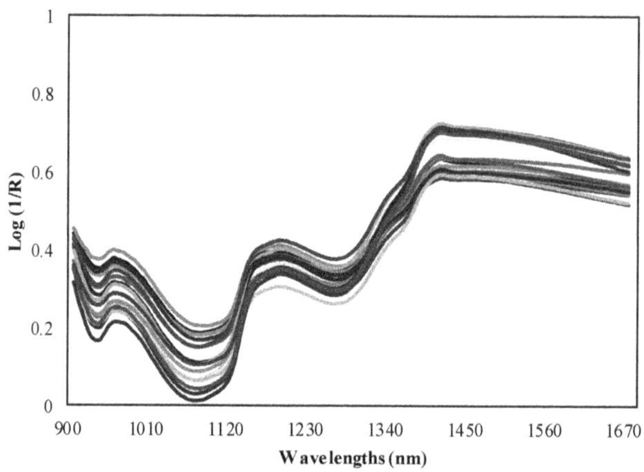

Figure 1. NIR spectra measured by the MicroNIR spectrometer.

2.4. Chemometric Analysis

First, the data spectral matrix was subjected to principal component analysis (PCA) to perform the spectral selection of samples, maintaining the spectral variability of the original matrix. The methods of spectra pre-processing included the mathematical procedures for correction and improvement of spectra, which were applied before the qualitative and/or quantitative interpretation of spectral data [31]. The fluctuations or drift of the spectral baseline were reduced through normalization procedures as well as spectra derivation [27]. The applied pre-treatments to eliminate spectral dispersion effects were multiplicative dispersion correction (MSC), standard normal variant (SNV), DeTrend (DT) or SNV-DT [32]. The calibration for the quality parameters was obtained after removing the samples for spectral reasons, according to the Mahalanobis distance (H criterion = 3) and chemical reasons (T criterion $\geq$ 2.5) [20]. The mathematical treatments were also used to develop NIRS calibrations considering a code of four digits (for example, 1,4,4,1). This encoding explains the first digit as the number of the derivative, the second digit as the interval over which the derivative was calculated, the third as the number of data points in an average or smoothing, and the fourth as the second smoothing. Then, the samples were selected by this procedure to establish the calibration set, and the best treatment was chosen later to calibrate each quality parameter independently.

NIR models were developed using 89 potato samples: 70 samples for the calibration group and 19 samples for the external validation group. Partial least squares (PLS) regression was used to obtain the models with the best prediction performance, taking into account the different spectral pre-treatments. Calibration equations were performed by modified partial least squares regression (MPLS) using the raw spectral data and testing the different spectral treatments, as well as allocating the corresponding reference values to each sample. During the processing of this method, the cross-validation is recommended in order to select the optimum number of factors and to avoid overfitting [33]. The group of calibration samples is divided into a series of subsets in order to perform cross-validation. Then, each subset is validated with calibration, which is developed on the other samples [33]. Finally, several statistics were considered to evaluate the predictive capacity of the equations obtained. The standard error of cross-validation (SECV) is considered a good estimate for the prediction capability of the equation [33]. The ratio performance deviation (RPD) is a non-dimensional statistic for the evaluation of a NIR spectroscopy calibration model [34,35], which is the relation between the standard deviation of the reference chemical values (SD) and the root mean square error of prediction (SEP) in the NIR model and the standard error of cross-validation (SECV). The statistics used to select the best calibration equations were multiple correlation coefficients (RSQ) and the standard error of cross-validation (SECV). The software WinISI II v.1.50 (Infrasoft International, LLC, Silver Spring, MD, USA) was used for chemometric processing.

3. Results

3.1. Quantified Reference Data on Tubers: Dry Matter and Reducing Sugars

The descriptive analyses (mean, minimum, maximum, and relative standard deviation) of the dry matter and reducing sugars, which are quantified in the tubers are summarized in Table 1, according to the potato cultivar. The data were presented according to the two groups established for the NIR treatment: 70 samples constituted the denominated calibration group, and 19 samples were used for the validation group.

Table 1. An overview of the samples by potato cultivar and distribution by the calibration and validation set.

		Mean	SD	Min	Max
Samples set by potato cultivar					
Kennebec (N = 48)	Dry matter (%)	19.88	1.63	16.00	22.10
	Reducing sugar (g/100 g)	0.23	0.09	0.15	0.49
Agria (N = 41)	Dry matter (%)	20.19	1.04	17.30	22.20
	Reducing sugar (g/100 g)	0.15	0.04	0.10	0.37
Calibration set (N = 70)					
	Dry matter (%)	19.67	2.07	16.0	22.0
	Reducing sugar (g/100 g)	0.19	0.08	0.10	0.49
Validation set (N = 19)					
	Dry matter (%)	19.89	1.25	17.80	22.0
	Reducing sugar (g/100 g)	0.20	0.09	0.12	0.43
Total sample set (N = 89)					
	Dry matter (%)	20.03	1.39	16.00	22.20
	Reducing sugar (g/100 g)	0.19	0.08	0.10	0.49

SD: Relative standard deviation; Max: Maximum; Min: Minimum.

The mean dry matter content was similar between the tubers of Agria and Kennebec ($p = 0.68$) (Figure 2), with a mean value of 20.19% and 19.88%, respectively (Table 1). The box and whisker plot showed a higher range in the dry matter content of Kennebec tubers, with values between 16.0% and 22.1% (Figure 2), and a relative standard deviation of 1.63%. Regarding the reducing sugars, Kennebec cultivar had a significantly higher content than Agria ($p < 0.0001$), with a mean value of 0.23 g/100 g, and maximum value of 0.49 g/100 g (Figure 2). Agria tubers had a mean value and maximum value of 0.15 g/100 g and 0.37 g/100 g, respectively (Table 1). Therefore, the greater relative standard deviation in reducing sugars content in Kennebec tubers was found (0.09 g/100 g).

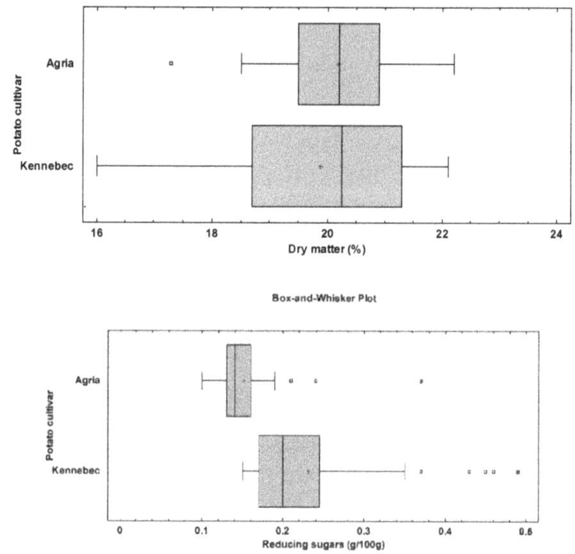

Figure 2. Box and whisker plot for dry matter (p-value = 0.68) and reducing sugars (p-value < 0.0001) by potato cultivar. The p-value according to the Kruskal-Wallis test by potato cultivar.

3.2. *Spectral Information and NIR Calibration Equation*

The PCA with the samples randomly selected in the calibration set was carried out (Table 1). The explained spectral variability was higher than 99.6% and between 4 and 10 principal components were required. In MPLS processing, the NIR residuals obtained after each factor and at each wavelength were calculated and standardized (dividing them by the standard deviations of the residuals at each wavelength) and then the next factor was calculated. The standardized method was conducted by dividing the NIR residuals with the standard deviations at each wavelength. Therefore, the obtained data of dry matter and reducing sugars, and the absorbance of the samples from 900 to 1700 nm were used to develop the calibration equations by this method.

The statistical parameters of calibration were obtained for each quality constituent after eliminating the samples using the spectral and chemical reasons. Between four and seven samples were eliminated to calibrate the dry matter, and seven and 10 samples for reducing sugars (Table 2). The optimal calibration equations for the determination of dry matter and reducing sugars were calculated based on the lowest SECV and the highest RSQ. The best NIR calibration models were shown in Table 2, indicating the best mathematical treatments, the range of applicability, the value of RSQ, and standard errors of calibration and cross-validation. For dry matter, the best equation showed a RSQ coefficient of 0.72 and a wide range of applicability (of the same order as the reference chemical method). Reducing sugars had a lower value of RSQ (0.55), and the marge of minimum and maximum values was acceptable. On the other hand, SEC and SECV were acceptable for both parameters. The RPD value was also taken into account to assess the predictive capacity of the models, with values of 1.90 and 1.48 for dry matter and reducing sugars, respectively (Table 2).

Table 2. Calibration descriptors of the best models obtained for each parameter by NIR.

Constituent	Math Treatment *	N	Mean	SD	Range of Applicability		SEC	RSQ	SECV	RPD
					Min	Max				
Dry matter	Detrend only 0,0,1,1	65	20.09	1.36	16.00	24.18	0.72	0.72	0.93	1.90
	Standard MSC 2,10,10,1	65	20.04	1.42	15.77	24.31	0.75	0.72	10.21	1.89
	None 2,4,4,1	65	20.17	1.25	16.41	23.94	0.68	0.71	0.98	1.85
	Standard MSC 2,4,4,1	63	20.15	1.28	16.30	24.00	0.70	0.70	0.96	1.84
	SNV only 2,4,4,1	66	20.07	1.43	15.76	24.37	0.79	0.70	10.31	1.82
Reducing sugars	SNV only 0,0,1,1	62	0.18	0.06	0.01	0.35	0.04	0.55	0.05	1.48
	Detrend only 2,8,6,1	61	0.17	0.04	0.06	0.28	0.02	0.51	0.03	1.42
	Standard MSC 0,0,1,1	63	0.17	0.04	0.05	0.29	0.03	0.50	0.03	1.41
	Detrend only 0,0,1,1	61	0.17	0.04	0.06	0.28	0.03	0.48	0.03	1.39
	None 2,4,4,1	60	0.17	0.04	0.06	0.27	0.03	0.48	0.03	1.39

N: Number of samples after removing the outliers; MSC: Multiplicative dispersion correction; SNV: Standard normal variate; SD: Standard deviation; Min: Minimum; Max: Maximum; RSQ: Multiple correlation coefficients; SEC: Standard error of calibration; SECV: Standard error of cross-validation; RPD: Ratio performance deviation. * In the math treatment, the first digit is the number of the derivative, the second is the gap over which the derivative is calculated, the third is the number of data points in a running average or smoothing, and the fourth is the second smoothing.

Cross-validation was carried out to evaluate the robustness of the obtained models. The set of calibration samples was divided into six subsets, of which five subsets were used for calibration and the other subset for the prediction set. This procedure was carried out several times, with the objective that all of the subsets were subjected to the calibration and prediction process. Subsequently, the resulting models for each parameter were validated and its predictive capacity was determined. The correlation between the values of reference (obtained by the reference method in laboratory) and the values predicted by NIR are represented in Figure 3. Internal validation showed better results for dry matter (higher RSQ and good errors of prediction) than reducing sugars. SEP, SEP (C), and bias indicated that the calibration models for the two quality parameters allows their determination. Therefore, the results reflected the prediction capacity and validity of the models. Although a larger number of samples including more types of potato cultivars, and more growing seasons could improve the predictions of these quality parameters.

3.3. External Validation and Prediction Capacity of the Models

The obtained calibration equations were validated with 19 new potato samples (validation set, Table 1) and the generated values were compared with the reference, according to the residual mean and root mean square error (RMSE) (Table 3). The predictive ability of models resulted as satisfactory, with RMSE of 1.17 and 0.07, and the mean residual values of 1.01 and 0.05 for dry matter and reducing sugars, respectively. The predicted values by NIR equations were compared with the reference data of samples that did not belong to the calibration equation using the Student's t-test for paired values. The null hypothesis is accepted and there is no difference between the reference values and the NIR method generated for each parameter ($p > 0.05$).

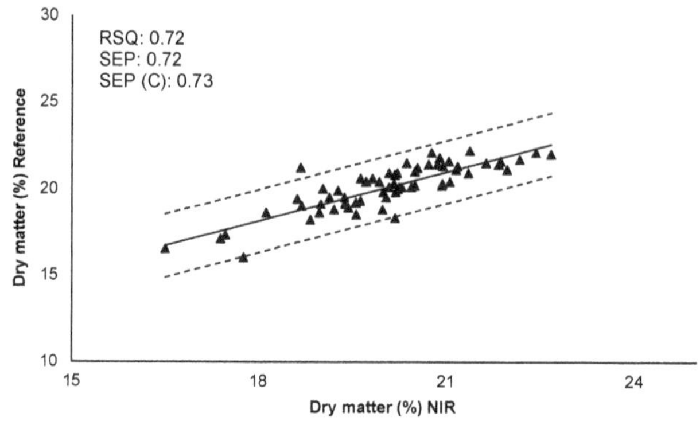

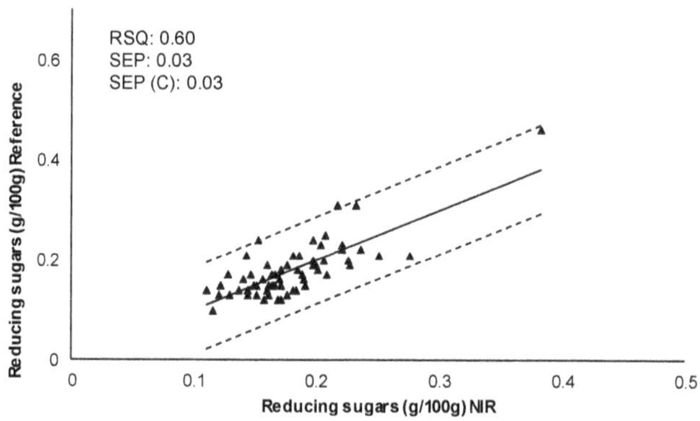

Figure 3. Internal validation. Comparison of reference values with predicted values by the NIRS model for each quality parameter. RSQ: Multiple correlation coefficient; SEP: Standard error of prediction; SEP (C): Standard error of prediction corrected by BIAS.

Table 3. External validation (19 samples) of potato quality parameters with the results of the NIR calibration.

Constituent	Mean Residual	RMSE	*p*-Value
Dry matter	1.01	1.17	0.22
Reducing sugars	0.05	0.07	0.16

RMSE: Root mean standard error. *p*-value: Level of significance calculated according to the Student's test.

4. Discussion

The demand for nutritional information and quality aspects of food by consumers makes it necessary to discover fast and safe methods that guarantee the safety and particularities of food. The precise assessment of potato freshness degree is a complex task. Food producers need techniques to evaluate changes in quality parameters, and non-destructive techniques, such as NIR technology, provide these advantages. These analytical methods are potentially useful tools to control the stability of the quality requirements in postharvest technology [3]. In this sense, the fresh and processed potato sector joins the challenge of offering quality products. Recent advances have shown good potentials of NIRS in real-time monitoring and modeling for different food processes. However, most of the studies have been carried out at a lab scale, while applications at industrial levels are still few [29], due to the difficulty of integrating scientific and industrial knowledge.

The sugars in potato tubers are very critical compounds for estimating the viability of processing, such as chipping and French frying [6,36]. In particular, glucose is responsible for the undesirable browning color of the frying process and it negatively affects the marketability of chips and other fried potato products [14]. PGI "Patata de Galicia" established values of reducing sugars content below 0.4% and dry matter above 18% for its industrial processing. Monitoring sugars in potato tubers before and during the storage has become a basic quality practice in the frying industry [6]. Therefore, it is of great importance for the processing industry that the quality of potato tubers meets their standards after storage [8].

In recent years, there have been investigations of the efficacy of NIR for determining dry matter and sugar contents in the potato processing industry. Benchtop NIR devices are generally built for experts in the laboratory, since they generate spectra that require interpretation and further data processing to generate a result. In response to an industrial demand, portable instruments can be designed for non-scientific personnel [28]. Portable NIR spectroscopic instrumentation and methods for spectral data analysis and interpretation are undergoing notable advancements. This has allowed for better optimization of the analytical procedures and the use of this technique directly on site [27].

NIR technology has been extensively studied for homogenized samples of potatoes, such as potato pulp, sliced potatoes, freeze-dried potato, and cooked potato mash [14,17,37,38], but to a lesser extent in intact potato [14,25,39,40]. Often, several pre-treatment methods are applied independently and the performance of the subsequent chemometric analysis was compared, in order to establish the best selection of the pre-processing set [27]. Regression analysis groups the methods used for the quantitative prediction of a physicochemical property for a large set of unknown samples. Then, the properly calibrated and validated regression model can be used significantly quicker and more efficiently compared with the conventional methods [27]. Therefore, chemometric methods are used to reduce the complexity of NIR spectral datasets and to build prediction models [17,20,37,41]. MPLS regression was used to estimate some chemical and quality constituents of potato tubers [14,17,25,40]. Specifically, models for glucose, sucrose, and soluble solids were built, with R^2 in sliced potato samples (around 0.96, 0.83, and 0.50, respectively) higher than whole tubers (around 0.90, 0.80, and 0.30, respectively) [14]. The reducing sugars and dry matter content of potato varieties for frying (Innovator, Lady Claire, and Markies) according to three types of preparations (unpeeled, peeled, and transversally cut tubers) were compared, resulting in whole peeled potato tubers as the obtained maximum accuracy of the models to predict the dry matter (around $R^2 = 0.84$) and reducing sugars (around $R^2 = 0.77$) [25]. The best results in the estimation of dry matter concentration in sliced potato samples ($R^2 = 0.95$) than the whole tuber ($R^2 = 0.85$) were also found [39]. R^2 around 0.98 was obtained for the prediction of dry matter content of the reported potato pulp [23]. In addition, lower values of R^2 for glucose and fructose content on intact potato were obtained (with values of 0.65 and 0.71, respectively), as well as an acceptable standard error of prediction [40]. However, potato processors are more interested in determining the quality on whole tuber, but the application on intact unpeeled tubers is less frequent.

The accuracy and goodness of the models were evaluated according to the statistic RPD. Some researchers considered that a RPD ratio of less than 1.5 indicates poor predictions and a value higher than 2 indicates a good calibration [20]. Although the results in the present study using the portable equipment were not excellent, in the case of dry matter they were acceptable (RDP = 1.98). The calibration was performed on unpeeled potato tubers, which complicates the model robustness of these quality parameters. In addition, it must be taken into account that products with a low moisture content, such as ground or dehydrated samples, usually presented high values of RPD. This is the case of the estimation of dry matter and total soluble solid content on peeled and freeze-dried potato by the NIR system, with a RPD greater than 2.8 [17]. For the prediction of glucose content on whole tubers, a RDP of around 2 was reported [14], and for reducing sugars a RPD lower than 2 was found [40]. Sugars are generally reported to yield a lower prediction performance by NIR technology [6,37,38]. Some researchers mentioned the effect of the skin on the lower performance of the models that varies depending on the cultivar [6]. Consequently, sorting potato tubers based on sugar content is a more challenging task than assessing sugars in ground, homogenized or even sliced samples. In particular, the peeled potato was determined as the most interesting in order to obtain precise models for sugar and dry matter contents, which improves the RPD values from 15% to 38% for reducing sugars and 35% for dry matter [25]. However, potato processors are interested in obtaining the prediction models for whole tubers, in order for the technology to be developed for application in potato processing lines.

The calibration models should be based on large datasets, which are obtained from different destinations, growing conditions, and operational conditions [18]. This is the first study to apply the MicroNIR directly to whole tubers in an area closely linked to the potato crop and with a notable economic impact on the agricultural sector of the Galician community. Rapid measurement devices, such as the MicroNIR, calibrated for working with multiple cultivars and different shapes make its application attractive for the potato industries. The portable equipment incorporated the analytical precision required for chemical identification and quantification with a spectral resolution, which is equivalent to the benchtop instruments [26,27]. These portable devices have the advantages of ease of transportation and the necessary flexibility for an analysis in an industrial environment. Undoubtedly, technological innovations in portable instruments have been increased by developing interesting advantages in their application in-line with respect to laboratory equipment.

5. Conclusions

Taking into account the industrial range defined for reducing sugars and dry matter parameters according to the PGI Patata de Galicia standard, we consider the MicroNIR as a useful portable device and with a promising performance for the sector. The efficacy of NIR spectroscopy was demonstrated as a rapid and non-destructive method for the estimation of dry matter and reducing sugars in intact potato tuber. Although the model calibrations were not excellent, as a first approach for the use of this technology on-site in the potato warehouses, it is attractive for the sector. The results are interesting for the use of MicroNIR as a tool for fresh potato quality control during in-line processing. The efficiency of the automation techniques of this type optimizes the management of industrial processing, guaranteeing the quality of the potato tubers. However, a close collaboration between the potato processors and researchers is very important in order to achieve these goals and for future improvements.

Author Contributions: Methodology, software, and validation, O.E. and L.M.; formal analysis, O.E. and M.C.S.; writing original draft preparation, O.E. and M.C.S.; writing review and editing, O.E., M.S.R.-F. and M.C.S.; supervision, O.E., L.M., M.S.R.-F. and M.C.S.; project administration, M.C.S. All authors have read and agreed to the published version of the manuscript.

Funding: This research was funded by the project included in Rural Development Programme 2014/2020—FEADER 2017/045B—Xunta de Galicia. Laura Meno is a beneficiary of the pre-PhD contract FPU 17/00267 granted by the Ministry of Education, Culture, and Sports.

Institutional Review Board Statement: Not applicable.

Informed Consent Statement: Not applicable.

Data Availability Statement: Data is contained within the article.

Acknowledgments: The authors want to thank the farmers, potato industries, and the Council regulation of Protected Geographical Indication Patata de Galicia for their close collaboration.

Conflicts of Interest: The authors declare no conflict of interest.

References

1. Burlingame, B.; Mouillé, B.; Charrondiere, R. Nutrients, bioactive non-nutrients and anti-nutrients in potatoes. *J. Food Comp. Anal.* **2009**, *22*, 494–502. [CrossRef]
2. Birch, P.R.J.; Bryan, G.; Fenton, B.; Gilroy, E.M.; Hein, I.; Jones, J.T.; Prashar, A.; Taylor, M.A.; Torrance, L.; Toth, I.K. Crops that feed the world 8: Potato: Are the trends of increased global production sustainable? *Food Sec.* **2012**, *4*, 477–508. [CrossRef]

3. Bianchi, G.; Scalzo, R.L.; Testoni, A.; Maestrelli, A. Nondestructive analysis to monitor potato quality during cold storage. *J. Food Qual.* **2014**, *37*, 9–17. [CrossRef]
4. Meno, L.; Escuredo, O.; Rodriguez-Flores, M.S.; Seijo, M.C. Prevalence of airborne fungal spores in two potato warehouses with different storage conditions. *Aerobiologia* **2021**, *37*, 309–320. [CrossRef]
5. Xunta de Galicia. Resolución de 9 de Junio de 2016, de la Dirección General de Ganadería, Agricultura e Industrias Agroalimentarias, por la que se da publicidad a la solicitud de modificación del pliego de condiciones de la indicación geográfica protegida Pataca de Galicia/Patata de Galicia. *Diario Off. Galicia* **2016**, *120*, 27028.
6. Rady, A.M.; Guyer, D.E. Evaluation of sugar content in potatoes using NIR reflectance and wavelength selection techniques. *Postharvest Biol. Tec.* **2015**, *103*, 17–26. [CrossRef]
7. Kasampalis, D.S.; Tsouvaltzis, P.; Ntouros, K.; Gertsis, A.; Moshou, D.; Siomos, A.S. Rapid nondestructive postharvest potato freshness and cultivar discrimination assessment. *Appl. Sci.* **2021**, *11*, 2630. [CrossRef]
8. Broćić, Z.; Dolijanović, Ž.; Poštić, D.; Milošević, D.; Savić, J. Yield, tuber quality and weight losses during storage of ten potato cultivars grown at three sites in Serbia. *Potato Res.* **2016**, *59*, 21–34. [CrossRef]
9. Seijo-Rodríguez, A.; Escuredo, O.; Rodríguez-Flores, M.S.; Seijo, M.C. Assessment of antioxidant potential of potato varieties and the relationship to chemical and colorimetric measurements. *Am. J. Pot. Res.* **2018**, *95*, 71–78. [CrossRef]
10. Hertog, M.L.A.T.M.; Tijskens, L.M.M.; Hak, P.S. The effects of temperature and senescence on the accumulation of reducing sugars during storage of potato (*Solanum tuberosum* L.) tubers: A mathematical model. *Postharvest Biol. Technol.* **1997**, *10*, 67–79. [CrossRef]
11. Abbasi, K.S.; Masud, T.; Qayyum, A.; Khan, S.U.; Ahmad, A.; Mehmood, A.; Farid, A.; Jenks, M.A. Transition in tuber quality attributes of potato (*Solanum tuberosum* L.) under different packaging systems during storage. *J. Appl. Bot. Food Qual.* **2016**, *89*, 142–149.
12. Helgerud, T.; Segtnan, V.H.; Wold, J.P.; Ballance, S.; Knutsen, S.H.; Rukke, E.O.; Afseth, N.K. Near-infrared spectroscopy for rapid estimation of dry matter content in whole unpeeled potato tubers. *J. Food Res.* **2012**, *1*, 55. [CrossRef]
13. Sun, X.; Zhu, K.; Liu, J. Nondestructive detection of reducing sugar of potato flours by near infrared spectroscopy and kernel partial least square algorithm. *J. Food Meas. Charact.* **2019**, *13*, 231–237. [CrossRef]
14. Rady, A.M.; Guyer, D.E.; Kirk, W.; Donis-González, I.R. The potential use of visible/near infrared spectroscopy and hyperspectral imaging to predict processing-related constituents of potatoes. *J. Food Eng.* **2014**, *135*, 11–25. [CrossRef]
15. Alamar, M.C.; Tosetti, R.; Landahl, S.; Bermejo, A.; Terry, L.A. Assuring potato tuber quality during storage: A future perspective. *Front. Plant Sci.* **2017**, *8*, 1–6. [CrossRef] [PubMed]
16. Sánchez, J.A.C. Using NIRS spectroscopy to predict postharvest quality. *CAB Rev. Perspect. Agric. Vet. Sci. Nutr. Nat. Resour.* **2012**, *7*, 1–14.
17. Escuredo, O.; Seijo-Rodríguez, A.; González-Martín, M.I.; Rodríguez-Flores, M.S.; Seijo, M.C. Potential of near infrared spectroscopy for predicting the physicochemical properties on potato flesh. *Microchem. J.* **2018**, *141*, 451–457. [CrossRef]
18. Nicolaï, B.M.; Beullens, K.; Bobelyn, E.; Peirs, A.; Saeys, W.; Theron, K.I.; Lammertyn, J. Nondestructive measurement of fruit and vegetable quality by means of NIR spectroscopy: A review. *Postharvest Biol. Technol.* **2007**, *46*, 99–118. [CrossRef]
19. Escuredo, O.; González-Martín, M.I.; Wells-Moncada, G.; Fischer, S.; Hernández-Hierro, J.M. Amino acid profile of the quinoa (*Chenopodium quinoa* Willd.) using near infrared spectroscopy and chemometric techniques. *J. Cereal Sci.* **2014**, *60*, 67–74. [CrossRef]
20. González-Martín, M.I.; Wells-Moncada, G.; Fischer, S.; Escuredo, O. Chemical characteristics and mineral composition of quinoa by near-infrared spectroscopy. *J. Sci. Food Agric.* **2014**, *94*, 876–881. [CrossRef]
21. Mehrübeoglu, M.; Coté, G.L. Determination of total reducing sugars in potato samples using near-infrared spectroscopy. *Cereal Foods World* **1997**, *42*, 409–413.
22. Van Dijk, C.; Fischer, M.; Holm, J.; Beekhuizen, J.G.; Stolle-Smits, T.; Boeriu, C. Texture of cooked potatoes (*Solanum tuberosum*). 1. Relationships between dry matter content, sensory-perceived texture, and near-infrared spectroscopy. *J. Agric. Food Chem.* **2002**, *50*, 5082–5088. [CrossRef] [PubMed]
23. Chen, J.Y.; Miao, Y.; Zhang, H.; Matsunaga, R. Non-destructive determination of carbohydrate content in potatoes using near infrared spectroscopy. *J. Near Infrared Spectrosc.* **2004**, *12*, 311–314. [CrossRef]
24. Haase, N.U. Prediction of potato processing quality by near infrared reflectance spectroscopy of ground raw tubers. *J. Near Infrared Spec.* **2011**, *19*, 37–45. [CrossRef]
25. Camps, C.; Camps, Z.N. Optimized prediction of reducing sugars and dry matter of potato frying by FT-NIR spectroscopy on peeled tubers. *Molecules* **2019**, *24*, 967. [CrossRef]
26. Sorak, D.; Herberholz, L.; Iwascek, S.; Altinpinar, S.; Pfeifer, F.; Siesler, H.W. New Developments and applications of handheld raman, mid-infrared, and near-infrared spectrometers. *Appl. Spectrosc. Rev.* **2012**, *47*, 83–115. [CrossRef]
27. Beć, K.B.; Grabska, J.; Huck, C.W. NIR spectroscopy of natural medicines supported by novel instrumentation and methods for data analysis and interpretation. *J. Pharm. Biom. Anal.* **2021**, *193*, 113686. [CrossRef] [PubMed]
28. Crocombe, R.A. Portable spectroscopy. *Appl. Spectrosc.* **2018**, *72*, 1701–1751. [CrossRef]
29. Grassi, S.; Alamprese, C. Advances in NIR spectroscopy applied to process analytical technology in food industries. *Cur. Opin. Food Sci.* **2018**, *22*, 17–21. [CrossRef]
30. VIAVI Solutions Inc. *MicroNIR Pro v3.0. User Manual*; VIAVI Solutions Inc.: Santa Rosa, CA, USA, 2019; p. 308.

31. Beć, K.B.; Grabska, J.; Huck, C.W. Biomolecular and bioanalytical applications of infrared spectroscopy—A review. *Anal. Chim. Acta* **2020**, *1133*, 150–177. [CrossRef] [PubMed]
32. Dhanoa, M.S.; Lister, S.J.; Barnes, R.J. On the scales associated with near infrared reflectance difference spectra. *Appl. Spectros.* **1995**, *49*, 765–772. [CrossRef]
33. Shenk, J.S.; Westhaus, M.O. *Analysis of Agriculture and Food Products by Near Infrared Reflectance Spectroscopy. Monograph NIR Systems. Infrasoft International, NIR Systems*; Silver Spring Inc.: New York, NY, USA, 1995; p. 488.
34. Williams, P. The RPD statistic: A tutorial note. *NIR News* **2014**, *25*, 22–26. [CrossRef]
35. Esbensen, K.H.; Geladi, P.; Larsen, A. The RPD myth .... *NIR News* **2014**, *25*, 24–28. [CrossRef]
36. Stark, J.C.; Love, S.L. (Eds.) Tuber quality. In *Potato Production Systems*; University of Idaho: Moscow, ID, USA, 2003; pp. 329–342.
37. López, A.; Arazuri, S.; García, I.; Mangado, J.; Jarén, C. A review of the application of near-infrared spectroscopy for the analysis of potatoes. *J. Agric. Food Chem.* **2013**, *61*, 5413–5424. [CrossRef] [PubMed]
38. Helgerud, T.; Wold, J.P.; Pedersen, M.B.; Liland, K.H.; Ballance, S.; Knutsen, S.H.; Rukke, E.O.; Afseth, N.K. Towards on-line prediction of dry matter content in whole unpeeled potatoes using near-infrared spectroscopy. *Talanta* **2015**, *143*, 138–144. [CrossRef]
39. Subedi, P.P.; Walsh, K.B. Assessment of potato dry matter concentration using short-wave near-infrared spectroscopy. *Potato Res.* **2009**, *52*, 67–77. [CrossRef]
40. Chen, J.Y.; Zhang, H.; Miao, Y.; Asakura, M. Nondestructive determination of sugar content in potato tubers using visible and near infrared spectroscopy. *Jpn. J. Food Eng.* **2010**, *11*, 59–64. [CrossRef]
41. Lei, T.; Sun, D.W. A novel NIR spectral calibration method: Sparse coefficients wavelength selection and regression (SCWR). *Anal. Chim. Acta* **2020**, *1110*, 169–180. [CrossRef] [PubMed]

sensors

MDPI

Article

Near-Infrared Sensors for Onsite and Noninvasive Quantification of Macronutrients in Breast Milk

Candela Melendreras [1], Sergio Forcada [2], María Luisa Fernández-Sánchez [1], Belén Fernández-Colomer [3], José M. Costa-Fernández [1], Alberto López [4], Francisco Ferrero [4,*] and Ana Soldado [1,*]

1 Department of Physical and Analytical Chemistry, University of Oviedo, 33006 Oviedo, Spain; uo257805@uniovi.es (C.M.); marisafs@uniovi.es (M.L.F.-S.); jcostafe@uniovi.es (J.M.C.-F.)
2 Department of Nutrition, Grasslands and Forages, Regional Institute for Research and Agro-Food Development (SERIDA), 33450 Villaviciosa, Spain; sforcada@serida.org
3 Service of Neonatology, Department of Pediatrics, Hospital Universitario Central de Asturias, 33011 Oviedo, Spain; bcolomer@gmail.com
4 Department of Electrical Engineering, University of Oviedo, 33204 Gijon, Spain; uo181549@uniovi.es
* Correspondence: ferrero@uniovi.es (F.F.); soldadoana@uniovi.es (A.S.); Tel.: +34-985-182552 (F.F.); +34-985-103583 (A.S.)

Abstract: Breast milk is an optimal food that covers all the nutritional needs of the newborn. It is a dynamic fluid whose composition varies with lactation period. The neonatal units of hospitals have human milk banks, a service that analyzes, stores, and distributes donated human milk. This milk is used to feed premature infants (born before 32 weeks of gestation or weighing less than 1500 g) whose mothers, for some reason, cannot feed them with their own milk. Here, we aimed to develop near-infrared spectroscopy (NIRS) measures for the analysis of breast milk. For this purpose, we used a portable NIRS instrument scanning in the range of 1396–2396 nm to collect the spectra of milk samples. Then, different chemometrics were calculated to develop 18 calibration models with and without using derivatives and the standard normal variate. Once the calibration models were developed, the best treatments were selected according to the correlation coefficients (r^2) and prediction errors (SECVs). The best results for the assayed macronutrients were obtained when no pre-treatment was applied to the NIR spectra of fat (r^2 = 0.841, SECV = 0.51), raw protein (r^2 = 0.512, SECV = 0.21), and carbohydrates (r^2 = 0.741, SECV = 1.35). SNV plus the first derivative was applied to obtain satisfactory results for energy (r^2 = 0.830, SECV = 9.60) quantification. The interpretation of the obtained results showed the richness of the NIRS spectra; moreover, the presence of specific bands for fat provided excellent statistics in quantitative models. These results demonstrated the ability of portable NIRS sensors in a methodology developed for the quality control of macronutrients in breast milk.

Keywords: breast milk quality control; chemometrics; handheld; spectroscopy

Citation: Melendreras, C.; Forcada, S.; Fernández-Sánchez, M.L.; Fernández-Colomer, B.; Costa-Fernández, J.M.; López, A.; Ferrero, F.; Soldado, A. Near-Infrared Sensors for Onsite and Noninvasive Quantification of Macronutrients in Breast Milk. *Sensors* **2022**, 22, 1311. https://doi.org/10.3390/s22041311

Academic Editors: Rafael Font Villa and Mercedes Del Río Celestino

Received: 14 January 2022
Accepted: 7 February 2022
Published: 9 February 2022

Publisher's Note: MDPI stays neutral with regard to jurisdictional claims in published maps and institutional affiliations.

1. Introduction

Breast milk is a complex and highly variable fluid that provides nutrients and bioactive components for the correct growth and development of infants. The composition of breast milk changes throughout the lactation period, adapting to the nutritional needs of the rapidly growing newborn. Depending on the time of lactation, three types of milk can be distinguished: colostrum, transitional milk, and mature milk.

Breastfeeding is necessary for the development of newborns because it protects them from infections and diseases such as diabetes, obesity, and or hypercholesterolemia. Due to the multiple benefits of breast milk for infants, in the middle of the 20th century, milk banks were established. Currently, the World Health Organization and national and international pediatric associations consider breast milk banks as necessary to guarantee adequate nutrition for premature infants (born before 32 weeks of gestation or weighing less than 1500 g) or sick infants who, for whatever reason, cannot be fed with their mother's milk.

Milk banks are a specialized service integrated into the neonatology units of hospitals. Their objectives are the promotion and support of breastfeeding; they are responsible for the selection of donors, and the storage, processing, analysis, and distribution of milk. To guarantee the safety of donated milk, it is subjected to a pasteurization process using the Holder method (62.5 °C for 30 min and then rapidly cooled in less than 15 min to 4 °C). In neonatal units, human breast milk contains 70 kcal/100 mL because it was obtained from women of term babies later in the lactation period. However, the amount of macronutrients in breast milk depends on different factors such as gestational age, feed, or sampling procedure [1] and no predictions can be established [2]. As such, when comparing preterm and term milk, the former one has less energy and protein than the latter. Considering these issues, one major concern when feeding neonates with breast milk from hospital banks is the nutritional adequacy of the milk in meeting the nutritional requirements of the neonate, because some nutrients such as protein are associated with neurodevelopmental outcomes [3].

To determine the quality of breast milk, nutritional analysis should be carried out. Nowadays, milk quality is controlled in laboratories using properly validated chemical reference methodologies such as Mojonnier, Folch, Gerber, or the Roese-Gottlieb method for fat quantification, and the Bradford method for protein elemental analysis. Another alternative used in some milk banks involves quantifying the referred compounds in milk analyzers using pre-calibrated medium or near-infrared (IR) instruments. To establish the quality of the results obtained by the IR technique, Fusch et al. compared eight different laboratory IR analyzers to quantify macronutrients in breast milk, concluding that research groups using these devices must be cautious about their measurements, finding that adequate sample preparation and instrumental calibration and validation are necessary following the Good Laboratory and Clinical Practices [4].

Given these considerations, the best method of guaranteeing the adequate nutrition of newborns is the use of onsite and real-time sensors that are able to quantify the macronutrients in breast milk. Moreover, due to the value and short supply of breast milk, a noninvasive analysis would be the best method to achieve final sample analysis. This type of analysis can be carried out just before consumption. Among the portable, noninvasive, and real-time analytical techniques for food analysis, near-infrared spectroscopy (NIRS) is a real-time, noncontaminating, and versatile technology capable of providing information on food quality attributes in situ. Moreover, no sample pretreatment or chemical reagents are required for the analysis, making it a waste-free technique, unlike traditional laboratory nutritional compositional analyses [5–7]. Nevertheless, there are some important limitations of NIRS analysis due to the strong absorption of water in the NIR region, which increases the background, preventing obtaining satisfactory results in quantitative analysis. Nowadays, the applications of NIRS have increased due to the development of aquaphotomics [8], a novel science and methodology that features water NIRS spectra in aqueous and biosystems for indirect analysis of components.

The drawbacks of using NIRS (Near Infrared Spectroscopy) with portable instruments are the sensitivity of the NIR signal, the narrow wavelength range of low-cost devices, and spectrum complexity. NIR spectra are difficult to interpret because the vibrations of different molecular bonds are involved in the same absorption band, resulting in overlapping information. However, to overcome these shortcomings and to extract information, it is necessary the use multivariate analysis [5,9,10]. By combining NIR spectra and appropriate multivariate analysis, it is possible to extract all the relevant information and to develop a fit calibration model that is able to quantify the macronutrients in breast milk both onsite and in real-time. This methodology will enable newborns to be fed with breast milk that is adequate for their development stage [11–13].

Qualitative models have been developed with portable instruments, which allow differentiation between colostrum, transition milk, and mature milk [14]; however, no information about breast milk quantification parameters or the effect of spectra chemometric pretreatments on the final calibration models with portable devices is available. NIRS methodologies were developed based on the use of handheld, portable NIRS, and were tested for the analysis of the three major components in cow raw milk (fat, protein, and nonfat solids) [15]. Because of this, the application of onsite NIRS technology to the study of breast milk can be an effective alternative for the characterization and control of the quality of donated breast milk received by milk banks. An important aspect to take into account, and studied by Kwan et al., is that quantitative NIRS studies on breast milk were carried out with high-performance laboratory equipment, that were acquired precalibrated, which can lead to errors due to a systematic displacement of the data or some inaccurate calibrations [16].

To the best of our knowledge, no studies have proved the effectiveness of low-cost, onsite, and easy-to-use handheld instruments in quantifying macronutrients in breast milk. The complexity of the analysis, the need for obtaining macronutrients quantification in real-time, and the small amount of available breast milk demand new methodologies able to meet all the above-detailed requirements.

As such, in this study, we constructed methodology based on the use of low-cost NIRS sensors and appropriate chemometric procedures for onsite, real-time quality control monitoring of breast milk. The aim was to provide the neonatology units and milk banks with a cheap and easy-to-use tool that is able to establish the quality of milk just before newborn consumption to enable real-time decision making and ensure adequate nutrient combinations for feeding newborns.

2. Materials and Methods

2.1. Milk Samples

A total of 17 samples from the Asturias Breast Milk Bank (University Central Hospital of Asturias, HUCA, Oviedo, Spain) at different stages (colostrum, transitional milk, and mature milk) were used in this study. These samples covered the variability in breast milk for feeding newborns. However, for the development of an NIR calibration procedure, the recommended calibration samples minimum for any quality parameter is about 50 [16]. To increase the number of samples and the variability in the macronutrients content, we prepared three other batteries of 17 samples each (17 × 3) mixing different breast milk samples in a 1:1 proportion or diluting them 1:1 and 1:3.5 with distilled water (Figure 1). The final range of concentrations of all the parameters is shown in Table 1.

Table 1. Statistics of macronutrients and energy in breast milk (N = 68).

	Calibration Set (N = 53)				Validation Set (N = 15)			
	Mean	Max	Min	SD	Mean	Max	Min	SD
Fat [1]	2.39	5.30	0.51	1.25	2.60	4.60	0.57	1.58
CP [1]	0.87	2.50	0.27	0.50	0.78	1.70	0.33	0.39
RP [1]	0.72	2.00	0.21	0.40	0.69	1.40	0.27	0.34
CH [1]	5.80	8.80	2.34	2.56	5.58	8.40	2.31	2.63
Energy [2]	49.30	86.00	15.60	22.06	50.09	81.00	15.90	26.36
TS [1]	7.42	14.60	3.27	4.02	8.04	14.50	3.27	4.17

Max: maximum, Min: minimum, SD: standard deviation, RP: raw protein, CP: crude protein, CH: carbohydrates; TS: total solids. [1] g/100 mL; [2] Kcal/100 mL.

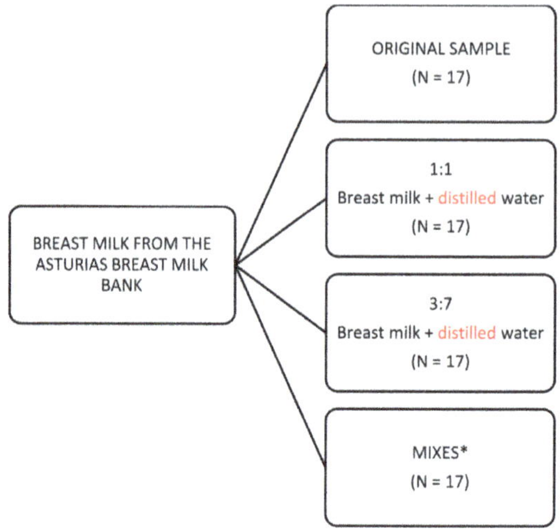

Figure 1. Scheme for battery sample preparation (N = 68). * Mixes (1:1) of following samples: 16 + 17, 10 + 4, 6 + 1, 7 + 1, 2 + 17, 3 + 16, 4 + 15, 5 + 1, 6 + 13, 7 + 12, 8 + 11, 9 + 10, 2 + 3, 5 + 8, 9 + 11, 12 + 13, and 3 + 15.

A total of 68 breast milk samples were involved in this NIRS research work. This initial set was separated into two groups: (i) a calibration set containing 53 samples because, when developing NIRS calibration models, a minimum of 50 samples is recommended [17]; (ii) an external validation set with 15 samples. Spectral data included in one or another set were randomly selected.

The statistics of the calibration and validation populations for all the parameters included for the quality control of breast milk are detailed in Table 1. The macronutrients analyzed were fat, crude protein (CP), real protein (RP), carbohydrates (CH), energy, and total solids (TS). As can be seen, we report two values for protein, crude protein, and true protein. The first is the protein content based on the total amount of nitrogen in breast milk; this value can include nonprotein nitrogen compounds, and true protein does not include these nonprotein nitrogen compounds [18]. All reference data were provided for the Asturias Breast Milk Bank.

All milk samples were stored frozen, allowed to thaw at room temperature before analysis, and then heated at 37 °C in a water bath. Once the samples were at temperature, they were homogenized by manual stirring and NIR analysis was carried out.

Each mother provided written informed consent for donating the samples for this study, which was approved by the institutional review board.

2.2. NIRS Spectra Collection

NIRS spectra were collected with a portable NIRS instrument (microPHAZIR Mod. 1624, Thermo Fisher Scientific Inc., Wilmington, MA, USA). This handheld instrument includes an electromechanical system (MEMS) and an incandescent tungsten light source for illumination, which is safe for users and ensures the integrity of the sample. It has a single, broad-spectrum InGaAs detector, which makes it a low-cost, energy-efficient device with a good spectral response. The scanning window or sampling area is 0.13 cm^2, and the wavelength range of the device is 1596 to 2396 nm, with an approximate interval of 8.7 nm. It is compact and easy to handle, has a gun shape, and an integrated reference for easy calibration. For sampling, we used a liquid cup (Foss IH-0397, Foss NIRSystems, Silver Spring, USA), 45 mm high and 25 mm wide, with an optical path of 17 mm. This cuvette had a quartz wall (the wall of radiation incidence) and a rear wall of aluminum that reflected the NIR radiation and allowed the radiation to pass through the sample twice. Spectral information was collected in transreflectance mode, from direct exposure on the cuvette, and each NIRS spectrum is reported as the average of 5 scans. To obtain a representative spectrum of each breast milk, the collection procedure was as follows: each sample was divided in three aliquots and one spectrum was collected for each one. The final signature for each sample is reported as the averaged spectrum of the three subsamples. In the global procedure, a total of 204 scans (68 samples × 3 scans per sample) were collected as log 1/R, where R is reflectance, to build the spectra database of the 68 breast milk samples.

2.3. Data Processing

NIRS spectra collected with the handheld instruments are defined by 100 points in a range of wavelengths between 1596 and 2396 nm. Unscrambler X software (version 10.1, CAMO Software, Oslo, Norway) was used for chemometric developments. The calibration set was centered prior to performing the regression models by principal component analysis (PCA) to detect potential spectral outliers, and the regression procedure employed to build the calibration models using the global spectrum (all the wavelengths 1596–2396 nm) was partial least squares (PLS) [19]. The models were optimized using a random cross-validation method included in the Unscrambler X software package, with 20 segments and 3 samples per segment. The optimal number of PLS factors was established considering the minimum residual variance.

With the aim of minimizing the scattering phenomenon, the standard normal variate (SNV) mathematical pretreatment was applied to raw spectral data. After that, different derivative pretreatments were applied to the spectral data to minimize unforeseen variations and to improve calibration. The pretreatments code in this chemometric software can be summarized using a four-digit notation, where the first digit (a) refers to the order of magnitude of the Savitzky-Golay derivative (SG) (0 = underived spectrum, 1 = 1st derivative, 2 = 2nd derivative, etc.); the second digit (b) indicates the polynomial order of the derivate; and the third (c) and fourth (d) digits indicate the size of the left and right intervals, respectively, expressed in nanometers, used for the derivative smoothing calculation. A total of 18 different models (6 parameters × 3 chemometric strategies) were developed using different pretreatments of the breast milk samples and PLS as the regression model. According the four-digit notation, the chemometric strategies assayed in this research work were 0 0 0 0, 0 2 2 2, and 1 2 2 2.

The best-fitting equations were selected by the statistical criterion for each parameter, based on the lowest standard error of calibration (SEC) and standard error of cross-validation (SECV), the highest coefficient of determination for calibration, (R^2), and coefficient of determination for cross-validation (r^2) [20,21].

The external validation was evaluated based on the lowest standard error of prediction (SEP) and the best Student's *t*-statistic for paired samples comparing the reference and NIRS method.

3. Results and Discussion

In order to understand the information in the collected NIRS spectra, Figure 2 plots the raw and after-derivation values of the averaged spectra for the calibration and validation sets. We observed no differences between the populations, with the water band (O-H interactions) at 1950 nm being the most intense. At this wavelength, the band was a result of multiple overlapping bands, and it was directly influenced by chemical interactions with other molecular species in the sample [5]. Other important bands that could be associated with macronutrients in milk located at 2300 and 2380 nm. These are described as protein and fat (oil) bands, respectively [22]. Based on the aquaphotomic principle (strategy of monitoring a spectral band associated with a specific parameter, such as water, fat, protein, etc.), the location of these specifics bands (protein and fat) could help improve the calibration models for the aforementioned macronutrients.

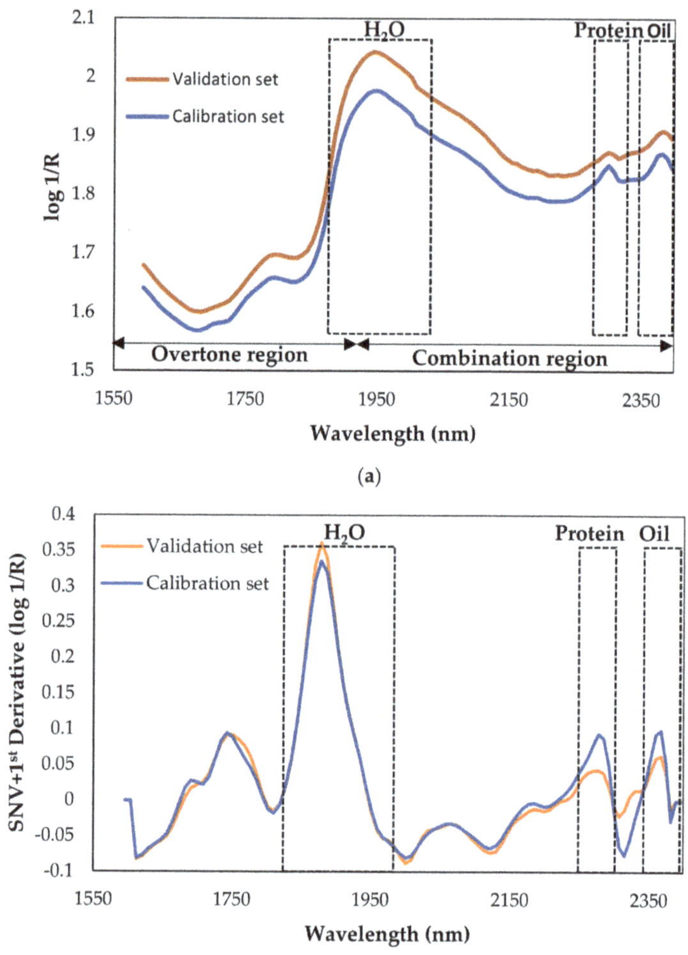

Figure 2. Average spectra of the calibration and validation sets: (**a**) raw spectra; (**b**) SNV + first derivative pretreatment.

After explaining the effect of vibrations associated with the macronutrients at different wavelength bands and their importance in the development of the NIRS procedure for the proposed parameters, we applied PCA with Hotelling's T^2 ellipse to raw NIR spectra of the calibration set to explore the spectra and detect outliers. As shown in Figure 3, the data outside the ellipse need to be checked because they are potential outliers. We can see that there are two samples located outside the ellipse: numbers 36 and 44. These samples were a mix of the originals 16 + 17 and 7 + 12, respectively. These results could be explained by operational error and not due to a compositional or spectral difference between these samples and all the others involved in this study. All original samples other than the mixed and diluted ones were satisfactory according PCA. Both outliers were deleted for the development of the final calibration models.

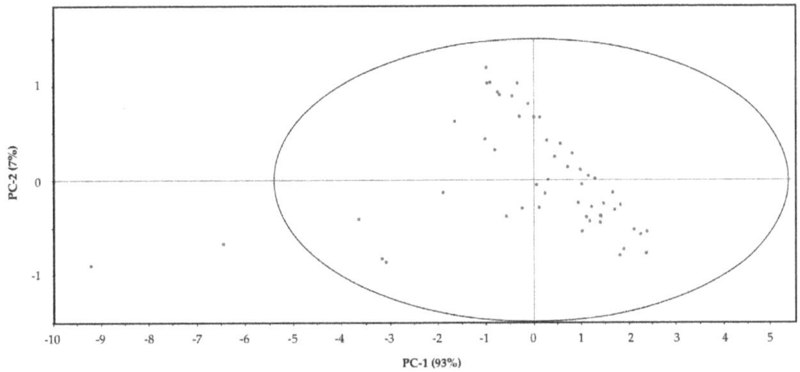

Figure 3. Principal component analysis with Hotelling's T^2 ellipse analysis for outlier detection in the calibration set.

The next step was to run the calibration models with the global spectrum using PLS as a regression strategy and cross-validation with random groups. As detailed in the Section 2, different pretreatment procedures were evaluated to obtain the best calibration statistics, ranging from no pretreatment to scattering correction (SNV) plus first derivative. As shown in Table 2, a total of 18 calibration models applying different pretreatments, both with and without scattering correction and derivatives, were used to quantify six parameters (fat, crude protein, raw protein, carbohydrates, energy, and total solids) for quality control of breast milk.

Table 2. Calibration and cross-validation statistics for breast milk samples using partial least squares regression.

Math Pre-Treatment	Parameter	R^2	SEC	r^2	SECV
0 0 0 0	Fat	0.910	0.37	0.841	0.51
	CP	0.782	0.19	0.508	0.30
	RP	0.797	0.14	0.512	0.21
	HC	0.874	0.91	0.741	1.35
	Energy	0.922	6.17	0.791	10.39
	TS	0.787	1.86	0.686	2.42
SNV 0 2 2 2	Fat	0.876	0.44	0.795	0.58
	CP	0.725	0.25	0.498	0.35
	RP	0.580	0.22	0.411	0.27
	HC	0.860	0.94	0.593	1.65
	Energy	0.835	8.96	0.756	11.32
	TS	0.709	2.13	0.529	2.77
SNV 1 2 2 2	Fat	0.826	0.52	0.779	0.61
	CP	0.796	0.22	0.524	0.36
	RP	0.787	0.16	0.482	0.25
	HC	0.894	0.83	0.699	1.42
	Energy	0.927	5.94	0.830	9.60
	TS	0.929	1.07	0.685	2.20

SNV: standard normal variate, N1N2N3N4: Savitzky-Golay derivative order, polynomial order of derivative, left, and right intervals for the derivative smoothing; R^2: coefficient of determination for calibration, SEC: standard error of calibration, r^2: coefficient of determination for cross-validation, SECV: standard error of cross-validation, RP: raw protein, CP: crude protein, CH: carbohydrates, TS: total solids.

As shown in Table 2, for all parameters, with the exception of raw protein (R^2 = 0.58), the R^2 calibration coefficient obtained was higher than 0.7, and the best performance was exhibited by fat, energy, and total solids with R^2 values all higher than 0.9. These results for fat could be explained due to spectrum bands: as shown in Figure 2, a specific band for fat is noted around 2380 nm, as was observed for the breast milk samples. The presence of this band allowed us to obtain excellent calibration results with raw spectra data, without any pretreatment (R^2 = 0.91). Notably, fat is one of the most important parameters when characterizing donated breast milk as it is the main source of energy for newborns. Furthermore, it provides essential nutrients such as fat-soluble vitamins and polyunsaturated fatty acids [23]. These results showed that the NIR spectra successfully captured quantitative variations in fat, showing the richness of the NIRS spectra.

Similarly, quantitative results could be obtained using the proposed procedure, because NIRS regression coefficients were obtained for energy and total solids, with R^2 values higher than 0.9 [5]. Precision values calculated as coefficient of variation (CV% = 100 × SEC/mean calibration set reference values) ranged between 10% and 15% for all the parameters involved in this study. Similar results were obtained in previous research [16] when comparing the results produced by different devices and reference methods. For these parameters, the best results were obtained after applying SNV and the first derivative to the raw spectra. For RP, CP, and CH, not many differences were observed when developing PLS regression with or without pretreated spectra. The R^2 value for protein data was higher than 0.75 and for CH, it was higher than 0.8. These values indicated that all the developed models can be used for quantitative analysis.

After studying the calibration statistics of the NIRS multivariate models in depth, the next step was to select the best chemometric models to quantify each macronutrient in breast milk. This choice was made based on the criterion detailed in the Section 2 (the highest r^2 value and the lowest SECV) and on the comparison between SEC and SECV, because a gap between SEC and SECV is related to large differences between calibration and prediction results, which indicate that the calibration model was not robust.

Although few differences were found when comparing the NIRS calibration and cross-validation statistics, the best results for fat (r^2 = 0.841, SECV = 0.51), proteins (RP, r^2 = 0.512, SECV = 0.21), and carbohydrates (r^2 = 0.741, SECV = 1.35) were obtained when no pre-treatment was applied to the NIR spectra. For energy (r^2 = 0.830, SECV = 9.60), SNV plus the first derivative obtained satisfactory results. As shown in Table 2, the best R^2 value for TS was obtained with a first-derivative pretreatment; however, the difference observed between cross-validation and calibration data was too large, indicating that the results were not robust and the validity of that model was limited. For these reasons, we determined that the best model for TS quantification was that without applying any pretreatment to the spectra (r^2 = 0.685, SECV = 2.42).

The energy and TS parameters showed the largest differences in SECV and SEC values. The reason for these results may be that both parameters are indirect. This means that they are not directly related to a molecular bond. However, the robustness of the calibrations could be improved by enlarging the sample size and the variability in the breast milk. An update of the calibration models is required when new samples are considered. However, no conclusions could be reached with this calibration and cross-validation data, as an adequate statistic test is needed to quantify the SEP (external validation).

After selecting the best treatments, to confirm the validity of the developed models, we then externally validated the models, predicting all the macronutrients by using the selected models. As detailed in the Section 2, 15 breast milk samples with reference data were selected for external validation and all the parameters were quantified. After quantifying all the breast milk parameters with the developed models, Student's *t*-test was applied to compare the results obtained when analyzing samples using the reference and onsite NIRS method.

As detailed above, the final acceptance or evaluation of the NIRS calibration models necessitated an external validation including samples not involved in the calibration procedure. In this work, 15 breast milk samples were included in the external validation set and quantified with the selected calibration models developed in this study. The selected methods and external validation statistics are detailed in Table 3.

Table 3. External validation statistics used for predicting nutritive parameters of breast milk (validation set, N = 15).

	Math Pre-Treatment	SECV	SEP	SECV/SEP	RPD	Accuracy %	$t_{student}$ Reference vs. Predicted
Fat	0 0 0 0	0.510	0.579	0.881	2.7	94	1.21
CP	SNV 1.2.2.2	0.359	0.426	0.843	0.9	114	0.57
RP	0 0 0 0	0.210	0.203	1.035	1.7	92	0.69
HC	0 0 0 0	1.347	1.630	0.826	1.6	108	1.36
Energy	SNV 1.2.2.2	9.603	11.757	0.817	2.2	94	1.74
TS	0 0 0 0	2.420	4.541	0.533	0.9	115	1.57

SNV: standard normal variate, N1N2N3N4: Savitzky-Golay derivative order, polynomial order of derivative, and left and right intervals for the derivative smoothing, SECV: standard error of cross-validation, SEP: standard error of prediction, RPD = standard deviation of the validation set/SEP; Accuracy %: 100 − ((reference value − predicted value)/reference value) × 100), *t*-critical value for 95% confidence and 14 degrees of freedom = 2.145, RP: raw protein, CP: crude protein, CH: carbohydrates, TS: total solids.

As shown in Table 3, we compared the SEP and SECV values for the selected models. The ratio between SECV and SEP ranged between 0.817 and 1.035 for all parameters except for TS (0.533). The similarity between SECV and SEP confirmed that no difference was found between the external and cross-validation predicted errors, indicating a sufficiently robust calibration for all parameters [24]. For TS quantification, an improvement in the models is required by including new samples and enhancing the multivariate model.

Another statistic that can explain model reliability in NIR spectroscopy is RPD (RPD = standard deviation/SEP). For this parameter, three categories can be defined: (1) excellent models, with RPD > 2; (2) fair models, with 1.4 < RPD < 2; and, (3) unreliable models, with RPD < 1.4 [25]. These values have been applied in NIRS studies; however, no statistical basis has been used to establish these thresholds. Moreover, researchers developed useful calibration models with RPD values lower than the proposed standard values [25]. Considering this statistic in Table 3, we the fat and energy models are categorized as excellent and the RP and HC models as fair. For TS and CP, the range of and variability in the samples should be increased to improve the NIR statistics and obtain a valuable model.

After determining the fit and quality of the calibration models, and taking into account the coefficients of determination and calibration or cross-validation errors, a statistical test including SEPs and reference data was conducted to evaluate the prediction errors of the multivariate NIRS models. In this work, we selected Student's *t*-test to compare paired samples. We applied a paired difference *t*-test to compare the results obtained when samples were analyzed by the reference and NIRS methods. Then, we used the set of differences to build the *t*-statistic using the mean and standard deviation of the differences. Student's *t*-test is a useful statistic strategy for comparing two data sets of quantitative results obtained with different analytical methods. In this study, we compared the reference and predicted data of all samples included in the validation set (N = 15). The results showed that there were no differences when considering a confidence level of 95% because, for all parameters, the calculated $t_{student}$ was is lower than 2.145 (theoretical value of t for 14 freedom degrees and 95% confidence level). Another parameter that we used to characterize the proposed methodology was accuracy. We calculated the accuracy for all the macronutrients involved in this study, and the best results were obtained for fat and energy with a value of 94%.

We could find no information about analytical methodologies for onsite and real-time quantification of the macronutrients in breast milk. Table 4 summarizes the most relevant studies conducted using NIRS technology, including the instrumentation employed to quantify macronutrients (high-performance laboratory instruments). Only dos Santos et al. [14] used a portable NIRS instrument to classify breast milk in colostrum, transition milk, and mature milk, which are the three stages of the lactation period.

Table 4. An overview on reported NIRS-based methods for breast milk analysis.

Reference	Device	Lab/Portable	Wavelength Range λ (nm)	Sample Size (N)	Analyzed Parameters
Corvaglia [26]	Fenir 8820, Esetek	Lab	700–2750	124	Fat and nitrogen contents
Sauer [27]	SpectraStar 2400 Near Infrared Analyzer, Unity Scientific	Lab	1200–2400	52	Fat, protein, and carbohydrates
Fusch [28]	SpectraStar 2400 Near Infrared Analyzer, Unity Scientific	Lab	1200–2400	1188	Fat, protein, and carbohydrates
dos Santos [14]	MicroNIR™ 1700, JDS Uniphase Corporation	Portable	910–1676	198	Qualitative (colostrum, transition milk, and mature milk)
Present study	MicroPHAZIR Mod. 1624, Thermo Fisher Scientific Inc.	Portable	1396–2396	68	Fat, crude protein, raw protein, carbohydrates, total solids, and energy

Lab: laboratory instrument; N = number of samples involved in the study.

Focusing on spectroscopic laboratory methodologies, previous researchers evaluated and compared the use of near- and mid-infrared instruments vs. reference methodologies using high-performance laboratory devices [28]. More specifically, the macronutrients quantified were fat, protein, and carbohydrates. The NIRS instruments employed were precalibrated NIRS laboratory devices, scanning in a range of 1200–2400 nm. No information about the statistics of calibration models were included in this work; however, external validation statistics can be compared with our results. The values of coefficients of determination detailed by Fusch et al. for validation were 0.76 for protein, 0.01 for lactose, and 0.79 for fat. In our study, employing a portable device and developing our own calibration models, the coefficient of determination of cross-validation was 0.52 for protein, 0.74 for HC, and 0.84 for fat MISSING [28].

Table 5 compares the external validation statistics obtained with laboratory instrument vs. the developed portable-device methodology by computing the linear regression of the reference and predicted NIRS data [28]. To evaluate the random error in the prediction based on regression results, the $S_{y/x}$ statistic (random error in the y-direction, y-direction indicates the prediction values and x-direction reference values) was calculated. Comparing these data with SEP values (Table 3) for fat, RP, and HC, we observed that the regression error ($S_{y/x}$) was equal or lower than the error obtained in external validation (SEP), confirming the validity of the developed models.

Table 5. Comparison of external validation statistics: Ref. [28] (laboratory instrument) vs. the proposed methodology (portable device).

	Portable	Laboratory	Portable		Laboratory
	Linear Regression		r^2	$S_{y/x}$	r^2
Fat	y = 0.69x + 0.72	y = 0.55x + 1.25	0.85	0.547	0.79
RP	y = 0.77x + 0.16	y = 0.55x + 0.54	0.67	0.154	0.76
HC	y = 0.95x + 0.60	y = 0.02x + 5.69	0.01	0.904	0.89

RP: raw protein, CH: carbohydrates, r^2: coefficient of determination, $S_{y/x}$: random errors in the y-direction, y: prediction values, x: reference values.

Another NIRS strategy for macronutrients milk analysis was carried out by de la Roza et al. [2], who developed in-house NIR calibrations with a high-performance laboratory instrument (working range 400–2500 nm) for fat, protein, and nonfat total solids in cow milk. The best result, with an R^2 value of 0.971, was obtained for the quantification of fat, applying mathematical SNV, detrend, and second derivative pretreatments. The R^2 value is similar to that obtained in this work (0.910). However, for total solids, the values of the calibration coefficient of determination for cross-validation were lower than 0.700

(r^2 = 0.612) [2], and better results, above 0.750, were obtained using our handheld portable instrument (r^2 = 0.787).

4. Conclusions

In this work we focused on the development of a real-time and simple methodology to quantify the macronutrients in breast milk. Notably, the implementation of this procedure requires the use of low-cost and handheld NIRS instruments. Moreover, expert personnel are not required for analyzing samples, facilitating the quality control procedure in the feeding of newborns in neonatology units.

In this paper, we demonstrated the feasibility of using a cheap and easy-to-use handheld NIRS instrument with a narrow scanning range, from 1596 to 2396 nm, and a small scanning window to control the quality of breast milk. By using a multivariate strategy and different pretreatments, we developed quantitative calibration models to determine the energy, fat, carbohydrate, and protein contents in this type of sample, with coefficients of determination for calibration (R^2) higher than 0.79 for all parameters. These results, obtained with a limited number of breast milk samples (68 samples), can be considered as a first step in the development of an appropriate method, even if more work needs to be carried out to improve the equations and to include more samples in the database. The interpretation of the results obtained for fat analysis demonstrated the richness of NIRS spectra because the presence of specific bands shows excellent potential for constructing useful quantitative models.

Prior to implementation, external validation was required. We tested the proposed methodology with breast milk samples not involved in the calibration procedure, and nonsignificant differences were observed when comparing the reference and portable NIRS methods. The accuracy obtained for the tested parameters was higher than 90%. Our results suggest that NIR sensor measurements of macronutrients are acceptable for clinical use in breast milk banks. The limitation of the proposed method is related to the range of values for each nutrient; however, further improvements can be achieved by including new samples to update this first calibration method.

Author Contributions: Conceptualization, F.F., J.M.C.-F., M.L.F.-S., A.S., and B.F.-C.; methodology, A.S., C.M., and S.F.; validation, C.M., S.F., A.S., and B.F.-C.; formal analysis, J.M.C.-F. and M.L.F.-S.; investigation, A.S. and C.M.; resources, A.L., F.F., J.M.C.-F., M.L.F.-S., and A.S.; data curation, S.F. and C.M. writing—original draft preparation, C.M.; writing—review and editing, A.L., F.F., J.M.C.-F., M.L.F.-S., A.S., and B.F.-C.; visualization, A.L. and F.F.; supervision, A.S.; funding acquisition, A.L., F.F., J.M.C.-F., and A.S. All authors have read and agreed to the published version of the manuscript.

Funding: This research was funded by Spanish Ministry of Science and Innovation (PID2020-117282RB-I00 and MCI-20-PID2019-109698GB-I00), by the Principado de Asturias GRUPIN IDI/2021/000081, and by "Fondo de investigación en Salud" (FIS) from the Instituto de Salud Carlos III in Spain (Project PI16/02084) co-financed by the Feder Programme of the European Union.

Institutional Review Board Statement: The study was conducted in accordance with the Declaration of Helsinki, and approved by the Ethics Committee of Hospital Central de Asturias (Proyecto n° 72/16; 05 May 2016).

Informed Consent Statement: Each mother provided written informed consent for donating the samples for this study, which was approved by the institutional review board.

Data Availability Statement: Data sharing not applicable.

Conflicts of Interest: The authors declare no conflict of interest.

References

1. Leghi, G.E.; Middleton, P.F.; Muhlhausler, B.S. A methodological approach to identify the most reliable human milk collection method for compositional analysis: A systematic review protocol. *Syst. Rev.* **2018**, *7*, 122. [CrossRef] [PubMed]
2. Fu, T.T.; Schroder, P.E.; Poindexter, B.B. Macronutrient analysis of target-pooled donor breast milk and corresponding growth in very low birth weight infants. *Nutrients* **2019**, *11*, 1884. [CrossRef] [PubMed]

3. Stephens, B.E.; Walden, R.V.; Gargus, R.A.; Tucker, R.; McKinley, L.; Mance, M.; Nye, J.; Vohr, B.R. First-week protein and energy intakes are associated with 18-month developmental outcomes in extremely low birth weight infants. *Pediatrics* **2009**, *123*, 1337–1343. [CrossRef]
4. Fusch, G.; Kwan, C.; Kotrri, G.; Fusch, C. "Bed Side" human milk analysis in the neonatal intensive care unit: A systematic review. *Clin. Perinatol.* **2017**, *44*, 209–267. [CrossRef] [PubMed]
5. Shenk, J.S.; Workman, J.J.; Westerhaus, M.O. Application of NIR Spectroscopy to Agricultural Products. In *In Handbook of Near Infrared Analysis*, 1st ed.; Burns, D.A., Ciurczak, E.W., Eds.; Practical Spectroscopy A Series; CRC Press, Taylor & Francis Group: Boca Raton, FL, USA, 2008; Volume 13, pp. 347–387.
6. Bertrand, D.; Dufour, E. *Infrared Spectroscopy and Its Analytical Applications*; Editions Tec & Doc: Paris, France, 2000.
7. Murray, I. Forage analysis by near infra-red spectroscopy. In *Sward Measurement Handbook*; Davies, A., Baker, R.D., Grant, S.A., Laidlaw, A.S., Eds.; NIR Publications: Chichester, UK, 1993; pp. 155–177.
8. Muncan, J.; Tei, K.; Tsenkova, R. Real Time Monitoring of Yogurt Fermentation Process by Aquaphotomics Near-Infrared Spectroscopy. *Sensors* **2021**, *21*, 177. [CrossRef] [PubMed]
9. Naes, T.; Isaksson, T.; Fearn, T.; Davies, T. *A User-Friendly Guide to Multivariate Calibration and Classification*; NIR Publ.: Chichester, UK, 2007.
10. Osborne, B.G.; Fearn, T.; Hindle, P.H. *Practical NIR Spectroscopy with Applications in Food and Beverage Analysis*, 2nd ed; Longman Scientific and Technica: Essex, UK, 1993; pp. 11–35.
11. Shenk, J.S.; Westerhaus, M.O. *Routine Operation, Calibration Development and Network System Management Manual for Near Infrared Instruments*; NIR Systems, Inc.: Silver Spring, MD, USA, 1995; PN IS-0119.
12. Shenk, J.S.; Westerhaus, M.O. *Analysis of Agriculture and Food Products by Near Infrared Reflectance Spectroscopy*; Infrasoft International: Port Matilda, PE, USA, 1993; 116p.
13. Shenk, J.S.; Westerhaus, M.O. Calibration the ISI way. In *Near infrared Spectroscopy: The Future Waves*; Davis, A.M.C., Williams, P., Eds.; NIR Publications: Chichester, UK, 1996; pp. 198–202.
14. Dos Santos, V.J.; Baqueta, M.R.; Março, P.H.; Valderrama, P. Human Milk Lactation Phases Evaluation Through Handheld Near-Infrared Spectroscopy and Multivariate Classification. *Food Anal. Methods* **2021**, *14*, 873–882. [CrossRef]
15. De la Roza-Delgado, B.; Garrido-Varo, A.; Soldado, A.; González, A.; Cuevas, M.; Maroto, F.; Perez-Marín, D. Matching portable NIRS instruments for in situ monitoring indicators of milk composition. *Food Control* **2017**, *76*, 74–81. [CrossRef]
16. Celia, K.; Fusch, G.; Rochow, N.; Fusch, C. Milk analysis using milk analyzers in a standardized setting (MAMAS) study: A multicentre quality initiative. *Clin. Nutr.* **2020**, *39*, 2121–2128. [CrossRef]
17. Windham, W.R.; Mertens, D.R.; Barton, F.E. Protocol for NIRS calibration: Sample selection and equation development and validation. In *Near Infrared Reflectance Spectroscopy (NIRS): Analysis of Forage Quality*; US Department of Agriculture: Springfield, MA, USA, 1989; pp. 96–97.
18. Czosnykowska-Łukacka, M.; Królak-Olejnik, B. Orczyk-Pawiłowicz, M. Breast milk macronutrient components in prolonged lactation. *Nutrients* **2018**, *10*, 1893. [CrossRef] [PubMed]
19. Mark, H.; Workman, J. *Statistics in Spectroscopy*; Academic Press: Cambridge, MA, USA, 2003.
20. Camo Analytics Welcome. AspenTech. Available online: https://www.aspentech.com/en/acquisition/camo-analytics (accessed on 3 November 2021).
21. Thompson, M.; Wood, R. Using uncertainty functions to predict and specify the performance of analytical methods. *Accred. Qual. Assur.* **2006**, *10*, 471–478. [CrossRef]
22. Burns, D.A.; Ciurczak, E.W. *Handbook of Near-Infrared Analysis*, 3rd ed.; CRC Press: Boca Raton, FL, USA, 2007.
23. Koletzko, B.; Rodriguez-Palmero, M. Polyunsaturated fatty acids in human milk and their role in early infant development. *J. Mammary Gland Biol. Neoplasia* **1999**, *4*, 269–284. [CrossRef] [PubMed]
24. Savenije, B.; Geesink, G.H.; Van der Palen, J.G.P.; Hemke, G. Prediction of prok quality using visible/near-infrared reflectance spectroscopy. Prediction of pork quality using visible/near-infrared reflectance spectroscopy. *Meat. Sci.* **2006**, *73*, 18. [CrossRef] [PubMed]
25. Bellon-Maurel, V.; Fernández-Ahumada, E.; Palagos, B.; Roger, J.M.; McBratney, A. Critical review of chemometric indicators commonly used for assesing the quality of the prediction of soil attributes. *Trends Anal. Chem.* **2010**, *29*, 1073–1081. [CrossRef]
26. Corvaglia, L.; Battistini, B.; Paoletti, V.; Aceti, A.; Faldella, G.; Capretti, M.G. Near-infrared reflectance analysis to evaluate the nitrogen and fat content of human milk in neonatal intensive care units. *Arch. Dis. Child. Fetal Neonatal Ed.* **2008**, *93*, F372–F375. [CrossRef] [PubMed]
27. Sauer, C.; Kim, J. Human milk macronutrient analysis using point-of-care near-infrared spectrophotometry. *J. Perinatol.* **2011**, *31*, 339–343. [CrossRef] [PubMed]
28. Fusch, G.; Rochow, N.; Choi, A.; Fusch, S.; Poeschl, S.; Ubah, A.O.; Lee, S.Y.; Raja, P.; Fusch, C. Rapid measurement of macronutrients in breast milk: How reliable are infrared milk analyzers? *Clin. Nutr.* **2015**, *34*, 465–476. [CrossRef] [PubMed]

MDPI
St. Alban-Anlage 66
4052 Basel
Switzerland
Tel. +41 61 683 77 34
Fax +41 61 302 89 18
www.mdpi.com

Sensors Editorial Office
E-mail: sensors@mdpi.com
www.mdpi.com/journal/sensors

www.ingramcontent.com/pod-product-compliance
Lightning Source LLC
Chambersburg PA
CBHW040838250926
51850CB00007BA/1348
9783036575018